Samuel Hughes

A Treatise on Waterworks for the Supply of Cities and Towns

With a Description of the Principal Geological Formations of England as Influencing

Supplies of Water

Samuel Hughes

A Treatise on Waterworks for the Supply of Cities and Towns
With a Description of the Principal Geological Formations of England as Influencing Supplies of Water

ISBN/EAN: 9783337139933

Printed in Europe, USA, Canada, Australia, Japan

Cover: Foto ©Andreas Hilbeck / pixelio.de

More available books at **www.hansebooks.com**

A TREATISE ON

WATERWORKS

For the Supply of Cities and Towns

WITH

A DESCRIPTION OF THE PRINCIPAL GEOLOGICAL FORMATIONS OF
ENGLAND AS INFLUENCING SUPPLIES OF WATER

BY

SAMUEL HUGHES, F.G.S.

CIVIL ENGINEER

NEW EDITION, REVISED AND CONSIDERABLY ENLARGED

Capio Lumen

LONDON

CROSBY LOCKWOOD AND CO.

7, STATIONERS' HALL COURT, LUDGATE HILL

1882

PREFACE.

In claiming for this little volume a modest place in Mr. Weale's now celebrated Rudimentary Series, I venture to offer a few words of explanation as to the mode in which I have been led to treat the subject.

After a brief allusion in the earlier pages to some celebrated works of antiquity and to the ancient modes of procuring water which were practised in the East, the second part of the book is devoted to a mixed geological and hydrographical examination of the surface of England.

This has been thought necessary in consequence of the extreme importance which physical structure exercises in every question of Water-supply—an importance which attaches alike to every source of supply, whether from springs, rivers, wells, lakes, or drainage areas. This branch of the subject is in itself so extensive that I can only pretend to have given a very meagre and imperfect sketch of those geological and other physical conditions which affect the water-yielding capacity of various districts. I trust I have said enough however to open up the subject to the attention of the young student, and to point out the right direction in which his investigations ought to be pursued.

The third part relates to the sinking of wells and borings, on which it has not been necessary to enlarge very copiously, inasmuch as the present series already contains a very interesting little volume by Mr. Swindell devoted especially to that subject.

In the fourth part, relating to pumping machinery for raising water, I have attempted to describe the principal varieties of engines and pumps used for this purpose both in this country and America, and have brought together a mass of facts and calculations relating to the duty and power of pumping engines of various kinds, and to the cost of pumping water, which I trust will not be altogether uninteresting.

The subject of raising large volumes of water by means of steam power engages unusual attention at this time, not only in reference to Waterworks, but also as bearing on the sewerage of towns, and especially that of the Metropolis. I have therefore dwelt at some length on the questions connected with actual engine-power, the nature and extent of the surplus or auxiliary power which ought to be employed, and the American system of using high-pressure engines for the auxiliary power. The calculations and tables which are given in this part of the Work show the great economy resulting from this practice.

The fifth part relates to Waterworks obtaining supplies from rivers, streams, and drainage areas, and contains some observations on the characteristics and flow of rivers, and on the cost and dimensions of embankments for impounding reservoirs. This part also contains a few particulars relating to filter-beds and service reservoirs.

The last part is devoted to the flow of water in open channels and pipes, and to the subject of gauging the flow of water under various conditions, as in rivers, pipes, and artificial open channels, also through orifices and over weirs.

It was originally intended to have extended the Work so far as to embrace the distribution of water in the streets of towns, but it was found very difficult to compress the preceding important divisions within a sufficiently small compass to admit of this. The subject of distribution therefore remains untouched; and as this division of the inquiry, with all its details of mains, pipes, services, standcocks, hydrants, etc. etc., will of itself be amply sufficient to fill a single volume of the series, it is proposed to treat this part of the subject in a separate form and at a future time, the distance of which will depend somewhat on the reception accorded by the public to the present humble attempt.

I avail myself of this opportunity to express the very sincere obligations I am under to many kind friends and professional acquaintances, who have aided me with advice and information in every department of the subject on which I have sought assistance.

14, Park Street, Westminster.

PREFACE TO THE SECOND EDITION.

The revised and considerably enlarged edition of this work bears a posthumous character, and I am certain that, independently of the recognised value of this volume, the circumstance that its revision was one of the author's last acts, will lend to it an additional and special interest.

I can bear testimony that the revision of these pages was completed within a few days of the fatal illness which closed a singularly active and useful life. Mr. Hughes's professional career had been unusually extended; it may be considered to have commenced at a time when railway enter· prise was being carried on with much spirit, and he was associated with some of the most distinguished engineers of the period in the projection of the principal schemes which have since developed our present admirable network of railways.

At a later period, and quite in another channel, the zealous efforts of Mr. Hughes to uphold the rights of public bodies in the promotion of economic and sanitary measures attracted a deal of attention. His views respecting large private undertakings, water companies, gas companies, &c., which he was of opinion should be vested under municipal control, and conducted solely for public advantage, aroused a powerful feeling of antagonism in many quarters; but there is little doubt that a few years will see this system developed to an extent that will do justice to his convictions.

It has been observed that he was often identified with extreme measures. This was to a certain extent true, because he initiated them; but, in reference to a subject to which he especially devoted much attention—the supply of gas —it is only just to record that the soundness of his views has been vindicated by the fact that most of his early suggestions have been gradually carried out, even in the face of considerable opposition.

The inhabitants of this metropolis are not likely to forget that in 1860, upon the occasion of the passing of the Metropolis Gas Act, Mr. Hughes, associated with Mr. James Beal, a gentleman whose assiduous devotion to public interests is so well known and appreciated, were the pioneers of a movement which secured the subsequent development for the public benefit of all the late improvements introduced in the supply of gas.

ARTHUR SILVERTHORNE, C.E.

1, WESTMINSTER CHAMBERS, LONDON,
      September, 1872.

CONTENTS.

PART I.

PART II.

PART III.

PART IV.

PART V.

PART VI.

PART VII.

APPENDIX.

ON THE VARIOUS MODES ADOPTED FOR COLLECTING SUPPLIES OF WATER.

THE sources from which water is commonly procured have, in reality, only one origin—namely, the supply given by nature in the form of rain and snow falling from the clouds. The water thus bestowed may be seized by man in various forms. It may be taken in a comparatively pure, distilled state, as it falls in the shape of condensed vapour from the clouds, free from admixture with any earthy salts, or infusion of any other ingredients except those which it meets with in passing through the air. It may be collected from drainage areas, where the quantity of rain happens to be greater than that which evaporates and sinks into the earth; again, it may be taken from rivers, streams, or lakes, which are themselves supplied chiefly from drainage areas: or, lastly, it may be taken from wells or springs, where the water has accumulated after passing through strata and rocks of various kinds, and become much changed from its original state of purity by the absorption of gases and by holding in solution earthy salts and other bodies which it has met with in its subterranean passage.

With respect to the first or primitive state of water, however well adapted this may be from its quality of softness, for certain commercial and culinary purposes, it is confessedly not palatable for drinking. It fails entirely in that agreeable taste imparted to spring water by the gases, and even by the mineral matters held in solution. At the same time pure rain water is so valuable for many purposes, and would answer so well for

B

many others where quality is of no consequence at all, that the inquirer may at first sight ask with some surprise, how it is that means are not taken to collect the rain water as it falls from the clouds instead of extracting it often by a difficult and expensive process from the bosom of the earth. Let us examine this question a little more in detail. It is evident, until mankind had made considerable progress in the arts, there were no roofs of houses or other buildings from which rain water could be collected, and therefore if any attempt were made to procure it otherwise than by embanking across valleys, and collecting a certain portion of the rainfall, as practised at the present day, it must be done by making an artificial impervious surface, from which the water, falling in the shape of rain could flow into cisterns placed to receive it. Let us suppose an acre of surface so constructed: the expense of this at the present day would probably not be less than one shilling per square yard, or £242 an acre, independently of the value of the land. Now the interest of this sum alone, at 5 per cent., would be more than the value of the water which could be collected on an acre of surface, assuming an available rainfall of 30 inches, which is probably as much as could be obtained in dry years on the average of England, after deducting such evaporation as would be unavoidable in spite of all precautions. The whole quantity of water which could be collected on an acre of ground, from an available rainfall of 30 inches per annum, is under 681,000 gallons, which, at sixpence per thousand gallons (a price at which water can not only be collected, but conveyed and distributed in towns, even where pumping to a high level is necessary), would amount to little more than £17, or about the interest of the money to be laid out in constructing the gathering ground and purchasing the land alone. So much for the collection of perfectly pure rain water by an artificial surface. But it may be said, we possess in the roofs of houses already built in every town, the means of collecting rain water of tolerable purity, except when the atmosphere is polluted by smoke, as in the metropolis. We shall find, on examination, that the quantity which could be so collected is wholly inadequate for the wants of any given population. Let us assume that the roof-surface of any town or large group of houses is equal to 60 square feet for each individual, an assumption which is probably much in excess of the real fact, then the annual quantity of water which would be collected on this surface on the same supposition as before, with respect to available rainfall, would be 935 gallons, or less than 3 gallons a day for each individual. Now the common allowance for towns in England, including every kind of use, both domestic and public, is not less than 20 gallons per head per day, so that the rainfall alone will only give one-seventh of the quantity actually required ; or, in other words, in order to obtain the required quantity we must assume an annual rainfall of 140 inches, which is scarcely yielded in any part of the world.

Professor Leslie calculates, in his Elements of Natural Philosophy, that the roof of a lofty house in Paris, containing on an average 25 persons, might deliver annually 1800 cubic feet of rain water, which would furnish to each individual daily the fifth part of a cubic foot, or rather more than one gallon.

We find, however, notwithstanding the inadequacy of the quantity, that Venice and many other continental cities have been for many years supplied with rain water, both from public and private reservoirs, which are commonly constructed underground, for the purpose of receiving the water from the roofs. Matthews, in his Hydraulia, describes one of these public reservoirs, which is situate under the court of the ducal palace at Venice. Here the underground cistern is provided with a sand filter, through which the water passes, and flows into a covered well in a clear and transparent state.

In speaking of a surface to collect rain-water from which surface it is to flow into a reservoir, it may be necessary to notice, in passing, that a reservoir is absolutely necessary, in order to prevent the loss of all the water by evaporation. For instance, if a man were to make for himself a tub or cistern large enough in surface to catch all the rain-water he required for his own use, calculating its area simply from the known rainfall, he would find that evaporation would carry off all the water as fast as it falls, and a great deal faster. From the very accurate observations made at the Highfield House Observatory, near Nottingham, during the last year, we learn that, whilst the whole rainfall was only 17.3 inches, the evaporation was equal to 41 inches, or more than double the amount of rainfall. As, however, the evaporation is in direct proportion to the area of the surface exposed, it follows that the rainfall of a very large drainage area may be collected in a small and deep reservoir with a comparatively small loss from evaporation.

There being such obstacles in the way of collecting pure rain-water, we find that mankind in the very earliest ages have turned their attention to the stores or reservoirs which nature has provided in the shape of springs, rivers, streams, and lakes.

ANCIENT MODES OF OBTAINING WATER FROM WELLS.

We shall not attempt to follow Mr. Ewbank in his minute details of the earliest processes adopted by mankind for procuring water, by first kneeling down at the side of a river and drinking from the surface, after the manner of the inferior animals, and then gradually advancing to use the hollow of the hand, and the concave cases or coverings of fruits for the purpose of lifting and containing the water. All this may be interesting in an antiquarian point of view, but is not within the scope of our present more limited inquiry. The first works of primitive nations which interest us, and require attention, are the ancient wells, the remains of which are scattered over all the first inhabited countries, and the origin of which probably goes back as far as the world before the flood. The earliest wells were probably mere drainage pits, dug in moist spots of ground to allow of the infiltration of the surrounding water. Such are the small square wells, dug or scooped out of the earth, and discovered by travellers at the present day in parts of Africa, New Holland, and other uncivilised countries. Many of them are so shallow that they are emptied every day, and only supplied by the water which trickles into them during the night.

Mr. Ewbank is of opinion, that amongst the first people in the world these wells were superseded soon after the seventh generation from Adam, about which time the discovery of metals took place, and consequently the power of digging and penetrating through rocks. In fact, in the very earliest records which have been handed down to us from the most remote antiquity, as for instance in the writings of Moses, we have mention made of wells; and modern travellers have not hesitated to point out the sites of some of the most ancient wells as discoverable at the present day. Many of these wells, whose origin dates from the very earliest period, have passed through both rock and quicksand, and therefore exhibit a knowledge of workmanship and mode of dealing with mechanical difficulties which has not always been associated with such remote antiquity. Mr. Ewbank in fact asserts, that to the constructors of ancient wells in the East we are indebted for the only known mode at present adopted of sinking deep wells through quicksands by the employment of a curb, which settles and sinks down as the excavation proceeds.

Among the most ancient wells in the world are those which bear the name of the Patriarch Jacob and his son Joseph; the former situate near Sychar (the Shechem of the Scriptures) and the latter near Cairo in Egypt. Jacob's Well has been visited by pilgrims in all ages, and has been minutely described by Dr. Clarke, in his Travels. It is 9 feet in diameter and 105 feet deep, made entirely through rock; and when visited by Maundrell it contained 15 feet of water.

Joseph's Well at Cairo is 297 feet in depth, and is altogether a much more elaborate work than the other, and indeed more so than most modern wells. The mode of raising the water was by means of an endless rope, carrying earthern pots or buckets and working over a wheel at top and bottom, similar to the buckets of the modern dredging engine, only that the chain of pots moved vertically, instead of working in a sloping direction.

The endless rope carrying the pots was put in motion by oxen walking round in a circle ; and as the depth of the well (nearly 300 feet) was too great to be worked in one lift, it was divided into two separate shafts by a compartment large enough for the oxen to work in, at a depth of 165 feet below the surface of the ground. Herein arises the great peculiarity of this well ; the upper shaft having a section of 24 feet by 18, with a spiral passage winding round it from top to bottom, of sufficient dimensions to allow the oxen to pass from the surface to the working chamber at the lower extremity of the upper shaft. The spiral passage is 6 feet 4 inches wide and 7 feet 2 inches high, and is made with so gradual an inclination that persons ride up and down upon asses or mules. The lower shaft goes from the bottom of this chamber to a further depth of 132 feet. This lower shaft is not in the same line as the upper one, but a little on one side, and is smaller in dimensions, being 15 feet by 9. The oxen working in the chamber between the two shafts raised the water into a reservoir immediately at the bottom of the upper shaft, through which it was again raised to the surface by another chain of pots, worked by oxen at the top of the well.

The extraordinary skill displayed in the construction of this well has excited the admiration of all travellers who have visited it. The spiral passage surrounding the upper shaft is executed with the utmost precision, a very thin portion of the rock (only about 6 inches) being left between the passage and the cavity of the well. Semicircular openings, or loopholes, are formed at intervals, by which the spiral passage is dimly lighted from the interior. Many curious conjectures have been hazarded as to this remarkable well and its peculiar oblong form. This latter has been attributed, with some show of reason, to the necessity for lighting the interior, and to the fact that this form would admit the light of the sun during more hours of the day than a square or circular section. Conflicting opinions are also entertained as to the date of the well. The common people of Egypt ascribe it to the patriarch Joseph. Many antiquarians, however, are inclined to imagine the well to be of much more recent date ; some ascribing it to the famous Saladin of the Crusades, whose real name was Yussef (or Joseph) ; while others believe it was made by a vizier named Joseph about 800 years ago.

The celebrated well of *Memzem* at Mecca dates, according to popular tradition, from a higher antiquity even than Jacob's Well, being in fact venerated as the well from which Hagar nourished the ancestor of the Arabian people ; the construction of the well itself being attributed to Abraham and Isaac.

The oasis of Thebes contains Artesian wells, which have been noticed by Arago. The practice of the ancient inhabitants was to dig square wells 60 to 80 yards in depth, through the superficial vegetable soil, clay, and marl, down to the limestone rocks, in which borings 4 to 8 inches in diameter were then made. These holes were fitted with a block of sandstone, which was furnished with an iron ring, and was used to stop the supply when necessary.

OTHER ANCIENT MODES OF COLLECTING WATER.

Although the very earliest people appear to have procured water chiefly from wells of various kinds, and sometimes from reservoirs and cisterns formed at spring heads to collect the water, it is evident that in process of time these means would become inadequate, and the wants of increasing populations concentrated in towns would require other modes of supply. Hence we find that aqueducts were used for conveying water into towns from a very remote period of history.

It is probable that the great canals and lakes of which traces still exist in Egypt were used as reservoirs for storing the waters of the Nile in times of flood, in order that they might be used for irrigating the land during dry seasons. The great aqueducts, reservoirs, and other hydraulic works of ancient

Egypt are perhaps not less wonderful, and certainly were far more useful, than their pyramids and colossal sculptures. Savary, in his "*Lettres sur l'Egypte*," describes some works of extraordinary magnitude for raising water into high reservoirs near Cairo, and observes that in other parts of the country the Egyptians conveyed the water to the tops of hills, where immense cisterns were hewn in the rocks to receive it, and whence it afterwards flowed among deserts, which this great people transformed into fertile fields.

Many parts of Syria contain the remains of aqueducts which are undoubtedly of great antiquity. Those in the neighbourhood of Tyre and of Jerusalem are attributed to Solomon, who lived 1000 years before the Christian era. The ancient aqueducts of Antioch and Hamah have also been noticed by many modern travellers.

The earliest account of any aqueduct for conveying water is probably that which is given by Herodotus (who was born B.C. 484 years). He describes the mode in which an ancient aqueduct was made by Eupalinus, an architect of Megara, to supply the city of Samos with water. In the course of the aqueduct a tunnel, nearly a mile in length, was pierced through a hill, and a channel 3 feet wide made to convey the water.

The first aqueduct of ancient Rome was that constructed B.C. 331 years, by the censor Appius Claudius, after whom it was named the Appia Claudia. Before the construction of this work, which brought in the water from a distance of about 11 miles, the inhabitants derived their supply from the Tiber, and from wells and springs in the immediate vicinity of the city.

About 100 years later the celebrated aqueduct, called the Aqua Martia, was commenced by Quintus Martius. This began at a spring 33 miles from Rome, the length of the aqueduct itself being upwards of 38 miles, or about equal to that of the New River. The works, however, are of a much more gigantic character, there being a series of nearly 7000 arches, some of them 70 to 100 feet in height. Many aqueducts were constructed during the reign of the Emperors, as the Aqua Virginia, in the reign of Augustus, the aqueduct for conveying the Anio in the reign of Nero, and the Aqua Claudia in the reign of Caligula. This last work commenced at the distance of 38 miles from Rome, and was formed under ground for 36¼ miles, about 3 miles of which consisted of tunnelling. The supply of water to ancient Rome was computed by Professor Leslie, on the authority of Sextus Julius Frontinus, who was inspector of the aqueducts under the Emperor Nerva, and who has left a valuable treatise on the subject, at 50 million cubic feet per day for a population of one million souls. This gives the immense average per head of 50 cubic feet, or 312 gallons, a consumption quite unequalled in modern times, except in the city of New York, where it is said to have amounted nearly to this quantity.

The Roman aqueducts were mostly built of brickwork, and consisted of nearly square piers carried up to a uniform level, allowing for the fall of the water, and connected by semicircular arches, on which the conduit for carrying the water was placed. The conduit had a paved or tiled floor, side walls of brick or stone, and a roof formed by an arch turned across it, or by a flat coping of stone. Various opinions have been hazarded by modern writers to account for the peculiar practice adopted by the Romans of conveying their water by means of a uniform channel through hills and over valleys, instead of adopting the modern system of following the undulations of the country by passing over hills and descending to the bottoms of valleys. A writer in Lardner's Cyclopœdia thus expresses himself : "*Ignorance of the principle by which liquids return to. their level* is shown in the construction of aqueducts by the ancients, for supplying water to towns." This supposition is grossly inconsistent with the skill and acquaintance with the laws of nature displayed by the Romans and other great nations of antiquity, as exhibited in the very aqueducts which this writer condemns. Professor Leslie says nothing can be worse founded than the opinion that the Romans were unacquainted with the laws of hydrostatics, and, therefore, with the method of conducting and raising water through a train of pipes. "The ancient writers," he observes, "who either treat of the subject or incidentally mention it, are clear and explicit in their remarks, while many vestiges of art still attest the accuracy of these statements. Pliny, the natural historian, lays down the main principle that 'water will invariably rise to the height of its source.' He subjoins that leaden pipes must be employed to carry water up to an eminence." Palladius, Vitruvius, and other Roman writers describe the method of conveying water in leaden and earthenware pipes.

Some of the Roman aqueducts were uncovered or open channels, and of course had to be formed with a uniform inclination, or the water would not have flowed through them at all. In other cases, where the aqueduct was covered, it was a question of expense between making the conduit sufficiently strong and watertight to resist the great pressure of water to which it would have been subjected in the valleys, or, on the other hand, raising it a sufficient height to make the flow uniform and reduce the pressure to a minimum. It was not, in fact, till the introduction of cast-iron pipes, strong enough to withstand the pressure of a column of water equal to several hundred feet of vertical height, that the moderns were able to adopt the mode of crossing hills and valleys by following the undulations of the ground. In the construction of the New River aqueduct, only 260 years ago, no other principle was thought of than that of a uniform channel for the water; the only difference between this and the Roman aqueducts being its excessively timid character; for, instead of boldly passing through hills and over valleys, it winds around the former and creeps up the latter in order to diminish every artificial work to its least possible dimensions. Yet who ever thinks of decrying the New River, or treating contemptuously the skill displayed by its constructors? Considering the period of its execution, it was as great a work as any which distinguishes our own times.

Some of the Roman aqueducts were remarkable for being

built in tiers or arcades one above another, and several of those which supplied the imperial city brought in the water of separate streams at different levels one over the other.

Nor did the genius and enterprise of the Roman people rest content with the embellishment and improvement of their own immediate territory by means of these magnificent works of art. Throughout all the continents of the old world, wherever the Roman eagles have penetrated for conquest and civilisation, there we find the remains of their gigantic aqueducts. Not only Italy and Sicily, but the more distant lands of Greece, France, and Spain, contain abundant traces of these great works.

Both the Moors and the Romans have left remains of many magnificent hydraulic works in Spain. An embankment 66 feet wide and 150 feet high, stretches across a gorge on the road from Alicante to Xativa, and dams up the mountain torrents into an immense reservoir.

The Roman aqueducts of Segovia and Seville still supply those Spanish towns with water ; while the noble aqueducts of Nismes, Metz, and Lyons, in the south of France, are imperishable monuments of Roman renown during their possession of ancient Gaul. The Pont du Gard, that wonderful structure consisting of three tiers of arcades placed one over the other to the height of 150 feet above the valley, still exhibits the sectional area of the Nismes aqueduct. M. Genieys, formerly engineer-in-chief to the municipality of Paris, calculates the quantity of water supplied by this aqueduct at nearly 14 million gallons per day.

The ancient works executed under the later Roman emperors for the supply of Constantinople, combine the system of aqueducts with the collection and impounding of water by means of reservoirs at the head of the aqueduct. The impounding reservoirs are situate about 12 miles from the city on the slopes of a range of mountains, which form the south-eastern prolongation of the great Balkan chain. There are four principal aqueducts, one of which conveys the water collected by three separate reservoirs, while the other three are each supplied by its own reservoir. Besides these extensive provisions for securing water to the city there are immense subterranean reservoirs, one of which, now in ruins, is called the Palace of the Thousand and One Pillars, not because this is the precise number supporting the roof, but because the number is a favourite one in the expression of eastern hyberbole. This great subterranean cistern is supposed to have been made by the Greek emperors for the purpose of storing water in case of a siege or similar calamity. Although originally of great depth, it is now nearly filled up with earth and rubbish. It is singular that in the nineteenth century we are reviving in our covered reservoirs, for the purpose of storing water in a state of freshness and uniform temperature, the practices which were followed nearly 2000 years ago by nations whose modern descendants are half barbarians.

The works of the ancient Peruvians, constructed for the purpose of irrigating vast rainless districts of country, are not inferior to those which have been executed by other nations. The necessity for water in South America will be well understood when we reflect that along the coast there are districts of 2000 miles in extent in which no rain ever falls. Some very interesting notices of the ancient aqueducts and wells of Peru are contained in Garcillasso's Royal Commentaries of Peru, which were translated into English by Sir Paul Ricaut and published in London, 1688.

He relates that the ancient Incas devoted much of their paternal care and attention to the improvement of the country, and amongst other great works constructed numerous aqueducts for conveying water from the hills to fertilise the otherwise dry and desert parts of the country. Instances are related of insignificant streams being conveyed a distance of 60 miles for the purpose of irrigating a few acres of land.

One of the Incas made an aqueduct 120 leagues in length, to convey the water of certain springs, which rose near the summit of a high mountain between Parcu and Picuy. The aqueduct was 12 feet in depth, and watered a tract of country more than 50 miles in breadth. Another great Peruvian aqueduct, 150 leagues in length, traverses the whole extent of one large province, and irrigates a vast extent of dry and arid pasture land. The Peruvians do not appear to have practised the method of traversing valleys by bridging them over with arches, but conveyed their water round the mountains, following such a contour as gave them a proper inclination for the water-course— something in the same style as that in which the New River was originally brought from Chadwell Springs into London. It appears from the account of Garcillasso, that the Peruvians had numerous wells, from which water was raised for irrigation and other purposes; but at the time of the Spanish conquest many of these wells were used as hiding-places to conceal treasure, and being filled up with earth, in order more effectually to hide them, the sites became obliterated, and all traces of the wells were destroyed.

ON THE ORIGIN AND NATURE OF SPRINGS, AND THE WATER PROCURED FROM SPRINGS.

Of the water which falls from the clouds in the form of rain and snow a certain proportion runs off the surface, and is received by rivers and open water-courses; another portion enters into union with vegetable substances, a third portion is evaporated from the surface, and a fourth portion sinks into the soil, and passing through strata which are more or less porous, forms the subterranean reservoir which yields, under certain favourable circumstances those fresh, cool, and delicious springs, that are met with in nearly every part of the world. To determine the proportion of the whole rainfall which actually sinks into the ground to be again yielded by the earth in the form of springs, is one of the most interesting problems presented to the study of the hydraulic engineer. Nor is it less important to determine the proportion which will naturally flow off the surface into collecting or impounding reservoirs, this being the question which has usually to be solved in considering the supplies of water to be obtained from gathering grounds or drainage areas. It is obvious that the proportions vary much according to circumstances; they depend greatly upon climate, and upon the geological formation of the district, and they vary also according to the season of the year. Thus in warm weather everywhere, and in tropical climates throughout the year, the evaporation must be very much greater than in colder seasons and climates, so that, supposing the rainfall to be equal, and all other things alike except the temperature, the quantity of water sinking through the surface to form springs, and the water running off to form rivers, will be much less in the warm than in the cold climate. Again, with respect to geological structure, it is evident that a district of retentive impermeable clay will carry off much more water in its streams and rivers, will admit of greater evaporation, and will allow less to sink into the ground than a district composed of permeable gravel, sand, or porous rocks. The physical shape and configuration of a country has also much to do with the proportions in which the rainfall is disposed of. Thus, all other things being alike, a very hilly or undulating country, full of steep slopes and declivities, will evidently pass off more water on the surface, admit of less evaporation and of less sinking below the surface than one with a more level surface. It is a subject of regret, that few observations of a trustworthy nature have been made to determine the proportion of the whole rainfall, which sinks into the ground in different districts. Such observations as we do possess have frequently been made by partizans in order to support some view or theory of their own, and in other cases have been made, rather with the collateral object of comparing one district with another, than to determine the absolute quantity which penetrates in any one given district. Dr. Dalton found, in the course of three years' experiments on the new red sandstone soil of Manchester, that 25 per cent. of the whole rainfall penetrated to the depth of 3 feet. Mr. Charnock in his experiments during five years, on the magnesian limestone soil of Ferrybridge, in Yorkshire, found that only 19.6 per cent. of the whole rainfall filtered through to the depth of 3 feet, while Mr. Dickinson, having observed the infiltration during eight years through the sandy gravelly loam which covers the chalk in the valleys around Watford, found that as much as 42·4 per cent. of the whole rainfall penetrated to this depth. It is considered, that observations made with Dalton's rain-gauge, which indicates the quantity penetrating to the depth of 3 feet, may be depended on as correct, to determine the yield of subterranean water in a given district, because it is probable, that at and beyond the depth of 3 feet, no evaporation takes place into the atmosphere. Mr. Dickinson's experiments are very interesting, as they indicate not only the quantity which penetrates annually, but the varying proportion in each month. The following table, showing the result of Mr. Dickinson's observations during the eight years, from 1835 to 1843, has been published in a paper by Mr. Parkes in the Journal of the Royal Agricultural Society. Vol. V., p. 147.

	Mean Fall.	Mean Infiltration.	Per Cent.
January	1·847	1·307	70·7
February	1·971	1·547	78·4
March	1·617	1·077	66·6
April	1·456	0·306	21·0
May	1·856	0·108	5·8
June	2·213	0·039	1·7
July	2·287	0·042	1·8
August	2·427	0·036	1·4
September	2·639	0·369	13·9
October	2·823	1·400	49·0
November	3·837	3·258	84·9
December	1·641	1·805	100·0
	26·614	11·294	42·4

It appears from these observations that in the first three months of the year the quantity which penetrates is about 70 per cent. of the whole rainfall, that in the winter months of November and December nearly the whole rainfall sinks into the earth, while in the four summer months, from May to August inclusive, the quantity which sinks is exceedingly small amounting on the average to little more than 2 per cent. of the whole rainfall.

Mr. Stephenson, in his report on the spring water from Watford, calculates the area or watershed draining into the rivers Verulam and Colne at $113\frac{1}{2}$ square miles, and he assumes the annual rainfall of the district at 20 inches. The total quantity of rain falling on this surface would thus be equal to $14\frac{1}{2}$ millions of cubic feet in 24 hours. Mr. Telford found that in a dry season the rivers which drain this district carried off 30 cubic feet per second, or about $2\frac{1}{2}$ million feet in 24 hours, which is equal to nearly $3\frac{1}{2}$ inches of rainfall. Dr. Thomson, however, calculates the quantity carried off by the streams and springs at 4 inches, and Mr. Stephenson adopts this quantity in his calculations. He then assumes that the evaporation in a chalk district, together with the quantity absorbed by animal and vegetable life, is equal to one-third of the whole rainfall.

The following table therefore represents the results which Mr. Stephenson arrived at for the chalk district round Watford :—

	Millions of Cubic feet per day.	Inches of rain per annum.
Quantity carried off by rivers and streams . .	3	4
Quantity evaporated and absorbed by animal and vegetable life	5	$6\frac{2}{3}$
Quantity sinking into the ground to form springs .	$6\frac{1}{2}$	$9\frac{1}{3}$
Total rainfall	$14\frac{1}{2}$	20

Proportion of rainfall which sinks below the surface, 44·8 per cent.

On referring to Mr. Dickinson's table at page 30, it will be seen that the quantity which he determined from actual experiment as penetrating below the surface was 11·3 inches out of a total rainfall of 26·6 inches. This is equal to about 42·4 per cent., or rather less than Mr. Stephenson's calculation. Other observers who have written on this subject have calculated, roughly, in formations less absorbent than the chalk, that streams and rivers carry off one-third of the total rainfall, that another third evaporates and enters into animal and vegetable life, while the remaining third sinks into the earth to form subterranean sheets of water, and breaks out again in the form of springs. Mr. Prestwich, an able and practical geologist, who has distinguished himself by numerous papers of great value, and who has devoted himself to a very interesting examination of the tertiary and cretaceous formations surrounding the metropolis, has given,* as the result of close examination and experiment, the following table, to show the probable amount of infiltration into three of the principal water bearing strata which surround the metropolis :—

	Total mean annual rainfall.	Amount of infiltration.	
	Inches.	Inches.	Per cent.
Lower tertiary sand . .	25	12	48
Upper green sand . . ,	28	10	36
Lower green sand . . .	26½	16	60

Springs which break out on the surface of the ground are caused in various ways : sometimes by simple superposition of a porous stratum on one which is impervious to water, and sometimes by the action of faults.

The annexed figures represent one of the most simple conditions under which springs are met with, where c is a cap of

Fig. 1.   Fig. 2.   Fig. 3.

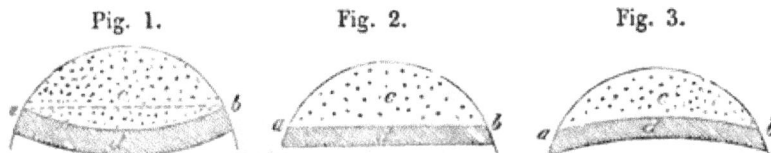

* Prestwich on the Water-bearing Strata of London, p. 120. John Van Voorst, 1851.

porous sand or gravel, resting on *d*, an impermeable mass of clay. In this case a portion of the water falling on the porous surface of the covering will penetrate downwards until it is stopped by the clay, and will break out at *a* and *b* in the form of springs. Where the underlying clay has a basin-shaped or hollow form the water will accumulate in the lower part of the sand to the level of the horizontal line *a b*, and below this line the sand will be permanently saturated with water. A great many of the shallow springs around London arise from the water lodging in depressions filled with porous gravel, which rest on the thick beds of London clay beneath. Hampstead Heath is another example where a mass of porous sand rests on a thick bed of impermeable London clay. At a number of points all round the heath, the water escapes from the sand in springs, and finds its way over the surface of the London clay. There are a great many towns situated in clay districts all over England, where the water is procured, either from springs arising in the drift-gravel lodged in basins and hollows, or from wells sunk into this drift-gravel to a point below the line of saturation. Many of the surface wells of London drew their supply from this source, the water being derived from what were termed land-springs, to distinguish it from the deep well-water lying below the London clay. Many towns on the new red sandstone, as Leicester, Nottingham, Wolverhampton, &c., had shallow wells under the same circumstances. The water procured from wells of this description is commonly very inferior to that drawn from deeper levels.

Figure 4, taken from Dr. Buckland's Bridgewater Treatise,

Fig. 4.

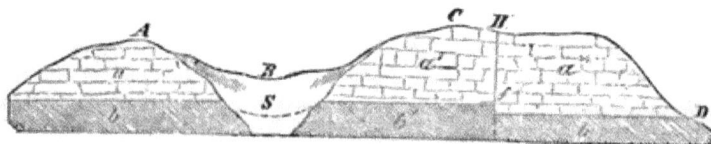

exhibits the origin of two kinds of springs. The valley B in this figure is one in which springs are caused by contact of a permeable with an impermeable stratum, and the hill c is one on the sides of which, as at *f*, springs are caused by the action of a fault. The following description is in the words of the distinguished author of the Bridgewater Treatise. "The hills A, c, are supposed to be formed of a permeable stratum, *a a′ a″*, resting on an impermeable bed of clay, *b b′ b″*. Between these two hills is a valley of denudation B. Towards the head of this valley the junction of the permeable stratum *a a′* with the clay bed *b b′* produces a spring at the point s; here the intersection of these strata by the denudation of the valley affords a perennial issue to the rain-water which falls upon the adjacent upland plain, and percolating downwards to the bottom of the porous stratum *a a′*, accumulates therein until it is discharged by numerous springs, in positions similar to s, near the head and along the sides of the valleys which intersect the junction of the stratum *a a′* with the stratum *b b′*.

"The hill c represents the case of a spring produced by a fault H. The rain that falls upon this hill, between H and D, descends through the porous stratum *a″* to the subjacent beds of clay *b″*. The inclination of this bed directs its course towards the fault H, where its progress is intercepted by the dislocated edge of the clay bed *b′*, and a spring is formed at the point *f*. Springs originating in causes of this kind are of very frequent occurrence, and are easily recognised in cliffs upon the sea shore. Three such cases may be seen on the banks of the Severn, near Bristol, in small faults that traverse the low cliff of red marl and lias on the N.E. of the Aust passage. In inland districts the fractures which cause these springs are usually less apparent, and the issues of water often give to the geologist notice of faults, of which the form of the surface affords no visible indication."

Other conditions under which water occurs, are illustrated in figs 5, 6, and 7. In fig. 5 A is an impermeable cap of clay, resting on a porous bed, B, which in its turn rests on an impermeable stratum, c. The water which falls on the surface of B, and perhaps some of that which falls on A, will sink into

Fig. 5.

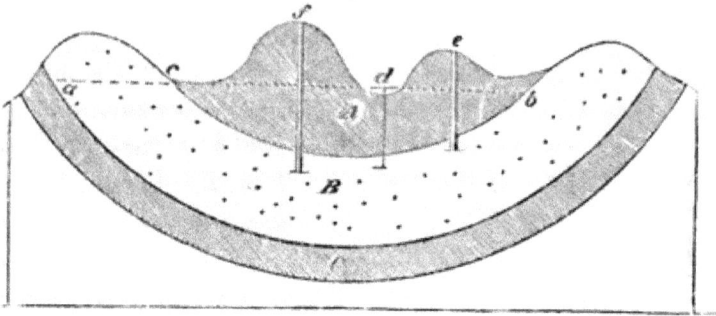

the porous stratum B, and accumulate nearly to the level of
ab, at which level it is drained by springs, breaking out at
c. In wells sunk at e and f, the water will rise to the level
of the line ab; also, in borings made at d, the water will
probably rise through the bore-hole and overflow the surface,
forming what is called an overflowing Artesian well. It is
evident, if the mass A covered the permeable strata to a higher
level than c, namely, to as high a level as the edges of the bed
c, then the line of saturation would correspond with that upper
level—a distinction which will be sufficiently understood by
inspection of fig. 5, without the aid of another diagram.

Fig. 6 represents the case of a basin drained by a river

Fig. 6.

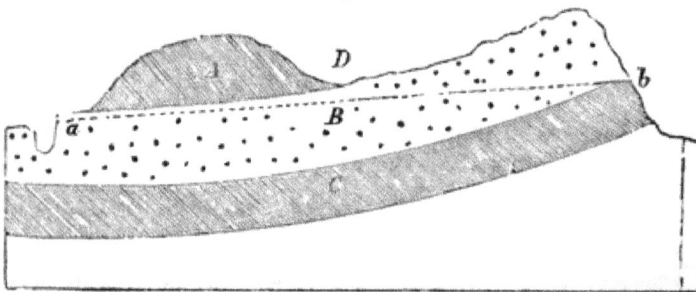

and having an inclined line of saturation. Here A, B, and C,
represent the same succession of strata as in fig. 5. At a

is a river, where the water lodged in B finds the means of escape; and hence the line of saturation and the height to which water will rise in wells becomes the line *a b*, drawn from the outcrop of c to the mean level of water in the river at *a*.

It is evident, if any part of the surface of B should lie below *a b*, then we may expect to meet with springs breaking out on the surface; and so, if any part of the surface of A should lie below *a b*, then we may expect to find overflowing Artesian wells, as in fig. 5.

It is probable that the line of saturation, *a b*, is not invariably a straight line, but in dry seasons is depressed into a hollow curve beneath the straight line, while in wet seasons it swells into a convex curve above the straight line. If we conceive it to swell in wet seasons to such an extent as to cut the surface D at any point to the right of the mass A, we shall have for a time a spring flowing at that point. This is one mode of accounting for intermittent springs, some of which will be hereafter noticed in speaking of chalk districts.

Fig. 7 shows an arrangement of strata which often prevails

Fig. 7.

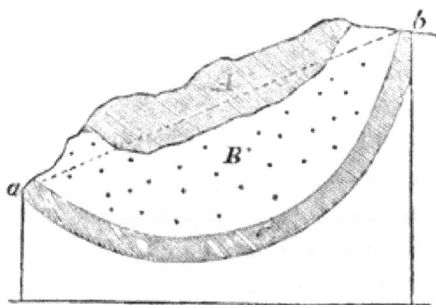

in nature, the impervious mass c cropping out at very different levels, *a* and *b*. Here the line of saturation also will be inclined from *b* to *a*, and at this level the water will stand in wells sunk between *a* and *b*. This explanation is somewhat at variance with that given for fig. 1, plate 69, of Dr. Buckland's Bridgewater Treatise, where in a diagram somewhat similar to this the water level is described to be at the level or a horizontal line drawn through *a*. This would perhaps be the case theoretically, if we conceive the stratum B thoroughly and completely porous, and offering no resistance whatever to the passage of the water. In other words, if we suppose B to be a liquid mass, of course the water will stand no higher than the level of *a*. This is in reality so far from being the case that in all valleys, such as that represented in fig. 7, the water stands at different levels, these being higher as we approach the elevated outcrop and lower as we descend. The observations of Mr. Clutterbuck, with reference to the water-level in chalk wells between Watford and London, completely confirm this view.

SPRINGS CAUSED BY FAULTS.

Figs. 8 and 9 show one of the most common modes of occurrence where the fault x has caused a dislocation of the

Fig. 8

x

Fig. 9.

x

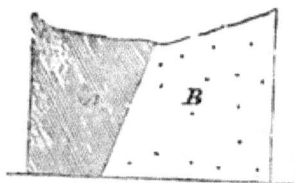 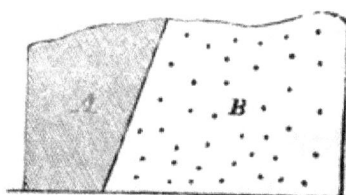

strata, and brought down the impermeable bed A in contact with the porous stratum B. Fig. 8 shows the spring breaking out in the valley at x, but the same effect sometimes takes place near the tops of hills or on high table land as at x, fig. 9, especially if the beds in B dip towards x.

Another class of springs is frequently caused by the fault in pervious rocks being filled up with clay or other matter impermeable by water. Such are many of the faults in limestone districts, as in the carboniferous limestone of Gower, where the faults are commonly filled with clay, which acts as a perfect dam, and throws out the water at the surface. It has been observed by geologists that the occurrence of springs in limestone districts is one of the best indications of the existence of faults. In the carboniferous district of Gower the limestone is traversed by a succession of nearly parallel faults, which range across the limestone at right angles to the coast line. The lines of these faults are invariably marked on the surface by a series of springs breaking out at different levels from that of the sea, up almost to the summit of the country. The lower springs are far the most copious, and some of those near the level of the sea never cease to flow, while those at the higher levels are readily affected in dry seasons, and often cease for months together to yield a drop of water.

Springs arising from faults, unlike those caused by alternations of strata in valleys of denudation, are by no means confined to combes or valleys. On the contrary, they often appear on table lands and other high elevations. The great boundary fault of the Dudley coal field, in the neighbourhood of Wolverhampton, where the magnesian limestone and red sandstone marls are brought down in contact with the coal measures, gives rise to numerous springs almost at the summit of an elevated district along the margin of the coal field. Many of these springs burst up in an almost vertical direction, and may be seen in several cases breaking through the hard surfaces of roads and flowing over into the gutters.

There are numerous other conditions connected with the juxtaposition of strata which give rise to springs. Some of these will be noticed in speaking of the principal water-bearing formations of this country. In a work of this description, however, it would evidently be impossible to go into all the details of the subject. Hence we have been obliged altogether to omit the phenomena of springs arising from many peculiar cases of alternating strata, as well as those arising from the unconformability of rocks and other stratigraphical arrangements, the consideration of which would be more suitable for a purely geological treatise

ON THE SPRINGS OF CHALK DISTRICTS.

The chalk is a formation of great extent, not only in England, but all over Europe, and there is every reason to suppose that at a former period—of course long before the age of history—it overspread nearly two-thirds of the whole of this continent. Its extraordinary composition, due in a great measure to the exuviæ and other remains of entomostraca and other microscopic beings, mingled with white mud, like that of tropical seas, and the *débris* of coral islands, and abounding with large forms of marine organic life, clearly point out the great conditions under which the chalk of England has been formed, namely in the deep and tranquil seas of a former world, swarming with all the rich and varied fauna peculiar to such conditions. To-day we find this great ocean bed of white mud hardened, consolidated, and raised up into dry land, occupying many thousand square miles of territory, and presenting every kind of undulation and irregularity of surface. Even when we come to the outside or escarpment of a chalk district, we find a broken truncated outline, which shows clearly that this great formation once extended much further than its present limits, and tells its own tale in language as plainly as the two opposite chalk cliffs of France and England show that the chalk has formerly been continuous between the two countries. Again, when we see the chalk lost beneath the tertiary sands and clays which cover it everywhere in the neighbourhood of London and in Hampshire, and see it reappear on the other side of the basin, we know perfectly well, independently of the evidence of borings and wells, that the chalk is continuous beneath the overlying strata—these simply reposing in a basin or depression of the chalk, which has not been raised so high as that which appears at the surface. Considering the great extent of the chalk formation, and the numerous towns placed within its limits, comprising the metropolis of Great Britain itself, besides many other populous places of minor consequence, it is evident that the hydrographical conditions of such a formation must be of great importance. We shall therefore discuss at some length the phenomena of springs and wells in the chalk, as these have an important bearing both on questions of present and future water supply.

Mr. Prestwich, in his valuable work on the Water-bearing Strata of London, states the area of the chalk district immediately surrounding the tertiary basin of London at 3794 square miles, but this is by no means the extent of the chalk formation in England. If we take the whole area of the chalk country which extends almost without interruption from Flamborough Head in Yorkshire to near Bridport in Dorsetshire, and with few slight exceptions, and except where covered by tertiary or newer strata, occupies the whole area between this line and the coast, we shall find the great chalk basin of England occupy an area of not much less than 15,000 square miles, or nearly one-third the whole surface of England. A considerable part of this area is, no doubt, covered by tertiary and other deposits; but as the chalk extends beneath these, and even influences materially the hydrographical phenomena which accompany them, it is usual to include the whole space so covered as part of the chalk basin. It is true also that throughout a considerable part of Kent and Sussex there is a protrusion of older rocks coming up to the surface, from which the chalk has been denuded, so as to leave an abrupt truncated escarpment along the whole line of the north downs, and a corresponding one along the line of the south downs. Even deducting this interposed area of older rocks, however, which is termed the Weald of Kent and Sussex, we shall still have for the great chalk formation an area of nearly one-fourth the whole of England. According to a rough calculation the figures would be as follows:

	Sq. Miles.
Whole area of the chalk basin, including the Weald of Kent and Sussex	14,707
Less area of the Weald	2,134
	12,523

The whole area of England, according to the last census, is 50,782 square miles, and that of England and Wales 58,320.

CHALK BASIN OF LONDON.

The hydrographical conditions connected with the great chalk basin of London formed a frequent subject of discussion at the Institution of Civil Engineers in 1842. The Rev. W. Clutterbuck, a geologist of some eminence, residing at Watford, took up the case of the millowners in opposition to Mr. Stephenson's project of conveying spring water from Watford to London. In one of his papers Mr. Clutterbuck describes the line of country through which the river Colne flows. Part of this district, he observes, is covered with gravel through which the rain water percolates and finds its way into the chalk, where it accumulates until it rises and finds vent by the small streams of the Ver, the Gade, the Bulbourn, and the Chess, which are tributaries of the river Colne. Another portion of the Colne Basin is covered by the London and plastic clays, on the surface of which the rain flows in open channels into the Colne, rendering it subject to sudden floods. "In the upper or chalk portion of the district," says Mr. Clutterbuck, "a periodical exhaustion and replenishment of the subterranean reservoir are continually going on." This he has traced through a series of wells, and found to be exactly in proportion to the distance from the river or vent. Mr. Clutterbuck first drew attention to the effect of pumping from the deep London wells sunk into the chalk. He stated that the effect of pumping during the week was to reduce the level of the water to the extent of 5 inches, and that the original level was resumed on Monday morning, owing to the cessation of pumping during Sunday. The alternations of level are somewhat varied by heavy falls of rain, or by extraordinary cessations of pumping; but Mr. Clutterbuck assumes, as a general rule, that the relative heights of water in the wells at some distance from London pointed out and corresponded with the metropolitan holidays.

Mr. Clutterbuck, in another paper, referred to a statement in Conybeare and Phillips' Geology, (Book I., Chap. IV., page 35,) in which there appeared an anomaly in the height to which the water rises in certain wells on the north side of London. Thus, at Mile End, the water stood at the level of high water mark in the Thames; at Tottenham, 60 feet; at Epping, 314 feet; and at Hunter's Hall, 190 feet above that level. Rejecting the Epping well, for the reason which will be given presently, Mr. Clutterbuck stated, that if a straight line be drawn on a vertical section from the level of the water at Hunter's Hall to mean tide level in the river Thames, it will cut the water level in the other wells, and give nearly the average inclination of 10 feet per mile, or something less than that which he had ascribed in his former paper to the water level in the wells between Watford and London. The author rejects the Epping well, because he ascertained on the spot that the water was derived from land springs, and not from the sands of the plastic clay as in the other wells.

In order to prove the depression caused by the London pumping, the author draws a straight line to mean tide level in the Thames from a point 3 miles south of the Colne and 170 feet above high water mark, which is the level of the Colne at Watford.

This line, which would be fourteen miles in length, and would have a total fall of 180 feet, or about 13 feet per mile, cuts the water level at the point where it is drawn at Hendon Union Workhouse, and at Cricklewood between Hendon and Kilburn, from which fact Mr. Clutterbuck infers that up to this point there is no depression of level caused by the pumping from the London wells. At Kilburn, however, the water level is considerably depressed below the straight line so drawn; and in fact the water is known to have stood here some years ago 20 feet higher than at present, so that he attributes this depression to the exhaustion of the water by pumping under London.

Mr. Dickinson, an extensive millowner residing near Wat-

ford, has kept for some years an ordinary rain gauge, and also
one on the principle suggested by Dr. Dalton. This latter
shows the quantity of rain falling on the surface, which sinks
in so far as to be beyond the reach of evaporation, and which
therefore may be calculated to reach the interior reservoir of
the chalk formation, from which its springs and rivers are fed.

Mr. Dickinson has published the results recorded by these
two gauges for a period of eight years (see page 15), namely,
from 1835 to 1843. It appears from his observations that the
average rain fall in these years, as indicated by the ordinary
rain gauge, was 26·61 inches, while the Dalton gauge gave
only an average of 11·29 inches as penetrating beyond the
reach of evaporation. It further appears in these observations, that from April to August inclusive, scarcely any of the
rain descended below the reach of evaporation, whilst the
greatest quantities recorded by the Dalton gauge were usually
in the months of October and November. It is remarkable
how nearly the average absorption shown by the Dalton gauge
agrees with that assumed by Mr. Robert Stephenson in his
report to the Watford Spring Water Company. Mr. Stephenson assumed that 6½ million cubic feet sank into the earth
out of a total rain fall of 14½ millions. This quantity on a
rain fall of 26·61, the average by Mr. Dickinson's gauge,
would give 11·50 inches for the quantity absorbed, whereas
we have seen that Mr. Dickinson's quantity is 11·29 inches.
It must be observed, however, that according to Mr. Dickinson
the *minimum* quantity absorbed in certain years falls very far
short of this amount. For instance, in the year beginning
September 1840 and ending August 1841, the total rain fall
was 25·58 inches, and the proportion recorded by the Dalton
gauge was only 4·67 inches, or less than ⅕ of the total rain
fall. Several of the other years give the proportion of rain
absorbed at less than ⅓, while in four years out of the eight,
the proportion was rather more than half.

If these observations of Mr. Dickinson are to be implicitly
relied on, it would not be safe to calculate on an absorption of

more than 4 inches on the chalk surface in certain dry years, even when the total rain fall is quite equal to an average; and the result of course would be still less if we conceive an extremely warm average temperature combined with a small rain fall of 20 ins., which is by no means uncommon in this part of England.

Mr. Clutterbuck preferred the indications of wells to that of the Dalton gauge, and pointed out several objections and alleged errors in the results recorded by the latter.

The streams which flow off the Chiltern hills, as the Ver, the Gade, the Bulbourn, and the Chess, have their origin in the water which sinks into the gravel beds overlying the chalk, and which being upheld by retentive beds in the chalk, seeks a vent and flows off by those valleys. Mr. Clutterbuck describes the district of these valleys as a reservoir dipping towards the south, at an average inclination of nearly 300 feet in 14 miles, or about 21 feet per mile.

It appears to be the general opinion of Mr. Clutterbuck, the late Dean of Westminster, and other eminent geologists who have studied the subject, that the surface of the subterranean reservoir in the chalk corresponds roughly with a line drawn from the river Colne, at Watford, to mean tide level in the Thames, below London. They assume that in wells sunk to the chalk in the neighbourhood of London, the water will rise to somewhere about the level of this line. There are facts however connected with the variable rainfall at different seasons, and the irregular pumping from the London wells, which materially affect this water level. Thus a considerable replenishment of the reservoir takes place between December and March inclusive, and during these months the water accumulates in proportion to the distance from the vent below London.

During these months the general level of the water line rises above its usual height, and the streams make their appearance at higher points in the valleys. In the season of exhaustion again, which usually takes place between April and November, the water level is depressed, and the streams break out at lower points. This variation of level at different seasons sometimes amounts in the higher districts to 50 feet of vertical height.

The pumping from the London wells produces not only a gradual and permanent general depression of the water line, but even causes periodic changes corresponding almost with the daily extent of pumping. Mr. Clutterbuck asserts that the level is gradually reduced by the pumping during the week from Monday morning till Saturday night; that the cessation of pumping on Sunday is marked by the increased height on Monday morning, and that holiday times, such as Christmas, Easter, and Whitsuntide, may be distinguished. "Thus," says Mr. Clutterbuck, "the measurement of a chalk well in London would show the days of the week and the great festivals by the daily variations; the seasons would be indicated by the average difference in the height of the level at different periods of the year; and the changes of the weather, by the falling of the rain, would also be shown."

Dr. Buckland says Mr. Dickinson's rain gauge showed that during two-thirds of the year the rain which fell rarely sank 3 feet into the earth; but in November, December, January, and February, it passed down into the chalk in proportions which accorded so constantly with the greater or less amount of rain falling in these four wet months, that Mr. Dickinson had been accustomed to regulate the amount of orders undertaken to be executed in his paper mills during the following spring and summer, by the indications on this rain gauge of the quantity of water that descended more than 3 feet in the preceding winter.

Mr. Dickinson said the quantity of rain which penetrated the chalk in the four months from November to February, varied from 6 to 17 inches, the former quantity being sufficient to cause the flow of the principal springs.

ON THE WATER LEVEL OR LINE OF SATURATION IN THE CHALK.

The inclination of this line in different parts of the chalk basin appears to vary with the dip of the stratification, and sometimes presents anomalies which are probably caused by some phenomena of a merely local nature. Mr. Clutterbuck has shown tnat the water line in the district between Watford and London has an inclination of about 13 or 14 feet per mile, and that the inclination in the district north of London is only about 10 feet per mile. Mr. Prestwich quotes some observations made by Mr. W. Bland on the height at which water stands in two lines of wells about 6 miles apart, traversing the chalk district between Sittingbourne and Maidstone. Reducing these observations, Mr. Prestwich finds that on one line the inclination of the water line is 47 feet per mile, and on the other 45 feet per mile, or nearly the same. All these inclinations, both those of Mr. Clutterbuck and Mr. Bland, appear to correspond roughly with the dip of the chalk strata in the respective districts. In some other observations by Mr. Bland, where a difference occurs in the water level of 93 feet and 102 feet in distances of less than a mile, Mr. Prestwich says that these probably arise from some local cause.

DEPRESSION OF WATER LEVEL IN LONDON WELLS.

Mr. Clutterbuck stated, in 1843, that the depression in the centre of London amounted to 50 feet; at the Hampstead Road 30 feet; and at the Zoological Gardens, 25 feet. Mr. Davidson, from observations made in 1822, said the water in ten of the principal wells in London then stood at the level of Trinity high water mark. In 1843, the water did not rise to within 50 feet of that level, thus showing a depression of more than 2 feet per annum.

CHALK WELLS.

The circumstances under which water is met with in sinking near Watford are thus described by Mr. Stephenson, in his report on the water supply. The valley of the Colne is covered by a bed of alluvial gravel (? diluvial) about 20 feet in thickness, and on sinking through this about 5 feet into the chalk, abundant springs of water are met with, which increase in magnitude and force as we descend.

Mr. Stephenson then describes two experiments which he

caused to be made at a well purposely sunk in Bushey meadows. The first set of experiments was intended to show whether the springs which are met with immediately on sinking into the chalk derived their supply from the river, and the result of accurate gauging of the river showed that no effect whatever was produced when the water was repeatedly pumped out of a well 34 feet in depth. He inferred from this that all direct communication between the river and the springs was cut off by the clay bed in which the former flowed.

In the Tring cutting on the London and North-Western Railway, springs were met with which yielded a million gallons per day, and although this part of the country is many feet higher than Watford, the chalk was found to be so saturated with water that it was extremely difficult, and even impossible by ordinary means to sink wells a depth of 60 feet.

Mr. Prestwich is of opinion that water does not circulate through the chalk by general permeation of the mass, but chiefly through fissures. He observes, that if a shaft be sunk into chalk to a depth of say 30 feet below the level at which the water stands, and the water be pumped out, it will be replaced by the abstraction of water from the communicating fissures, and the distance to which these will be affected depends on several circumstances, but chiefly on the head of water on or above the level of the point of draught. In the higher chalk districts the fissures are soon exhausted by pumping, and Mr. Bland mentions an instance in the high chalk district between Maidstone and Sittingbourne where the pumping at one well drained another nearly a mile distant. At lower levels in the chalk district, however, and especially along the boundary of the tertiary area, as at Watford, the head of water is usually much higher than the surface of the ground, and wells sunk here can draw their supply from the whole mass of upland chalk lying beyond, and at a higher level. Hence the effect of pumping in this situation is much less sensibly felt than in the higher chalk districts. Mr. Paten, who was produced as a witness in favour of the Watford supply, before the Committee which sat in 1840, made some special experiments to determine the effect of pumping from contiguous wells in that neighbourhood.

His first set of experiments consisted of sinking twenty borings in a part of the valley at Watford, comprising two miles in length and three-quarters of a mile in breadth. These borings were each sunk to the depth of 134 feet, and the witness declared that he obtained an equal quantity of water from each boring. He then sunk a 12 inch boring in Bushey meadows to the same depth of 134 feet, and this boring yielded 480,000 gallons in 24 hours. His next experiment was made by sinking a shaft 20 feet in diameter and 34 feet deep. This shaft reached the chalk at 21 feet deep, and was sunk 13 feet into the chalk, the last 8 feet being solid chalk. At this depth of 34 feet the shaft yielded two million gallons of water per day of 24 hours. He then made four borings, each of 5 inches diameter, in the bottom of this shaft. The borings being 100 feet deep, the whole depth from the surface was 134 feet, and the whole yield of water was then three million gallons per day.

He further states that there are many wells in the neighbourhood of Watford yielding 400,000 gallons per day, and quoted four borings, which together yielded $1\frac{1}{2}$ million gallons. He seemed to be of opinion that in the neighbourhood of Watford the subterranean sheet of water is so abundant as to be inexhaustible, and that borings do not affect each other even when made at very short distances apart.

The great mass of evidence which has been brought forward of late years goes to show that a large supply may undoubtedly be obtained at those levels in the chalk district where springs now break out in abundance, and where sinkings are made below the line of saturation. At the same time there are many instances of very deep wells being sunk into the chalk through the tertiary strata, and proving either entire failures, or yielding a very small quantity of water. Mr. Prestwich mentions a well at Saffron Walden, which passed through 1000 feet of chalk before meeting with a sufficient supply of water, and it was not till the whole thickness of the chalk was traversed that water was obtained. Most of the large wells at the breweries in London are sunk from 200 to 300 feet into the chalk, and at this depth few of them yield more than a hundred thousand gallons a day. The well in the Hampstead Road, sunk by the New River Company, the well at Southampton, the famous artesian boring at Grenelle, and more recently the one sunk by M. Mulot at Calais, have all failed in procuring any large supply of water from the chalk. At the same time it has been stated, that the aggregate yield of all the wells sunk into the chalk in the neighbourhood of London is not less than 10 million gallons a day. Some doubts have been expressed whether the water of the chalk be distinct from that of the overlying sands of the plastic clay.

Mr. Clutterbuck appears to think there is no real distinction between them, and that the water is all derived from the chalk, from which it rises up into the sands.

Mr. Simpson, on the other hand, considers the waters of the two formations are distinct, as they rise to different levels. He observes, that previous to 1830, there were many overflowing artesian wells on the west side of London, and he believes that in the majority of these the water came from the chalk. (See his paper presented to the Institution, giving an account of 67 wells.)

Fig. 10, which is a section from Tring in Hertfordshire, to Seven Oaks in Kent, will serve to illustrate the principal hydrographical conditions of the London chalk basin. In this section A is the gault clay, an impervious stratum, which underlies the firestone, chalk marl, and chalk. The firestone and chalk marl are not shown in the section, because it is believed in this chalk basin the water penetrates through both of these, and is really not stopped till it reaches the gault. In the south downs, however, it is otherwise, for there Lydden Spout, and other copious springs are thrown out by the chalk marl. B is the great mass of chalk 800 or 1000 feet in thickness. The tertiary or Thanet sands, resting on the chalk, are marked c,

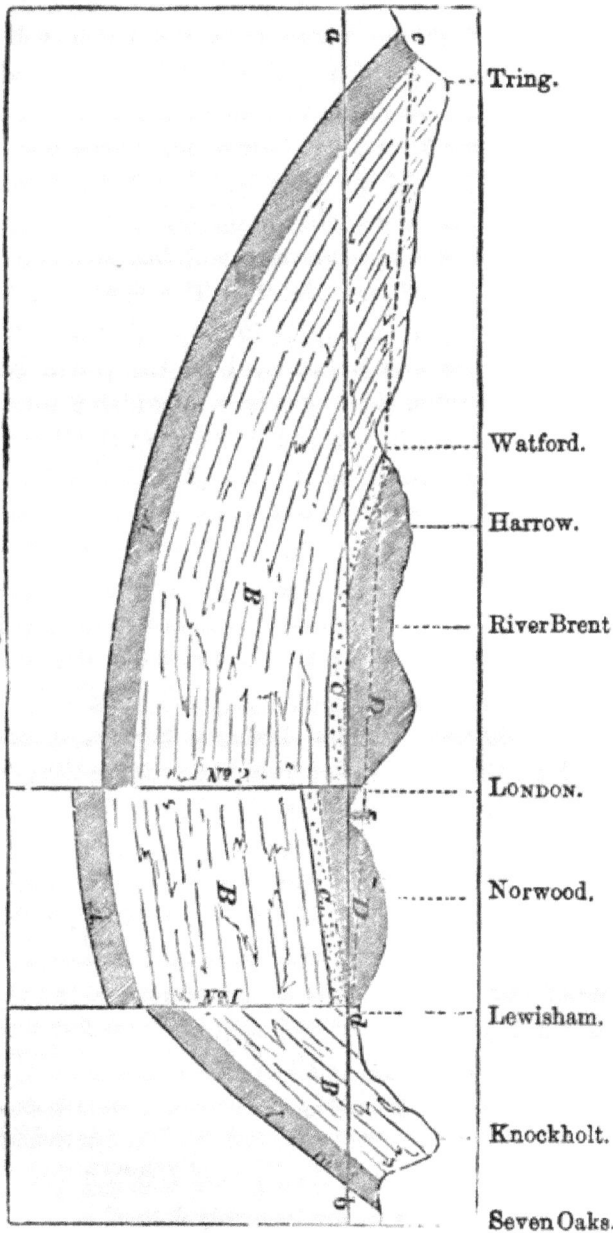

Fig. 10.

Tring.

Watford.

Harrow.

RiverBrent

LONDON.

Norwood.

Lewisham.

Knockholt.

Seven Oaks.

and the impervious mass of London clay is marked D.   The

level of high water mark in tne Thames is represented by the
horizontal line *a b*, and the presumed line of saturation
or height to which the water from the chalk will rise at
any point between London and Tring, is represented by the
inclined line *c d*, drawn from the top of the gault below Tring
to tide level in the Thames at Lewisham, where the chalk is
exposed in the basin of the Thames. It will be observed that
in this section I have not shown the beds in one continuous
uninterrupted basin-shaped arrangement, but intersected by two
faults, marked No. 1 and No. 2. The fault No. 1, which
brings down the London clay D to the level of, and in contact
with, the chalk, is clearly exhibited on the North Kent and
London and Brighton Railways, both of which it intersects at
New Cross, in a north-east and south-westerly direction. It
has been well described by the late Mr. De la Condamine,
in a paper read before the Geological Society of London, in
June, 1850.* With respect to the fault No. 2, although not
exposed at the surface, we have good evidence of its existence from well sections. Thus, the depth to the chalk below
Trinity high water mark at Gray's Inn Lane, the Hampstead
Road, Tottenham Court Road, and the Regent's Park, varies
from 80 to 100 feet; while at Trafalgar Square, Wandsworth,
and Chelsea, the depth varies from 250 to more than 300 feet,
which shows either a fault or a very great curvature of the
strata. Mr. Prestwich believes that this fault or axis of elevation, whichever it be, passes along the valley of the Thames, in
an east and west direction. It is clear that this fault as well
as the one at Lewisham, No. 2, would be intersected by
the line of our section. The main drainage of the chalk formation is not so much interfered with by these faults as might
be supposed at first sight. The line *c d* shows the height to
which the chalk is probably saturated with water, according to the views first promulgated by Mr. Clutterbuck, and
afterwards corroborated by the Dean of Westminster and other

* Published in Vol. vi. of the Quarterly Journal of the Geological
Society, page 440.

eminent geologists. The drainage of the chalk will still take place at *d*, notwithstanding the faults, because the communication between the separate masses of chalk is still uninterrupted, the fault being probably not filled up with impermeable clay and made into a puddle dyke, as happens in some districts. According to the views of Mr. Clutterbuck, the water will rise in wells between Tring and London, to the level of *c d*, and he has found by measurements of numerous wells intermediate between the two places, that the water stands at, or nearly at this height. It will be observed that the ground at Watford lies below the line of saturation *c d*, and this accounts for the numerous springs which break out in the meadows there, and for the fact, that every excavation, made only a few feet in depth, is immediately filled with water. Again, it will be observed, that a part of the London clay district, close to the metropolis, lies below the line of saturation. This is precisely the condition under which artesian wells may be expected to yield a stream of water that will overflow the surface. On boring down through the London clay, D, to the chalk on either side of the fault No. 2, we come to water which is acted on by the pressure from Tring or Knockholt as the case may be, and which, as soon as the boring is effected, rushes up through it and rises above the surface namely to the line *c d*. This is the explanation of many overflowing artesian wells in the neighbourhood of Fulham, Brentford, and other places in the valley of the Thames.

The section, fig. 10, differs somewhat from that given by the Dean of Westminster, in his celebrated Bridgewater Treatise, where the line of saturation is drawn horizontally at the level of the low ground at Watford: and in the text of that work it is said that the water of artesian wells will rise to this level. I prefer, however, the view taken by Mr. Clutterbuck, and which I have here ventured to follow, and I do this without implying the smallest disrespect for the opinion of so eminent a geologist as the Dean of Westminster, because I understand that in the discussions which took place in 1842—3, on the views and opinions of Mr. Clutterbuck, the Dean expressed his agreement with him on this point.* The faults which I have ventured to add in the section have only been brought to light within the last few years.

FAULTS AND DISTURBANCES OF THE CHALK DISTRICT, INFLU-
ENCING THE PHENOMENA OF SPRINGS AND THE HEIGHT OF
WATER IN WELLS.

The chalk basin of London appears to be intersected by two principal lines of disturbance, one of which has nearly an east and west direction, and follows the course of the Thames from the Nore to Deptford.† The other is supposed to be nearly at right angles to this passing across the Thames near Deptford, and ranging nearly north and south up the valley of the Lea, towards Hoddesdon. The North Kent and London and Brighton Railways cross these lines of disturbance near their intersection at Lewisham, and exhibit a curious series of small faults in a direction from north-east to south-west, or nearly in a diagonal direction with the main lines of disturbance.

The general effect of these faults is to bring down the tertiary sands, and sometimes even the London clay, to the same level as the chalk. This will be understood from fig. 11, which represents a section taken across the fault from west to east, in the neighbourhood of Lewisham or New Cross. A similar section, from north to south, across Blackheath, shows the chalk below Blackheath and Greenwich Observatory, in contact with the London clay under Greenwich marshes.‡

Fig. 11 explains the influence of this fault, on the height at which the water will stand in wells sunk down through the

---

* I find the Dean saying, in the discussion at the Institution of Civil Engineers, May 31st, 1842, "Mr. Clutterbuck's repeated observations upon wells along the line in question must be considered to have proved the existence of this *inclined* line."

† Prestwich, p. 40.

‡ See a very interesting paper by the Rev. H. M. De la Condamine, M.A., on the tertiary strata and their dislocations in the neighbourhood of Blackheath. Vol. vi. Quarterly Journal of the Geological Society, p. 440.

London clay. A B is a line representing the fault, on the west side of which the impervious mass of the London clay is thrown down in contact with the porous beds of tertiary sand and chalk, so that the water in these sands will stand at the level F on the east side of the fault, and not higher than D on

Fig. 11.

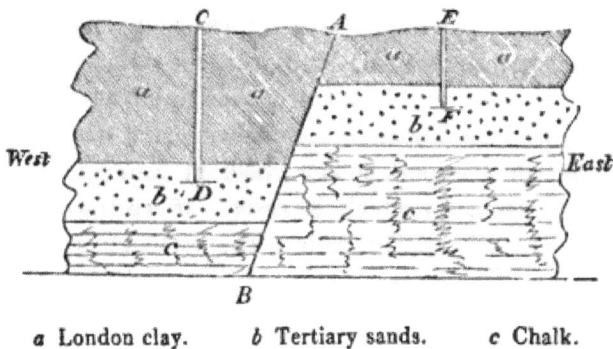

a London clay.   b Tertiary sands.   c Chalk.

the west side: the difference between the depths of E F and C D will show the different level to which the well will require to be sunk in the two cases.

The wells in Essex, even on the east side of the fault, are commonly of great depth, but this is principally owing to the general elevation of the country, which renders it necessary to penetrate through a great thickness of the London clay, some-times nearly 400 feet, before the same sands can be reached which, at Tottenham and Mile End, are found from 50 to 100 feet below the surface, and which at Lewisham, Charlton, and Woolwich, are actually at the surface.

At Bow, Stratford, and West Ham, the water stands in wells at about the height of high water mark in the river Thames, while in many parts of Essex it stands as much as 330 feet above that level. It is not probable that the whole of this great difference in elevation is caused by the fault, although some portion of it may be so accounted for. The tertiary sands are not uniform in composition over extensive areas, and it is possible that the flow of water through them may be im-peded by interposed lenticular masses of clay, and thus many of the anomalies in connection with the Essex wells may, in some degree, be explained.

In addition to the fault which has been already alluded to as traversing the London district in a north and south direction, there is another line of elevation passing east and west up the valley of the Thames, which appears either to have disrupted and broken through the strata beneath the London clay, or to have bent them into a saddle-shaped form, and brought them up within a comparatively small depth from the surface. This line of disturbance, however, does not appear to have influenced the hydrographical conditions of the chalk district nearly so much as the position occupied by the chalk in connection with the impervious strata on which it rests. If we trace the whole line of the chalk formation, both in the north and south downs, in Wiltshire, Hants, and again, in Hertfordshire and Cambridgeshire, on the north side of London, we shall find a remarkable difference in its hydrographical aspect. For instance, along the whole range of the north and south downs, if we except the insignificant streams of the Ravensbourne and the Wandle, the uniformly arid nature of the surface is not varied by a single streamlet, nor is a spring to be found any where on the surface of the chalk, except where the lowest marly beds give rise to springs at a level far below the general surface of the downs. But in the great development of chalk in Wiltshire and Hants, and, in fact, more or less throughout the whole of the chalk country on the west and north sides of the London basin, streams of all sizes rise from the chalk and flow over its surface.

This remarkable difference is due, no doubt, to the relative elevation of the chalk and of the gault formation which passes entirely beneath it, and is only separated from the lower chalk by a comparatively small thickness of firestone or upper green sand. Whatever may be the retentive capacity of the lower marly beds in different varieties of chalk, and at different parts of the range, it may be laid down as a general rule that water, in any considerable quantity, is never met with much above the level of the gault, which serves, in fact, as a barrier or dam to keep up to its own level the water that has penetrated the chalk. Hence, if the arrangement of these strata in any district be such that a large surface of chalk is greatly elevated above the gault, then we shall find a dry and parched surface on which the water no where seeks an open channel, until it arrives at the level of the gault. Below this level the porous beds of the chalk and the upper green sand are already saturated with water, and therefore the new supplies furnished by rain and snow, after having filtered through the chalk down to this level, will either overflow or force out the water which lies beneath, and already fills up the crevices and hollows of the chalk. Now, in the elevated ranges of chalk, the position in which this effect takes place is not less than from 300 to 400 feet below the general level of the downs, so that in such districts the area of chalk country lying below the level of the gault is in extent quite insignificant, consisting of a few hundred yards in breadth on the outcrop side of the chalk summit, and of a few narrow valleys which run up from the tertiary surface into the chalk country. Throughout the greater part of the elevated ranges of chalk, these are the only portions which are irrigated by natural streams. The great mass of the chalk, composing both the north and south downs, in Surrey, Kent, and Sussex, is therefore quite dry, and it is found that borings in the higher part of the range have to pass through many hundred feet before a drop of water can be obtained.

But in the less elevated districts of chalk, as in the counties of Bucks, Herts, and Cambridge, the level of the gault is not remarkably below the general level of the chalk, which in this part of England is nowhere distinguished by the bold mountainous features which prevail in the south-eastern counties. The chalk of Hertfordshire, in particular, is traversed by numerous valleys which are watered by copious and constantly flowing streams; a peculiarity which no less than its main physical features distinguishes it in the eyes of the agriculturist from the dry and barren downs of the south. The valleys in which these streams flow are usually not much below the level at which the gault crops out, but are probably points at which a line joining the outcrop of the gault and the permanent line of chalk drainage cuts the surface of the ground, as explained in the description of fig. 10; on the north and west sides of London it appears to be the gault which keeps up the water, but in the south downs the same office is performed by the chalk marl.

It is remarkable that the streams which rise in the chalk of Herts and Bucks, as the rivers Verulam, Colne, Gade, &c., together with those of the Chesham, Amersham, and High Wycombe valleys, all flow in the same direction as the dip of the chalk, whereas the small streams which rise in the Surrey and Sussex downs flow off the chalk and in a direction contrary to its dip.

It will appear obvious that borings, which penetrate the chalk of Hertfordshire to a very small depth, namely, to the level of the valleys in which streams already exist, will meet with water in abundance; for in such borings the water will evidently stand at the same level as that of the streams which now rise in the country. In the neighbourhood of Watford and Bushey main springs of a very copious character are met with on sinking a few feet below the surface; and at higher levels the water is found, even before sinking to the level of the main springs or point of saturation, because the water is intercepted in the course of flowing towards that level of saturation. In some works recently executed at Watford for supplying the town with water, under the direction of Mr. Pilbrow, C. E., a well of 9 feet diameter has been sunk about 30 feet into the chalk, and a considerable pumping power was found necessary to keep down the main springs during the progress of the works.

Stephenson's experimental well at Watford yielded 1,800,000 gallons per day. And the excavation of the London and North-Western Railway at Cow Roost, which cut through the gault and thus gave a passage to the water from the chalk in a north or north-west direction, had, during the execution of the work, a considerable stream flowing at the foot of the slope on each side of the cutting. These streams, which are now arched over and conducted in culverts, are said to have yielded 1 million gallons per day.

Streams and rivers in a chalk district are commonly more equal and uniform in volume than those in clay districts, and are not so liable to be swollen by floods as the latter. It is in consequence of this, and of the more absorbent quality of the chalk, that the streams are comparatively much fewer in chalk districts. Not only are the streams much less in number, but the bridges and culverts made to carry off water in the chalk districts, present a most insignificant area compared with those in clay districts of similar extent. This fact, which I believe was first noticed by the Dean of Westminster, was used with very good effect by Mr. Homersham in preparing a sort of popular comparison between districts of clay on the one hand, and chalk on the other.

According to Mr. Homersham's measurements, it was found that the water-way of arches in clay districts varied from 8 to 17 superficial feet for each square mile of drainage area, while in chalk districts the water-way varied from $\frac{1}{3}$rd of a foot to 2 superficial feet per mile of drainage area. Mr. Homersham gives one example, namely, a part of the river Blackwater having a water-way at Coggeshall, in Essex, equal to 3 square feet per mile of drainage area. This, however, is not a fair sample of a clay drainage, as this river flows over a considerable extent of the crag formation, where it very thinly covers the chalk, and some part of its upper course is even through chalk. Mr. Homersham has similarly quoted against himself the water-way of several rivers whose course is partly in chalk and partly in clay. On a careful review of his table, the least water-way which I can find in a pure clay drainage is about 8 feet per mile, ranging up to 13 and even 17 feet; while the water-way in chalk districts varies from $\frac{1}{3}$rd to $1\frac{1}{2}$ square feet, except in the case of the river Beane, the drainage area of which contains a large proportion of impermeable drift. In this case, the water-way has an area of 2 square feet per mile of drainage.

It is probable that the most copious single springs of the chalk do not yield more than from 3 to 500 cubic feet per minute, as will be seen more particularly from a table of these springs in the Appendix. At the same time, it is certain that streams flowing through chalk districts receive a gradual accession of water, which can only be due to invisible springs breaking out in the beds of the rivers, and therefore not capable of being seen or separately measured. The facts in the following statement of gaugings, relating to the river Lea, are on the authority of Mr. Beardmore, by whom they were supplied to Captain Vetch, and given in his evidence before the Board of Health in 1850, and before Sir James Graham's Parliamentary Committee in 1851.

Tables showing the increase of the river Lea from invisible springs in the bed of the river.

	Cube feet per minute.
Gauging at Field's Mill after deducting the proportion brought down by the Stort and the Ash	6,444
Gauging at Ware Mill, 5½ miles above Field's Weir after deducting the quantity abstracted by the New River	5,344
Increase from springs between Ware Mill and Field's Weir	1,100

Again, the four rivers which meet at Hertford, namely, the Lea, the Beane, the Rib, and the Mimram, were all gauged at points very little above the common meeting place, the most distant being the gauging of the Mimram at Panshanger, 2 miles above its junction with the Lea.

	Cubic feet.
The gauging of the Lea, after receiving all the above streams below Ware Mill, is	6,594
The separate gauging of the four streams	6,159
Increase from springs principally between Panshanger and Ware Mill	435

The above gaugings, which, I presume, were all accurately

made without any interruption in point of time, show that between Panshanger and Field's Weir, a distance of about $7\frac{1}{2}$ miles, there is an increase of 1535 cubic feet per minute from invisible springs, in addition to from 3 to 500 feet, the volume of the Chadwell spring, which breaks out at a sufficient height above the river to be measured with accuracy.

DISCHARGE OF CHALK STREAMS.

Mr. Telford, in 1834, found the volume of the Verulam at Bushey Hall, a little above Watford, equal to 1800 cubic feet per minute, and that of the Gade equal to 2520 cubic feet.

The volume of the Wandle, with a drainage area of 35 square miles, as stated in the evidence before the Parliamentary Committee in 1852, was 2693 to 3000 cubic feet per minute; but this is the volume at Wandsworth near its junction with the Thames.

Captain Vetch, who has investigated the subject of chalk rivers with some attention in reference to the supply of London, caused a contour line of levels to be traced at a height of 148 feet above high water mark, and proposed to abstract the streams at this height, with a view of bringing them into London at a sufficient elevation to supply the high service.

The following table gives the result of Captain Vetch's gauging of chalk streams at this level :—

	Miles in length flowing through chalk.	Cubic feet per minute.
The Verulam	19	2080
The Gade	14	2770
The Chess	9	1390
The river Darenth, near Shoreham, including the springs at Orpington		2080

Mr. Beardmore has gauged with great care all the chalk streams uniting to form the Lea at Hertford, and has added the drainage area of each as follows :—

	Cubic feet per minute.	Square miles of drainage.	Cubic feet for each square mile of drainage.
Lea proper at Horris Mill	.	2096	112	18·71
The Beane at Molewood	.	1483	83	12·42
The Rib at Ware Park	. .	959	61	14 34
The Mimram at Panshanger	.	1532	29·3	52·39

The following gauges are by others :—

		Cubic feet per minute.	Square miles of drainage.	Cubic feet for each square mile of drainage.
Lea at Lea Bridge	. . .	8880	570	15·58
Wandle below Carshalton	.	1800	41	43·9
Verulam, Bushey Hall	. .	1800	120·8	14·9
Gade, Hunton Bridge	. .	2500	69·5	8·19

There is something very remarkable in the great discharges of the Mimram and the Wandle, which amount to three times as much per square mile as the other chalk rivers. This has not been satisfactorily explained, nor does it appear clear whether this disposition is due to the drainage areas of these streams being less absorbent and more covered with impermeable beds than usual, or whether it be due to faults or lines of disturbance traversing the chalk districts. The Wandle flows entirely through a clay country, and can only be called a chalk stream, in consequence of the supply which it receives from the chalk springs at Croydon, Beddington, and Carshalton.

There seems to be evidence of a peculiar line of disturbance throughout the whole chalk district between Saffron Walden and Maidenhead. A line joining these two places will coincide with some very remarkable reaches of valley which lie in this north-east and south-west direction. I may instance that part of the river Rib between Widford and Hertford, the peculiar course of the river Lea, between Hatfield and Ware, and the whole course of the river Colne from Hatfield to Watford, in which Captain Vetch states there are many swallow holes. This subject requires more minute investigation than it has yet received, and may possibly have something to do with holding up the water in the valley of the Mimram, and producing the extraordinary discharge at Panshanger.

Fig. 12.

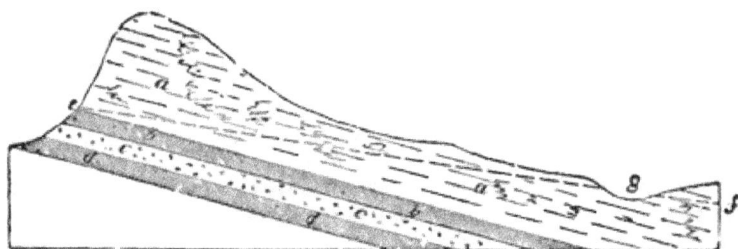

Figure 12, representing a section across the chalk escarpment as at Merstham, in Surrey, shows one set of conditions connected with the chalk springs. Here *a* represents the chalk, *b* the chalk marl, which throws out powerful springs at Lydden Spout, and many other places along the south-coast; *c* is the upper green sand or firestone, *d* is the gault clay, and the dotted line *e f* shows the line of drainage or subterranean flow of the water, held up by the chalk marl and saturating the lower beds of the chalk. In such a case as that represented by this diagram, no springs will be found at *e*, because the water will simply have its subterranean course in the direction of *f*. Springs will probably break out however at *g*, which represents a point where the water line cuts the surface of the ground, or, in other words, where the surface of the ground is below the water line. Few great chalk basins have their edges or circumference entire, but are commonly interrupted and broken through by seas or great rivers, which drain alike the waters on the surface and those which flow in subterranean sheets. It follows from this, that the line of saturation *e f* is not horizontal, as in an ordinary basin, like that in which rests the sand of Hampstead Heath, but has an inclination towards the level at which it is drained into the sea. The inclination of this line has been carefully ascertained on the north side of London by the Rev. Mr. Clutterbuck and others, who have determined, from observation of the wells, &c., and the heights at which the water stands in these, that the *water level*, as it has been latterly improperly termed, or the line of saturation

has an inclination of about 1 in 400, or 13 feet per mile.
In other parts of the chalk district, as shown in the preceding pages, the inclination of this line is probably not
less than 40 feet a mile. Referring again to the diagram,
the point *g* marks the position of numerous places both
in this country and in France, where the finest springs break
out in great abundance. It is probable that the springs of
Croydon and Carshalton, which feed the river Wandle on
the south side of London, as well as those of Watford and
other places on the north and west, have their origin under
these circumstances. The towns of Guines and St. Omer, in
the Pas de Calais, have springs of great power and abundance,
rising in the same manner and under the same circumstances.
In nearly all parts of this country the chalk presents a section

Fig. 13.

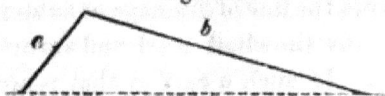

of the kind shown in fig. 13, *a* being the short slope or escarpment side, and *b* the long slope or direction in which the
strata dip underneath, and the ground naturally inclines on
the surface. A great number of towns, situate on the long
slope of the chalk, derive an abundant supply of spring
water, because they occupy the position represented by *g* in
fig. 12. Amongst these in the London chalk basin may be
mentioned St. Alban's, Hemel Hempstead, Chesham, Watford, Amersham, High Wycombe, and others north and west
of London. In a similar position are Andover, Winchester,
Croydon, and other places situate on the long slope of the
north downs. In the chalk wolds of Lincolnshire we have
Louth and Alford occupying a similar situation, and possess-
ing similar copious springs; while along the line of the south
downs we find Fareham, Emsworth, Lavant, Chichester, and
other places equally remarkable for copious chalk springs.

Fig. 14.

Fig. 15

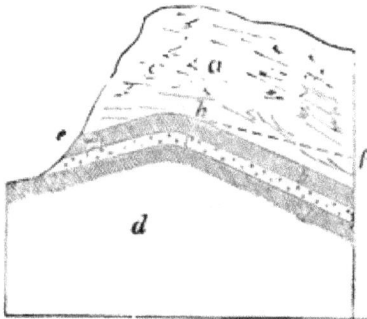
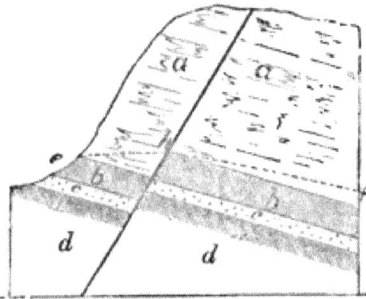

Figs. 14 and 15 represent sections across a chalk escarpment, illustrating the mode in which springs occur on the short or escarpment slope of the chalk. In fig. 14, the chalk marl is shown bent upwards at h, in such a manner that the water falling on and sinking into the porous chalk a, will flow backwards and break out in springs at e. In fig. 15, the chalk marl is faulted at h in such a manner that the water will be retained in the water-tight depression between e and h until it accumulates and breaks out in springs at e, as before. Wherever springs break out under the chalk escarpment, at the top of the chalk marl, as at Lydden Spout, Cheriton, and many other spots along the line of the south downs, they are probably caused by the configuration shown in one or other of these sections.

The same effect of a spring at e will be produced if we conceive the chalk marl to be simply compressed laterally, as shown in fig. 16. The ridge there shown at x, provided it be at a higher level than e, will produce a spring at e in the same manner as the actual rupture of the strata exhibited in fig. 15.

INTERMITTENT SPRINGS.

There is yet another class of springs in chalk districts, which it may be desirable to notice. These are the intermittent or varying springs, such as the Ravensbourne and others of the Surrey downs.

D

This stream of the Ravensbourne breaks out about six miles south of Croydon, in the valley along which passes the high road from Croydon to East Grinstead. The point where it first appears is in a flat part just below Birchwood House. This point is situated between the Half Moon Inn at Caterham Bottom and the inner entrance to Marden Park. Its appearance at the source is indicated by a series of small jets, none of them more than a quarter of an inch in diameter, but so numerous that in the course of 20 yards the water from these jets is sufficient to form a stream; and this increases in size till, at Caterham Bottom, it may almost be called a river. The point at which the stream first appears is about 350 feet above Trinity high water mark, and it falls towards Croydon at the rate of about 36 feet per mile.

The Bourne broke out in the early part of 1837, and on the 16th February, 1840; on the latter occasion with unusual force, filling not only its usual channel in the Riddlesdown valley, but overflowing the adjacent land for some distance. It appears that in the neighbouring quarries south of the Bourne source, water mostly appears in the autumn previously to the eruption of the Bourne, so that from the state of their quarries the workmen confidently predict the flowing of the Bourne. It is said the works on the Brighton Railway, especially the tunnel driven through the chalk at Merstham, four miles west of Caterham valley, were much retarded by water at the time the Bourne stream broke out.

Mr. Prestwich explains the phenomena of the Bourne, the Lavant in Hampshire, and other intermittent springs, by a section of the kind shown in fig. 16, in which the chalk marl is bent into an undulating form, having a saddle back at x, and a depression between x and y. When the depression between x and y becomes filled up, the water will flow over the summit x and be discharged at s, which opening acts like the longer leg of a syphon. The discharge will continue till the reservoir between x and y is exhausted, when the flow of the spring will stop till the reservoir is again replenished.

Fig. 16.

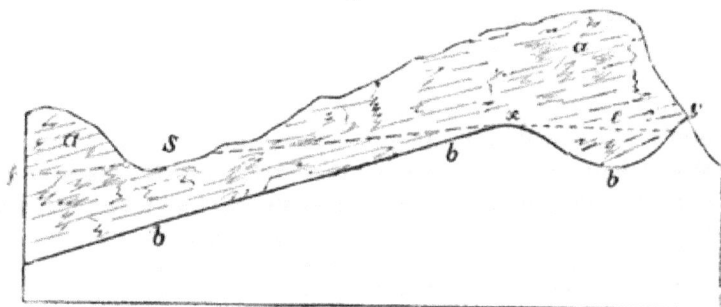

In fig. 16, b represents the top of the chalk marl, and $e f$ the line of saturation, when the basin between x and y is full, that is, when the spring is flowing. It will be observed that, in order to produce a spring at S, the level of the ridge at x must be below the level of y, the outcrop of the chalk marl. If it were otherwise, namely, if y were the lowest, we should have a spring at y, as in fig. 14. A fracture of the strata, as shown in fig. 15, would produce the same effect of a spring at y, if the point x were higher than y, but not otherwise.

It is probable that along the range of the chalk escarpment the various conditions, exhibited in figs. 11 to 16, are all to be met with at different points, as, for instance, in adjacent valleys, where they influence the hydrographical phenomena each in its own peculiar mode.

It is remarkable that nearly all the rivers which flow across the chalk district, whether they rise on the chalk itself, or whether they rise beyond and break through the escarpment, have their course in the same direction as the dip of the chalk beds. The only exceptions are to be found in some branches of the Cam, and other small streams in the north-eastern district. These, however, will be found to flow on a part of the chalk where the drift accumulations are very thick, and where the chalk is entirely obscured by these.

The general course of the streams in chalk districts being in the same direction as the dip, has given rise to the expression, that the subterranean flow of water follows the same direction as that of the surface. This is not true of all formations, however, the fact being frequently the reverse; for instance, in the mountain limestone, the rivers generally flow through great chasms in a direction opposite to the dip of the beds, which are commonly very highly inclined. In other words, the beds usually dip up stream instead of down. Magnificent examples of this are to be seen in the defile of the Avon below Bristol, and on a smaller scale, in the streams of the Gower district, beyond Swansea. I may mention a stream at Bishopton, in the tract of Gower, which rises in the coal measures, and, after flowing some miles in a winding course, comes suddenly up to a limestone district, in which the beds pitch as steeply as the roof of a house, and appear at some distance to oppose a solid barrier. On a nearer approach, however, a large cleft or chasm is visible through which the stream dashes with great rapidity. Thence it pursues its course through a deep hollow glen, all the way to the sea in a southerly direction, the beds all the time dipping to the north. It is evident when rivers run in a direction opposite to that of the dip of inclined strata, that the subterranean flow of water must be in an exactly opposite direction to that of the surface flow.

UNIVERSALITY OF THE CHALK FORMATION.

Viewed in relation to the entire surface of the earth it is held by geologists that the chalk has at least covered all those parts lying within the extreme limits at which it now appears. In fact, we seem to have little more left either in this country or the Continent, than the broken irregular edges of great basins which were once continuous with each other.

In France we have the edges of what has been called the Pyrenean Basin, the Mediterranean* Basin, and the Anglo-Parisian Basin. The chalk of England, from the coast of

* Cours Elementaire de Paléontologie et de Géologie Stratigraphique, par M. Alcide D'Orbigny, tome 2, fasciculus 2nd, p. 571.

Dorsetshire to the Humber, with its continuation into York-shire, forms the north western extremity of the Anglo-Parisian basin, while the chalk hills of the north and south downs are merely ridges inside the basin, raised up by the elevatory movement which has brought up the jurassic,* or neocomian strata of the Weald, within the area of the true chalk basin.

Besides its immense development in France, Spain, and the Mediterranean, in the three great troughs or basins which have been mentioned, particular members of the cretaceous formation are found in many other parts of the world. They occur in Belgium, Holland, Prussia, Westphalia, Hanover, Saxony, Bohemia, Poland, Sweden ; also in Mingrelia, Circassia, Georgia, the Caucasus, Bulgaria, Servia, Wallachia, Transylvania, Gallicia, Volhynia, and Podolia. Immense surfaces of chalk extend across Russia from Poland to the Ural Mountains. In North America the chalk formation extends over 35° of latitude, from Texas to the eastern part of New Jersey. In South America the chalk has been observed in New Granada, in Peru, in Chili, and the Straits of Magellan. In Asia it exists in Pondicherry and in Java.

In the western hemisphere, therefore, we find the chalk existing in New Jersey at 35° of north latitude, and extending to the farthest extremity of South America, or 53° of south latitude. In Europe it extends still further to the north, or nearly as far as 56° of north latitude ; thus clearly showing, by remains which still exist, that this great formation has been at one time almost universal over the whole surface of the earth.

---

* There is some difference of opinion among geologists as to whether the Weald clay and Hastings sand are to be considered as neocomian, and therefore part of the cretaceous series, or whether they are to be ranked with the jurassic or oolitic series.

DEPOSITS ABOVE THE CHALK IN THE LONDON BASIN.

These are, in ascending order, the lower tertiary clays, sands, and gravel beds, commonly called the plastic clay series, and next the thick mass of the London clay, above which are the sands of Bagshot and Hampstead Heath. The surface of the London. clay and that of the lower tertiary beds, is frequently overlaid by a thick stratum of diluvial gravel, which will be noticed hereafter.

The lower tertiary beds consist commonly of alternating series of sands, clays, and gravel, resting on the chalk all round the edges of the basin and passing beneath London with a thickness varying from 35 to 100 feet. In the western part of the basin, as about Reading and Uxbridge, this series of beds contains a large per centage of argillaceous matter, and is well entitled to the appellation of plastic clay. Beneath London, however, the mottled clays of the west are replaced in a great measure by sand and gravel beds, while at the eastern extremity of the basin, as about the Reculvers, nearly the whole mass consists of sands, which here assume such a considerable thickness that Mr. Prestwich proposes to confer upon them the distinguishing title of the "Thanet sands." Proceeding upwards from the chalk, Mr. Prestwich gives the following as an average section of the lowest tertiary beds beneath London :—

20 to 50 feet of light-coloured siliceous sands,
15 to 45 feet of sands, mottled clays, and pebble beds, very
  irregularly stratified,
1 to 3 feet of sands, pebbles and shells,
_____
36 to 98 ;

above this is the London clay.

Mr. Prestwich, who has taken immense pains to investigate the hydrographical conditions of the water-bearing strata around London, proposes to divide the tertiary basin into four distinct parts, by means of two lines crossing each other nearly in the direction of the disturbing lines which have been already mentioned as intersecting the chalk basin. One of these lines passes in a direction due north and south, a little on the east side of London, namely, in the valley of the Ravensbourne and through Hoddesdon and Waltham Abbey, in the valley of the Lea. The other passes in a true east and west direction immediately south of London, through Windsor, Brentford, and Woolwich, each line being continued to the outside of the basin or into the sea. These lines intersecting each other at Lewisham, divide the tertiary area into four parts, which Mr. Prestwich distinguishes by their position as the North-West, the North-East, the South-West, and South-East divisions. At a great many points in each of these divisions, Mr. Prestwich has examined and measured sections of wells, pits, cliffs, &c., showing the thickness and composition of the various beds composing the lower tertiary series. From these data he has made very accurate estimates of the relative thickness of clays and sand in every part of the basin. The outcrop of the beds has also been investigated with much care, as well as the area or surface of country which they occupy. The following table contains a summary of the information which Mr. Prestwich has collected on these heads :—

DIVISIONS.	Total extent of tertiary area.	Lower Tertiary Strata.		
		Extent of area.	Length of outcrop.	Thickness of permeable portions.
	Square Miles.	Square Miles.	Miles.	Feet.
North-Western or Watford* Division .	345	50	60	15
North-Eastern or Chelmsford* Division	1524	64	95	36
South-Western or Epsom* Division .	741	45	130	22
South-Eastern or Rochester* Division .	1524	64	95	36
	4134	223	380	

From this table it will be seen that although 4134 square

* These names of towns, situated nearly in the centre of each division, are merely added to aid the memory in identifying, without the aid of a map, the different divisions here indicated.

miles of country in the London chalk basin are covered with
tertiary strata, yet only 223 square miles consist of lower beds
partially permeable by water. Mr. Prestwich is of opinion
that the two great lines of disturbance which have been mentioned, prevent the free communication of water between one
part of the district and another, so that wells sunk in one division will not draw the water from the other. Taking then
the area of lower tertiary strata in any one district, we have still
to consider that only a certain part of this consists of sand, and
that a large mass of it in every part of the outcrop is composed
of clay. The surface also, even where the beds outcropping
consist of sand, is frequently covered with argillaceous drift
deposits, which impede and often altogether prevent the infiltration of water. It is probable, also, that the sands, even when
they crop out at the surface, may pass into clays, and thin
out into lenticular masses, through which water will permeate
with great difficulty. Mr. Prestwich points out several other
disturbing causes which will interfere with the flow of water
through these beds ; and on the whole it appears, looking at
their small area as compared with the immense surface occupied by the chalk, that these sands present much less prospect
of affording a large supply of water.

At the same time, owing to the considerable elevation at
which the sands crop out, especially on the northern side of
the basin, it is found on sinking wells into them through the
London clay, that the water bursts up with considerable force,
and rises to a considerable height in the well. The bottom or
basement of the London clay is frequently indicated by a hard
pebble bed, sometimes only a few inches and sometimes several
feet in thickness. In places where this bed is argillaceous and
serves to keep down the water, it frequently bursts up with
violence when the boring tools first penetrate through it. This
is also the case in passing through the beds of tabular septaria,
which are commonly found in the lower part of the London
clay. The same thing has been observed by borers in passing
through the bed of green coated flints which usually separates

the chalk from the tertiary beds; and, again, a similar upward rush of water is met with in breaking through the tabular masses of flint which occur in the upper white chalk.

The springs which break out from the chalk at Chadwell, Watford, Croydon, and other places, owing to the overflow of the line of saturation have been described at page 48. The tertiary sands when denuded of the London clay, yield springs of a similar description due to the same cause, namely the level of the ground at certain points being below the general line of saturation. The springs on the south side of Peckham Rye Common, and others which feed the Peckham branch of the Grand Surrey Canal, are due to this cause; and many similar springs are to be met with in various parts of the tertiary area. The springs which break out at the foot of the sand hills between Greenwich and Woolwich, and flow across the marshes to the river Thames, are probably caused by the argillaceous loamy alluvium of the marshes, abutting against the sand and keeping up the water till it breaks out and flows through the marsh ditches.

The river Thames has probably at one time flowed at the very base of these sand hills, but in process of time has altered its course and filled up the marshes with alluvial detritus. The water from the sand having now to find its way across these marshes appears in the form of springs, whereas formerly the river probably drained the sand directly and immediately without the action of water-courses and springs at their heads. All rivers naturally drain the lands through which they pass, whether these lands consist of the older stratified rocks or of mere drift deposits. When excavations are made near the margin of a river flowing in diluvium, it is sometimes difficult to say whether the water which accumulates is derived from the river or is due to its percolation through the drift in its passage towards the river. The latter origin is the most probable, as the water is generally found in the excavation at a higher level than the water in the river. This has often been observed in sinking the foundations for bridge abutments, in constructing subsiding reservoirs, filter beds, &c., for water-works and other purposes by the sides of rivers.

EXTENT OF THE TERTIARY SANDS IN THE LONDON BASIN.

They may be traced from the Suffolk coast, near Aldborough, extending in a narrow band seldom more than a mile in width, by Woodbridge, Ipswich, Hadleigh, Sudbury, Great Yeldham, Bishop Stortford, Hertford, Hoddesdon, Hatfield, Watford, Uxbridge, Windsor, Reading, and Newbury, nearly to Hungerford. This is the western extremity of the tertiary basin, the boundary of which then takes an easterly direction, passing by Kingsclere, Odiham, Farnham, Guildford, Epsom, and Croydon, into the valley of the Thames, which it occupies for a considerable breadth all the way to the sea, extending also under the Isle of Sheppy, by Rochester, Faversham, and Canterbury, to the sea between Ramsgate and Deal. The length of the outcrop, according to Mr. Prestwich, being 380 miles, and the area 223 miles—this gives an average breadth to the lower tertiary strata of less than two-thirds of a mile. Where the outcrop is intersected by valleys, however, the London clay is denuded and a greater breadth of the lower tertiaries exposed, as in the valley of the Lea between Hertford and Hoddesdon, in the neighbourhood of Watford, and other parts of the outcrop. Also in the valleys of the Thames and the Medway extensive denudation, combined with faults, has exposed a considerable breadth of lower tertiary sands. For instance, they extend almost uninterruptedly from Stratford to Croydon, occupying here a breadth of nearly 15 miles, broken only by a few high points, capped with London clay. Eastward of this line they somewhat diminish, but the zone continues to be several miles in breadth all the way to the German Ocean.

Mr. Prestwich states, that large supplies of water are derived from the tertiary sands throughout a district westward of the meridian of Greenwich, which is bounded on the north by Hertford and Watford, on the west by Uxbridge, and the south by Croydon. He computes that, in the valley of the Wandle, there are about 15 to 20 artesian wells, deriving a daily supply of from 800,000 to 1,200,000 gallons of water from the tertiary sands, and in the valley of the Lea from 20 to 30 such wells, deriving a supply of 120,000 to 200,000 gallons a day. The quantity yielded by single wells from these sands, would be utterly insignificant for purposes of public supply, except for very small towns. There are few instances of single wells which derive a water supply of more than 100,000 gallons a day from these sands.* It has been asserted, however, that many of the London wells which are sunk or bored into the chalk, really derive their supply from the tertiary sands ; and certainly the remarkable difference between the chalk water of the London wells and that of chalk water, for instance, from Watford or Ware, lends some countenance to the supposition.

Most of the water from the London chalk wells yield carbonate of soda and magnesia, and a comparatively small quantity of carbonate of lime, while, on the other hand, the water from Ware and Watford has a large proportion of carbonate of lime, and seldom any of the other carbonates. Again, the sulphates of soda and potash abound in the London well waters, but are absent from the pure chalk water. The chloride of sodium is also found extensively in the London well water, but is almost wanting in pure chalk water. See a valuable collection of analyses of river and well waters, published by Mr. Prestwich.

THE LONDON CLAY.

The London clay itself is, from its nature, destitute of water, although there are numerous wells in and about the metropolis deriving their supply from the drift gravel which covers it. Wells are exceedingly common all over England in drift

---

* Mr. Swindell states the yield of the well at Hanwell Lunatic Asylum. which terminates in the tertiary sands, at 100 gallons per minute, which is equal to 144,000 gallons in 24 hours. The first 30 feet of this well are 10 feet diameter the rest of it 6 feet.

gravel, especially where the gravel rests on an irregular surface of London clay, or on the marly clays of the new red sandstone formation. It is probable the water lies in hollows and troughs in the surface of the clay, and is frequently found at shallow depths beneath the surface. The water never overflows in such wells, and never spouts up as in an artesian boring. Its level is, however, affected by the dryness of the season, and it frequently happens that the highest wells in such districts become dried up, while those sunk deeper, or at lower levels, continue to yield a supply.

The drift gravel is exceedingly variable in thickness, and wells sunk into it commonly range from 10 to 50 feet in depth.* It is quite unfit for yielding large supplies for the use of towns, although in all parts of the country there are numerous private wells drawing a supply from drift gravel.

This is the case not only with such towns as Southampton, Portsmouth, and others on tertiary formations, but also with towns such as Leicester, situate on the clays or marls of the new red sandstone. Where drift gravel is overlaid by an argillaceous deposit, as in some parts of Essex, and wells are sunk at points lower than the outcrop of the gravel, the water will sometimes rise, as in artesian wells, and may be successfully obtained by means of a simple boring. It is far more common, however, to find the drift gravel merely forming the surface and not overlaid by any deposit sufficiently thick or impervious to keep down the water. In all such cases it is obvious the water in the gravel merely rests on the clay beneath at its lowest level, and can only be obtained by sinking wells into which the water will filter, but will not

---

* The expression "land springs," which is very common with the London well sinkers and borers, is applied to the shallow surface springs rising in the drift, either from the alternation of clay with the more porous gravel beds, or from the water held in the irregular hollows of the London clay beneath. Mr. Tabberner (quoted by Mr. Prestwich from the *Daily News*, 13 March, 1850) estimates the yield of the London drift wells at an aggregate daily quantity of about 3,000,000 gallons.

rise above the ordinary level of the water in its subterranean basin. From these wells the water must be pumped to the surface, and any attempt to procure water by simple artesian bores will of course be fruitless. Although, in London, and many other towns, the drift gravel formerly supplied numerous wells, it is found that the construction of sewers gradually drains these wells, rendering it necessary in some instances to sink them deeper, and in others entirely abstracting the supply and drying up the wells. Besides this, the burial-yards found in all large towns, and the innumerable impurities arising from gas-works and offensive manufactures being poured into the sewers, have in time so saturated the soil as to poison the water of all such wells and render it wholly unfit for use.

THE BAGSHOT SAND.

We shall see in a future page, that at a certain period in the history of London the springs breaking out from below the sands of Hampstead Heath were looked on with much satisfaction as a new source of supply of great value and importance. The whole cap of sand on these hills, however, does not probably exceed one square mile on which the whole rainfall, even if we conceive it all to be absorbed into the soil, does not amount to a million gallons per day, so that if all the springs on every side of Hampstead Hill were collected they would fall somewhat short of this amount, or less than $\frac{1}{50}$th of the whole supply required for the metropolis. With respect to all the springs on the west side of Hampstead, they would scarcely be worth the expense of the works necessary to collect and convey them; and, in point of fact, the water actually collected in the intercepting ponds of the Hampstead works may be considered as comprising all that can be collected with advantage. This quantity does not probably amount to more than about 200,000 gallons per day, a quantity perfectly insignificant in comparison with that required for the supply of London.

THE BAGSHOT SANDS OF SURREY AND HAMPSHIRE.

These sands extend almost continuously from Esher to Strathfieldsaye, with an extreme breadth from the north side of Virginia Water to the neighbourhood of Farnham. Estimated roughly, the length of the district may be taken at 30 miles, by an average breadth of 10 or 12 miles. The Bagshot sand is capable of three subdivisions. The first or uppermost is pure siliceous sand varying from 200 to 300 feet in thickness, and covering an area of 80 to 100 square miles. The upper sand attains its greatest thickness* in the north and east part about Bagshot Heath, Chobham Ridges, Romping Downs, Finchhampstead Ridges, and Hartford Bridge Flats. This is the upper Bagshot sand of geologists.

The middle division consists of a retentive stratum of white or pale yellow clay or marl, from 15 to 30 feet in thickness. These clays are extensively used for making bricks.

This middle division forms the Brackesham beds, or middle Bagshot of geologists.

Below this middle division is a lower series of light coloured sands, which like the upper beds consist of nearly pure siliceous matter. These are the lower Bagshot sands of geologists.

Professor Ramsay, in his letter to the Board of Health, points out that a covering of gravel, varying in thickness from a few inches to 20 or 30 feet, frequently obscures the surface of the sands; but, notwithstanding this, both the upper and lower sands admit very freely the percolation of water, so that little passes off by evaporation, and nearly the whole rainfall is absorbed. Many of the small streams rising from Bagshot and the neighbouring heaths, break out at the top of the middle or argillaceous portion, while the water falling on the lower sands is absorbed by them, and does not appear at the surface till thrown out by the underlying London clay.

---

* Report by Robert Austen, Esq., F.G.S., of Chilworth Manor, near Guildford, to the Commissioners of the General Board of Health.

PROPOSED GATHERING GROUND ON THE BAGSHOT SANDS.

About twenty years ago the Bagshot sands were proposed by the General Board of Health as an immense gathering ground for procuring the greater portion of the supply required for the metropolis. The project is first dimly shadowed forth in the Report on Supply of Water to the Metropolis, dated 28th May, 1850, and presented to both Houses of Parliament. The Report says, "The portion of this district (the Bagshot sands) to which our attention has been more immediately directed comprises an area of less than 100 square miles, lying east and west of a line from Bagshot to Farnham. The remaining district, which we have had under consideration, although of the same bleak and barren character, is of a different geological construction, consisting of the upper and lower green sands, and gault of the green sand formation, and constitutes the uncultivated sand districts draining into the east and west tributaries of the river Wey, situated south of the chalk ridge in the midst of which the town of Guildford stands." The report contains no details of the project—no statement of the mode to be practised for collecting the water, the levels at which it is to be conveyed and stored for distribution, nor, in fact, any of that kind of information which first occurs to an engineer as most essential in an inquiry of this nature.

After stating, however, as the result of numerous inquiries and investigations, that the daily supply required for the metropolis was 40 million gallons,[*] they set forth the following estimate of the yield of their gathering grounds from gaugings taken at the end of nearly six weeks of dry weather :—

[*] This quantity is already far below that supplied at the present moment by the existing Companies. In fact, the returns of 1853, only three years after the date of this report, show a much larger quantity than 40,000,000, and the supply now in April, 1870, actually exceeds 100,000,000 gallons a day, or $2\frac{1}{2}$ times the amount in 1850.

	Gallons per day.
From surface gathering grounds of sand, comprising specimens averaging from $\frac{1}{2}$ to 1 degree of hardness, and equal in quality to the water delivered at Farnham, from which district, and from streams derived from similar grounds, the average hardness may be estimated as under 3 degrees	28,000,000
From certain tributaries to the river Wey, containing some water from the chalk, but of a general quality of hardness, $\frac{1}{3}$ the average of the present supply of the Metropolis	60,000,000
From other tributaries to the Wey, of a harder quality, but only one-half the hardness of the present supply to the Metropolis	90,000,000

In the appendix to this report the Board of Health publishes an immense mass of medical evidence, the general effect of which is to show that the water from the gathering grounds is of very superior quality as to softness. Neither in the report nor in the evidence is to be found one word or one fact, as far as I am aware, in support of the above estimated yield of the gathering grounds.

The only other extract which I shall make from the report at present, is the following estimate of the gross estimated cost of this magnificent project, which I presume, from the wording of the estimate, and its including street and branch mains, services, &c., was intended entirely to supersede and replace all the works of the eight existing companies, and render all their plant and apparatus of every kind entirely unnecessary. Here then is the estimate in the very words of the report :—

"Storage reservoirs, and intercepting culverts on gathering ground; covered aqueduct thence to service reservoirs; covered service reservoirs and filter beds; principal mains from reservoirs, street, and branch mains, and services, &c., &c., over the whole district, including land for works and compensation, £1,432,000."

The estimate has one other item, namely, £710,000 for

sewerage, to be carried out in connection with the water supply. With this we have nothing to do at present.

It appears that the idea of collecting water extensively from the Bagshot sands, originated in the observation of some small works which had been executed at Farnham. This town is situated on the southern escarpment of the north downs, at a part where the chalk is extremely narrow, not more than half a mile in width, and immediately north of Farnham is the ridge of Tucksbury Hill, about four miles in length, and about one mile in breadth. This clay ridge is capped with Bagshot sand, a portion of which had been drained, and the water received into a shallow circular well about three feet below the surface, from which it flowed into the main supplying the town of Farnham with water. The population of Farnham being about 7000 persons, it is probable that at the very utmost the supply from this source did not exceed 100,000 gallons a day. This experiment, however, in which this comparatively small quantity of water was procured by means of tile drainage, seems actually to have given rise to the gigantic scheme of the Board of Health for supplying more than 50 millions of gallons a day to the inhabitants of London.

Let us now turn to the chemical evidence to see how the statements made in the report as to the quality of the water are supported. The evidence of Dr. Angus Smith contains the first experiments on the hardness of the water; and he appears to have tried 23 specimens of water from the Bagshot sands or district north of the Hog's Back, and 16 specimens of the lower green sand water from the district of Leith Hill, Hind Head, &c., south of the Hog's Back. The lowest degree of hardness which he found in any of the Bagshot waters appears to be that of 1° in the water of Aldershot Heath, while the hardness of the rest of the Bagshot sand water ranged from 1° upwards to 7·8. Is it on the evidence then, of Dr. Smith that the Board of Health ventured to put forth the statement, that they could procure from the Bagshot sands 28 million gallons of water daily, of a quality varying *from one-third to one degree in hardness?* Of the 23 specimens of Bagshot sand water analysed by Dr. Angus Smith, the average hardness is actually 4°.

Of the 16 specimens of lower green sand water, Dr. Smith finds the hardness vary from 4·75 to 14·6, the average being 7·9 degrees, yet this is the water of which the Board states the possibility of procuring 60 million gallons a day, of hardness equal to one-third of the average of the present supply to the metropolis. It must be observed, that the report of the Board of Health, and the evidence of Dr. Angus Smith were presented to Parliament at the same time, namely in 1850, so that the one must be taken to be founded on the other. It is true the Board afterwards published a report by the Hon. William Napier, containing statements entirely at variance with those of Dr. Angus Smith, but inasmuch as Mr. Napier's report is only dated January 1851, whereas the report of the Board bears date May 1850, it is obvious none of the statements by the Board could have been founded on Mr. Napier's report.

With respect to the other chemical evidence published by the Board as an appendix to their report, we find Professor Way saying that he analysed "Some water from a small well near its source, through which the water flows on its way to the town. The proportion of lime I found to be .168 grains in a gallon, which is equivalent to exactly $\frac{3}{10}$th of a grain of carbonate of lime in the gallon, or $\frac{3}{10}$th of a degree of hardness of Dr. Clarke's test."

Perhaps it was on this examination of a little pool of rain-water by Professor Way that the Board hazarded the magnificent assertion, which has been before quoted, that 28 million gallons a day, varying from $\frac{1}{8}$ to 1 degree in hardness, could be procured from the Bagshot sands.

I can find no analysis of the Bagshot waters by any of the other eminent chemists who were examined by the Board of Health. Among the witnesses so examined are Dr. Sutherland, Mr. Holland, Dr. Hassall, Dr. Gavin, Dr. Lyon Playfair, Mr. Spencer, Professor Clark, Mr. J. T. Cooper, Professor Hoffman, and Mr. R. Phillips, all exceedingly able as analytical chemists, and not one of these has given any but the most vague and general statements as to the Bagshot waters.

The Board probably felt on calmly reviewing their report of 1850, and the evidence by which it was accompanied, that there was an immense hiatus to be filled up in some way or other, and they accordingly deputed the Honourable William Napier to make a most comprehensive examination of the whole subject, namely to examine the whole of the drainage grounds both north and south of the Hog's Back, and to report fully both on the quantity to be collected, its quality, and the mode of conduction and delivery, and further, on the estimate of the whole works.

This gentleman, who appears to have been thus invested with the combined offices of engineer and chemist to the Board, first publishes a table of the yield of all the springs and rivulets occurring in an enormous district of about 400 square miles, in which he shows a total daily discharge of more than 39 million gallons. The Bagshot sand district, over which his investigations extended, appears to range from Pirbright and Chobham as far as Eversley and Bramshill, ten miles beyond Bagshot, and nearly as far west as Strathfieldsaye, while he also takes in an enormous tract of country called Easthampstead Plain and Bagshot Heath, on the north side of Bagshot. This it will be observed is materially extending the district which had been contemplated by the Board before making the report of 1850. This report (page 100) says, "the portion of this district to which our attention has been more immediately directed, comprises an area of less than 100 square miles, lying east and west of a line from Bagshot and Farnham." Now the distance from Bagshot to Farnham is about 11 miles in a straight line, so that the Board probably contemplated the drainage of a district about $4\frac{1}{2}$ miles wide on each side of this straight line. It therefore excites some surprise to find Mr. Napier wandering to Eversley and Bramshot 10 miles west of Bagshot, and ranging over Easthampstead Plain which lies north of Bagshot. But let us see the

quantity of water which he obtains in his 200 miles of
Bagshot sand area.

	Gals. per day.
From Chobham ridges, the district east of Bagshot, including Pirbright, and probably also Aldershot Heath, where Dr. Angus Smith's single specimen of 1° hardness was obtained, he finds the gaugings amount to .	3,020,086
From Easthampstead Plain and the district north of Bagshot	1,509,348
From the district west of Bagshot and Farnham, extending to Bramshill and Eversley, he obtains . .	7,712,090
Total from 200 square miles of Bagshot sands .	12,241,524

This last quantity of more than 7,000,000 contains 6,426,000
gallons from one stream called North Fleet, which from its
situation on the Ordnance Map appears to be considerably
outside the limits of the 100 miles described in the report of
the Board, so that the whole quantity of water gauged by Mr.
Napier in this 100 miles, clearly does not exceed 6,000,000
gallons. It will be remembered that the quantity estimated
by the Board from this area is 28,000,000—truly a singular
coincidence and one which entitles the statements of the Board
and their calculations to the confidence of the country !

Mr. Napier, then, having obtained only a little more than
12,000,000 gallons from the whole Bagshot sand district,
which Mr. Austin describes as having an area of 300 miles,
derives all the rest of his 39,000,000 from the green sand
districts of Hind Head, Blackdown, Leith Hill, &c. The
quantity of 27,000,000, said to be derived from about 200
square miles of green sand country, need not excite much
astonishment as the honourable gentleman seems not to have
been content with gauging springs and rivulets, but appears
to have taken in whole rivers without any regard to the levels
at which they were flowing, or the possibility of conveying
them to the top of Wimbledon Common without an enormous
pumping power. For instance, among the *springs and rivulets*
flowing from Hind Head and Blackdown, he finds one which
is pithily called Bramshot, yielding no less than 13,399,714

gallons a day, or nearly equal to the whole volume brought in by the new river. It would be amusing, though perhaps not very instructive, if the Honourable William Napier were to publish the details of these gaugings, and enlighten the public as to the mode in which they were taken.

So much for quantity; now for quality—bearing in mind, or referring to what Dr. Angus Smith has proved as to the hardness of these waters, one is fairly surprised to find that Mr. Napier states the whole of the Bagshot sand water, without the slightest exception or variation, at one degree of hardness. Yes, opposite to every *spring and rivulet*, every silver thread which this gentleman has visited stands the figure 1, in the column for hardness. Nearly the same uniformity prevails in the green sand waters, these being all either one or two degrees with the exception of Bramshot, which, being rather a large quantity, we may suppose has been examined with unusual care, and is accordingly marked one-and-a-half. In order to show the extreme patience and care with which Mr. Napier has investigated the tough subject of hardness, the following remarks are quoted from his table, as further explanatory of the degrees of hardness marked in the proper column. For instance, with respect to Holywater Spring, which is described as two degrees, the remark says, " will be *led* away at one degree of hardness." Bramshot water, marked one-and-a-half degree, " will probably be led away at half a degree of hardness." All the water of Hascombe Hills, marked one and two degrees, " will be taken away at half a degree of hardness." And again, the water of Leith Hills, marked two degrees, " will be led away at one degree of hardness." Mr. Napier's boldness is not exhausted by even all these grave remarks. He " can answer for at least ten millions more under two degrees of hardness." I am at a loss to understand the meaning of the following remark in a report such as this professes to be, " though these gaugings are only offered as an approximation, I consider they will eventually prove to be rather under than overstated."

Mr. Napier himself seems to have stood somewhat aghast at the difference between himself, and the only analytical chemist who examined these waters as to their hardness. Whether the following extract gives any very satisfactory explanation of the difference, I will not pretend to say :—

"Thus by gauging and testing the streams at their sources, instead of in their course and outfalls, we have the realization of the principle laid down by the Board; and this difference will go far to account for the variance of my results with those of Dr. Angus Smith."

As to the mode of executing the works, the information is very scanty. We are not told whether there are to be any collecting or impounding reservoirs, and not a word has been said about filtration, although it appears that many thousand acres of the gathering grounds consist of peat and moorland, the solutions of which would require the water to be filtered. We are told in a very off-hand sort of way that the water south of the Hog's Back is to be conveyed through the chalk ridge at Guildford; and that the water from the sands is to be brought in, in the direction of Woking. Not one word of information is given about levels. The river Wey at St. Catherine's lock, immediately south of Guildford, is only 92 feet above Trinity high water mark, and as the Guildford pass is to be used for the main aqueduct, and a great deal of the county north of Guildford is falling even below this level, it is probable the main aqueduct would not be higher than 100 feet above Trinity high water mark at Guildford. Allow only a loss of 5 feet per mile for friction, which would require an enormous main to carry such a volume of water, and we have 120 feet absorbed by friction in the 24 miles between Guildford and Wimbledon. This would destroy the whole effect of the presumed altitude at Guildford, and render the expense of pumping not one farthing less than that incurred by the Companies who now take the water from the Thames.

Mr. Napier's estimate, like his gathering ground, is some-

what different from that of the Board (less than half) Here
it is in his own words :—

Collection	£40,000
Conduction to service reservoir on Wimbledon Common in a double brick culvert, twenty-four miles at £7000	168,000
Covered service reservoir to contain four days' supply at 50 million gallons per day	80,000
Estimate of expense of mains to connect the reservoir with the present street pipeage.	200,000
Probable amount of compensation for millowners, irrigation, &c.	100,000
	588,000
10 per cent. for contingencies	58,800
Total	£646,800

Any comment on such an extraordinary document will surely
be unnecessary.

CONFIRMATIONS OF MR. NAPIER'S GAUGINGS.

Mr. Napier's report and the statements it contained probably astonished the Board of Health, nearly as much as they have subsequently astonished most other persons who have had patience to read them. We find, accordingly, Mr. Rammell and Mr. Quick sent down to test the gaugings in October and November, 1850. Both these gentlemen far out-Napier Napier! Mr. Rammell hands in his gaugings, amounting to 51 million gallons, or 12 millions more than Mr. Napier, and accompanies the statement by some very flowery observations on the surpassing qualities of the water. He glances at the geology of the district, the 300 or 400 miles of drainage ground, and reveals in a very *naïve* and amusing style his method of gauging, which, however, I have not time to notice at present. Mr. Quick modestly contents himself with producing his naked statement of gaugings, which amounts far beyond either of the others, namely, to

62 million gallons a day, and this, too, trom only 15 streams whereas it appears Mr. Napier gauged no less than 45 or three times as many. Well may the Board have been alarmed at the gradual increase announced by each successive investigator,—who shall pretend to say to what amount the gaugings might have increased had they continued to send one person after another in this way?

The gauging mania then seems to have slumbered for more than a year, when we again find Mr. Ranger gauging the streams from the 1st to the 10th of August, 1852, when he makes the volume about 4 million gallons a day less than Mr. Napier had made in the middle of summer.

About this time, Mr. Bateman seems to have been directed to examine the district and report on the subject, as I find his report to Lord Shaftesbury, dated 27th January, 1852, printed among the papers laid before Parliament by the Board of Health.

This report by Mr. Bateman seems effectually to have settled the question, and to have been the last act in this amusing farce. We have here the first traces of sound engineering judgment applied to the scheme. Mr. Bateman appears to state fairly enough the capabilities of the district, and the practicability of the scheme, always provided the money can be found. He is apparently not startled by the enormous sum at which he finds it necessary to estimate the cost, but coolly leaves it to the judgment and sense of the Board, whether they can find it practicable to bring forward such a scheme.

Mr. Bateman's gaugings are worthy of attention. The general result is, that he found 33,238,093 gallons per day against 39,407,324 by Mr. Napier, against 51,375,000 by Mr. Rammell, and against 62 millions by Mr. Quick. Mr. Bateman thus comments on these discrepancies. "Allowing for several evident errors in Mr. Napier's results, arising from *the streams having been gauged by him below mills which were working at the time, and using water which had been pre-*

riously stored, Mr. Foster's* measurements rather exceed those of Mr. Napier. Mr. Rammell's gaugings are generally considerably higher, but he has also been led into some errors by gauging below mills, and has, probably, not made sufficient allowance for the loss of velocity by the friction of the water on the bottom and sides of the channels in which he measured the streams." Mr. Bateman does not notice the gaugings of Mr. Quick or Mr. Ranger, as he was probably not in possession of these.

As well as I can understand Mr. Bateman's gaugings he appears to give about 8,000,000 gallons a day from the 200 square miles of Bagshot sands, and nine-tenths of this quantity he describes as having a hardness of $2\frac{1}{2}$ to 3 degrees. Of pure green sand water he gives upwards of 33,000,000 gallons derived from Hind Head, Hascombe, Hambleden, and Leith Hills, with an average hardness of $2\frac{1}{8}$ degrees. He also finds the Farnham branch of the river Wey, which is probably chall. water, yielding 10,600,000 gallons, *with a hardness of 14 degrees.* Mr. Bateman appears to think highly of the lower green sands as a gathering ground, but evidently is not much captivated with the prospects of collection from the Bagshot sands. He says, "the waters from Bagshot Heath, and those flowing from the sands and gravels north of Farnham, into the rivers White-water and Blackwater, form together not less than 6,000,000 or 7,000,000 gallons per day of excellent water; *but they are* distant, and could not easily be combined with a scheme for bringing the water of the green sands south of Guildford. I should prefer omitting both the Bagshot waters and the Farnham branch of the Wey, and consider the scheme as affording a daily supply of 39,000,000 or 40,000,000 gallons of pure soft water under three degrees of hardness."

This is very quietly and softly extinguishing the scheme of the Board of Health with reference to the Bagshot sands, and setting up in its place the conveyance of water from the green sand hills. This then is purely and exclusively Mr. Bateman's

* Mr. Foster is an assistant who gauged for Mr. Bateman.

scheme, and in this shape it is intelligible. He further speaks of the Wealden district, lying much more to the south, and instances the Hastings sand of St. Leonard's forest as a good gathering ground.

We hear little of this project of the Board of Health after the date of Mr. Bateman's report. It is not noticed in any Parliamentary inquiry subsequent to 1852, nor in the recent reports by Commissioners on the Supply of Water to the Metropolis. The failure of this scheme is not brought forward with any hostile feeling to the late Board of Health. On the contrary, one merit must not be denied to them—their proceedings have elicited some useful information, although mixed up with much that is merely speculative and hypothetical.

It is now generally admitted that immense gathering grounds of several hundred square miles are not adapted to furnish large concentrated supplies of water, in consequence of the numerous difficulties and expense of collecting it.

The district of Bagshot Heath presents little or no analogy with the scenery of the Palæozoic rocks, which have hitherto been resorted to almost exclusively as gathering grounds. Neither the shape of the valleys, the surface soil, nor the elevation of the ground affords any features of similarity.

Notwithstanding all the reports and other documents which have been published about the Bagshot sands, the actual proposed mode of collection has been very obscurely described. We are still almost ignorant of the absolute levels at which the springs were to have been collected, and the subject of impounding reservoirs is never once alluded to. Judging from the nature of the sands, it seems extremely doubtful whether impounding reservoirs could ever have been formed to hold water without lining or puddling the whole bottom, an expense that would have proved fatal to any such scheme.

Although the Bagshot sand district must be now regarded as totally inadequate for the supply of the metropolis, it is possible that many of the towns might advantageously derive a supply from the springs. Farnham is said to be already supplied from a sandy district which has been drained ; and no doubt the same may be practicable on a small scale, and for other towns not requiring so large a supply as the metropolis. Such towns as Windsor, Wokingham, Reading, Guildford, Woking, Weybridge, Kingston, and Staines, may not improbably find it advantageous to resort to the Bagshot sands for a supply of softer water than they can procure in their own neighbourhood. In the case of small supplies to be taken from springs, many of the difficulties with respect to large impounding reservoirs are avoided, and adequate collections of water may be made without the expense which would have to be incurred for collecting the water from 200 or 300 square miles of country.

BEDS ABOVE THE TERTIARIES.

About three-fourths of the whole surface of Norfolk and Suffolk consist either of a diluvial deposit covering the chalk, or of the crag formation, which is considered by geologists superior to the Bagshot sand. The diluvium consists generally of clay or loam with numerous fragments of chalk imbedded in it, and this is covered frequently by sand and other light soils. The diluvium is in many places as much as eighty feet thick.

The crag consists chiefly of thin layers of sand, gravel, and shells, resting sometimes on the chalk and sometimes on the London clay. Frequent sections of the coast seen in the low cliffs of the Norfolk and Suffolk coasts, from Aldborough to Cromer, commonly exhibit a succession of this kind—red loam at the base, gravel above this, and the gravel again covered by chalk rubble. The whole district over which these deposits extend is remarkably flat, and nowhere rises into undulating scenery. The supply of water is mostly derived from open wells, sunk into the gravel or through the loam into the chalk. The principal towns of the districts are Norwich, Yarmouth, and Ipswich, besides which there are a great number of secondary market towns scattered all over Norfolk and Suffolk.

There is also a remarkable tract of diluvial country, extending for several miles inland from the Humber to the Wash. The deposits here rest on the chalk which forms the Wolds of Lincolnshire. The deposit is mostly impervious, consisting of retentive beds of clayey or loamy gravel, and is remarkable from the great number of overflowing artesian wells sunk through the diluvium down to the chalk. The wells are simple borings, frequently 80 to 100 feet in depth, and are locally known as blow-wells. They abound in the neighbourhood of Louth, Alford, and Great Grimsby, and are commonly met with throughout a district about 50 miles in length from Wainfleet to Barton, on the Humber, with an average breadth from the coast of eight or ten miles. A similar district of blow-wells exists on the coast of Essex, between the mouth of the Stour and the estuary of the Thames. In these bore-holes, however, the water rises from the lower tertiary sands of the plastic clay formation. The quantity of water which they yield is very small, not more than from one to eight gallons per minute. Some of them, however, are very deep; those in Foulness Island, Mersea Island, and Wallasea, varying from 300 to 450 feet in depth.[*]

ON THE STRATA BETWEEN THE CHALK AND THE OOLITES.

These in descending order comprise :—

1. The upper green sand or firestone.
2. The gault clay.
3. The lower green sand or *neocomian* of the French geologists.

The upper green sand is an arenaceous or sandy formation, underlying the chalk with variable thickness, which probably ranges from a few inches to 150 feet in Wiltshire, where it attains its largest development. It extends from Filey on the

* See Dr. Mitchell's paper in Proceedings of Geological Society, vol. 3, p. 131.

coast of Yorkshire to near Wainfleet in Lincolnshire, where it is cut off by the estuary of the Wash. It again appears under the chalk of Norfolk, and extends with little interruption in a south-westerly direction through Cambridgeshire, Bedfordshire, Bucks, and Oxfordshire. The outline and breadth are then very irregular through Berkshire, Wiltshire, and Dorsetshire, where it terminates on the coast. Another range of green sand commences at Folkestone, and passing under the chalk of the north downs, ranges in a horseshoe form through Kent, Surrey, Hampshire, and Sussex, where it again comes to the coast beneath the chalk of Beachey Head. In a considerable part of its range, especially in the more northern district, it is often very feebly represented, and being sometimes only a few inches in thickness, and often covered with drift, frequently escapes observation.

It is, nevertheless, believed to exist in the shape of a sand bed or beds, more or less indurated into stone, beneath the chalk throughout the whole extent indicated. The breadth it occupies on the surface is insignificant, often not more than a few yards, except in valleys where the superior strata have been removed by denudation, and where, as in the valley of the Thames at Wallingford, and the vale of Pewsey in Wiltshire, it occupies a breadth of several miles.

With reference to the upper green sand surrounding the chalk of the London basin, Mr. Prestwich estimates the length of its outcrop from a point near Cambridge entirely round the horseshoe outline formed by connecting the vale of Pewsey with Farnham, and extending the outline to the English Channel at Folkestone, at 255 miles, with an average breadth of a mile and a half. Its thickness throughout the northern part of its course, as far south as Cambridge, is seldom more than 10 feet, and sometimes not more than a few inches. Throughout Oxfordshire, Wiltshire, Hants, and Surrey, Mr. Prestwich says, the average thickness is probably 75 feet, while in Kent and East Surrey its thickness is probably about 25 feet. With the exception of Watlington,

Wallingford, and Warminster,—the latter a celebrated locality for upper green sand fossils,—there are few towns, or even large villages, which can be said to be situate on the upper green sand, and even where collections of houses do exist on it, the surface is generally covered by a thick drift or diluvium.

There is in some cases considerable uncertainty whether the water falling on the chalk sinks through the chalk marl and penetrates the green sand. It certainly appears to do so in Cambridgeshire, where, as well as in the neighbourhood of Tring, it commonly penetrates down to the gault. Most of the chalk springs of the south downs, however, as Lydden Spout, Cheriton, &c., are thrown out by the chalk marl, and do not reach the upper green sand. Generally speaking, the green sand cannot be expected to yield a large supply of water, and few springs break out from it except in the district west of a line between Farnham and Petersfield. Here the gault of Holt Forest, Woolmer Forest, and Petersfield, throws out numerous small streams which, probably, have their origin in the upper green sand. In Conybeare and Phillips' "Geology of England and Wales," a well at Pottern, near Devizes, in the vale of Pewsey, is mentioned in the upper green sand. This is sunk 126 feet deep through the sand down to the gault, but the quantity of water which it yields is not mentioned. Mr. Gravatt (Trans. Inst. Civil Eng. vol. i.) mentions two borings made through the upper green sand at Tring in order to procure water for the Grand Junction Canal. The boring in each case appears to have passed through the chalk marl and upper green sand into the gault. The yield of one boring is not stated, while that of the other is said to be 1300 cubic feet in 24 hours. This, which is only a little more than 8000 gallons a day, is of course a very insignificant quantity, with reference to a supply even for a large village.

Selborne, the residence of the accomplished Gilbert White, who wrote a most amusing and interesting book on its natural history, is situated on the upper green sand, west

of Woolmer Forest. The wells at Selborne, which are probably sunk nearly, if not quite, down to the gault, are said
to average about 63 feet in depth. " When sunk to this
depth they seldom fail, but produce a fine limpid water, soft
to the taste and much commended by those who drink the
pure element, but which does not lather well with soap."

THE GAULT.

This is a deposit of stiff tenacious blue clay which lies
between the upper and the lower green sand. It will be
unnecessary to describe its range and direction, as it everywhere accompanies, and lies parallel to, the upper green
sand, the course of which has been already described. Many
streams and springs break out on the edge of this formation
and flow over it in a direction opposite to the dip. The volume
of these, however, is not considerable, and the valleys are not
of such a deep and capacious shape as to encourage the mode
of collecting water by storage reservoirs. Either on or closely
contiguous to that part of the gault formation which extends
from the Cambridgeshire Fens and the Isle of Ely into Wiltshire, are several towns of third or fourth rate importance,
among which may be mentioned Cambridge, Potton, Biggleswade, Shefford, Leighton Buzzard, Prince's Risborough,
Wantage, and Devizes. The towns of Petersfield, Dorking,
Wye, and Folkestone, are also situate on the gault of Hampshire and the north downs. The common mode of procuring
water for towns situate on the gault, is by sinking through
it down to the lower green sand. When a boring is made
down to the latter, the water generally rises nearly to the surface and sometimes overflows. Wells and borings through the
gault are common at Cambridge, Biggleswade, Shefford, and
all the line of flat clay country through Leighton Buzzard into
Wiltshire, and Mr. Prestwich describes the quality of water as
remarkably soft and pure. The artesian wells of Cambridge are
very numerous. They are commonly sunk through the gault,
and abundantly supplied with water from the lower green sand,

which probably derives its supply from the high hills of Bedfordshire. The supply from the wells in Cambridge was formerly esteemed of fair quality, although a considerable part of the inhabitants always availed themselves of water from the Cam, or from the Nine Wells' stream, which is derived from chalk springs. Latterly, however, the supply from all these sources has been found inadequate, and a Company has been formed, and works executed, for taking a supply of about 600,000 gallons a day from the Cherry Hinton stream and spring, which is also derived from the chalk. It appears that for some years after 1812, when the first artesian well was made in Cambridge, the water used to rise three or four feet above the surface. Owing to the increase in the number of borings, which are now probably between 500 and 800, the water stands, at present, from six to twelve feet below the surface. The hardness of two specimens of the Cambridge well water, as analysed by Mr. Warrington for Mr. Prestwich, was 8·8° and 11°.

Another famous locality for artesian wells penetrating through the gault to the lower green sand, is at Wrest Park, the estate of the Countess Cowper, near Silsoe, in Bedfordshire. Probably not less than 20 artesian borings have here been made through the gault, and in every case the water from the lower green sand flows over above the surface, furnishing a never-failing supply of water. Most of the wells have been furnished with a pipe which is bent over at the top, so as to discharge downwards a constant stream of water. The water tastes perceptibly of lime, and is highly esteemed as a chalybeate. Mr. Prestwich speaks favourably of the well water at Leighton Buzzard, Biggleswade, and other towns drawing from the lower green sand beneath the gault. Mr. Prestwich, Dr. Fitton, and other observers have stated the thickness of the gault at Folkestone at 126 feet; at Merstham 120 to 140 feet; between Guildford and Merstham somewhat thinner; at Devizes and Swindon about 100 feet; increasing in Cambridgeshire to about 150 or 160 feet.

Recent trials have shown 160 to 200 feet in thickness at borings on the Medway, near Maidstone; at Kentish Town about 130 feet; at Wrotham, Kent, 126 feet; at Eddlesborough, Bucks, 205 feet; at Baldock, Hertfordshire, about 170 feet; at Hitchin 214 feet; and at Harwich 61 feet.

Numerous wells and borings have lately been sunk in the gault district between Hitchin, Ampthill, and Biggleswade, in order to procure water for brickmaking, and for washing the coprolites which are extensively dug in the neighbourhood. These wells usually pass through about 200 feet of gault.

A well has recently been dug at the village of Arlesey, where the gault was found to have the same thickness, and at the well sunk for the Three-counties' Asylum at Arlesey the thickness of the gault was very accurately ascertained to be 204 feet.

Thus the gault seems to attain a maximum thickness of about 214 feet at Hitchin, and thence to diminish gradually in a westerly direction, till at Cambridge it does not exceed 160 feet, and where it finally ends at Hunstanton, in Norfolk, the bed of so-called red chalk which represents the gault is only 4 feet in thickness. The area occupied by the gault formation, extending from Cambridgeshire into Wiltshire, and thence under the north downs to the sea at Folkestone, is given by Mr. Prestwich at 340 square miles, with an average breadth of about a mile and a half.

THE LOWER GREEN SAND.

It is generally understood by geologists that this formation everywhere accompanies the gault, which is represented as resting on a mass of sand varying somewhat in mineralogical character and still more in thickness, and which in English geology is known as the lower green sand. North of the Humber, in the cretaceous district of Yorkshire, however, there are scarcely any traces of the green sand, and it is probably owing to faults or to thinning out that the Speeton clay, another name for the gault, is almost in con-

tact with the Kimmeridge clay and other members of the
oolitic series. Scarcely more visible is the lower green sand
skirting the Wolds of Lincolnshire, the chief place where it
attains any development in this county being the gently undulating district about Hagworthingham between Market
Raisin and Spilsby. In the great fen district of the Bedford
Level, in the counties of Norfolk, Suffolk, and Cambridge, the
lower green sand is entirely concealed by the alluvial deposits
of the fens, but a few miles north of Cambridge it begins to
appear and thence extends in a south-westerly direction, forming a well-marked zone of considerable breadth. Its general
range here is parallel to that of the gault, but projections of
the sand hills frequently jut out westward to a considerable
extent, thus breaking the general smoothness of the outline
and giving to this formation a marked and peculiar character
which is altogether wanting in the gault and the upper green
sand. With the exception of these projecting eminences, the
breadth of the green sand (about two miles and a half) may be
considered as tolerably uniform between Cambridge and
Leighton Buzzard, a little south of which it contracts rather
suddenly, and continues to Abingdon with an average breadth
of little more than a quarter of a mile. From this point it is
very slightly developed all along the western border of the
chalk. It appears however in insulated masses or outliers of
great extent, capping all the hills in the western part of Dorsetshire and the neighbourhood of Chard, Axminster, and
Honiton. In Devonshire it forms the picturesque and highly
varied scenery of the Blackdown Hills, which owe much of
their peculiar character to the deep valleys and alternating
ridges of this sand. In the Blackdown Hills the green sand
overlaps in succession the edges of all the formations between
itself and the red sandstone, and actually rests on the latter
throughout most of its western and southern boundary.

Smaller and still more detached outliers extend much further westward in Devonshire, in fact considerably to the
south-west of Exeter, where, as at Chudleigh, the lower green

sand is found in contact even with the Palæozoic rocks, proving the enormous area originally covered by the greer sand before its denudation.

The chalk of Hampshire and the north and south downs is accompanied throughout by a prominent zone of lower green sand, which preserves a parallelism to the gault and has the same general horseshoe shape encircling the Weald of Kent and Sussex. About one-half of the whole surface of the Isle of Wight is also occupied by the lower green sand.

The formation attains its chief prominence in the range of hills stretching almost in a westerly direction from Reigate, by Dorking, and Godalming to Haslemere. In this part of the range are the towering eminences of Leith Hill, Hind Head, and the Blackdown Hills of Hampshire, several of which are 1000 feet high above the sea, and considerably over-top the neighbouring chalk hills.

A considerable number of towns are situate either on the lower green sand, or in such contiguity to it, that this formation may not improbably, either now or at some future time, be resorted to for a supply of water. Among these towns are Cambridge, St. Neots, Potton, Biggleswade, Bedford, Shefford, Buckingham, Fenny Stratford, Leighton-Buzzard, Aylesbury, Thame, Oxford, and Abingdon. Similarly situated are the towns of Wellington, Chard, Honiton, Axminster, Sidmouth, Colyton, Axmouth, and Charmouth, in the west of England.

Within the influence of the green sand which encircles the Weald of Kent and Sussex, (to say nothing of the metropolis itself,) we have Folkestone, Sandgate, Hythe, Maidstone, Seven Oaks, Tunbridge Wells, Reigate, Dorking, Guildford, Godalming, Farnham, Alton, Petersfield, Petworth, Arundel, Eastbourne, Brighton, and Lewes.

The lower green sand in the tract between Cambridge and Leighton Buzzard, occurs in the form of grey or brown sand, but "chiefly," says Dr. Fitton, "as a coarse ferruginous compound of quartzoze sand cemented by hydrate and oxide of iron, and more or less indurated. At the top, however, is some green sand, as appears from the first discharge from the borings through the gault, after the rod has passed the clay ; the water subsequently obtained, depositing an ochreous matter of the colour of the Woburn sands." The whole mass, however, does not consist of sand as there is a considerable thickness of fullers-earth interposed in the sands of Woburn, which are extensively dug for the purpose of procuring the fullers-earth. This mineral has also been observed in the lower green sand of Norfolk. A great deal of the green sand district, west of Woburn, particularly in the neighbourhood of Leighton Buzzard, is much covered by drift gravel which conceals the sand except where it rises into hills. According to Greenough's map there is alluvium no less than 600 feet thick covering the green sand at Elsworth, west of Cambridge.

The green sand of the Blackdown hills rests usually on the red marl of the new red sandstone, and sometimes on the lias. Dr. Fitton describes the sandy surface as barren, but the marl, which forms the base of the hills for about two-thirds of their height, is usually fertile and presents a great contrast to the barrenness of the sand. The Blackdown hills are celebrated for the scythe stones which are obtained from sandy concretions occurring in these hills. Everywhere around the Blackdown hills are found the sources of the principal rivers of Devonshire, Somerset, and Dorset. The Culm, the Tone, the Parret, the Axe, and the Otter, all derive large supplies from the water of these sand hills thrown out by the marls on which the sand reposes.

The lower green sand of Sussex and Hampshire has been described in great detail by Dr. Fitton, in his admirable paper on the strata below the chalk (2nd series, Geol. Trans., Vol. IV., p. 103). He divides the lower green sand into three distinct groups, which may be characterised as follows :

 a. The upper division consisting principally of sand, white, yellowish, or ferruginous, with concretions of limestone and of chert frequently in false stratification. This division immediately underlies the gault rising up above the valley of the latter, and bearing a dry barren soil. Its thickness at Folkestone is about 70 feet.

b. The second member is described as a retentive stratum abounding in green matter and containing little stone. The water falling on the upper group *a*, in the neighbourhood of Folkestone, does not penetrate to the base of the lower green sand, but is thrown out in springs above the retentive stratum *b*. Many of the wells in Folkestone derive their supply from water which is upheld by this middle sandy clay. Thickness, near Folkestone, from 70 to 80 feet.

c. The third or lowest division, which rests on the Weald clay, is more calcareous in its composition than either of the others, and contains the principal beds of stone bearing the name of Kentish rag. These indurated beds commonly form a steep ridge or escarpment overlooking the valley of the Weald. Thickness, probably, 200 feet.

A part of the supply of water to Sandgate is derived from the springs which break out on the surface of this middle bed, as described in Mr. Blackwell's Report to the President of the Board of Health.* The water from this part of the green sand has been so bad during the last year, namely so full of gritty sand, and so impregnated with iron, that Mr. Blackwell recommends the supply to be taken from the same springs which are used for Folkestone, namely the chalk marl springs near Cheriton.

Dr. Fitton believes, from the general information he has received, and the observations he has made in Surrey and Hampshire, that the same sub-divisions exist in the lower green sand all round the Wealden area which it encloses.

The lower green sand of Sussex occupies about an average

---

* Report by Thos. E. Blackwell, Esq. to the President of the General Board of Health, dated 21st February, 1855. Parliamentary Paper, Session 1355.

breadth of a mile and a half from Pevensey as far west as the valley of the Arun. Here it expands to a much greater width, and continues through Hampshire and part of Surrey, as far as Dorking, with an average breadth of not less than five miles. At Dorking it contracts in width, while it accompanies the steep ridge of chalk called the Hog's Back ; but again expands to a width of several miles in the neighbourhood of Seven Oaks and Maidstone, and continues through Kent, still with a breadth of several miles, to its termination on the coast at Hythe and Folkestone.

The most prominent parts of the lower green sand in Surrey are the elevated crests of Hind Head and Leith Hill. Dr. Fitton attributes the great development of the lower green sand in this part of the country to flexures and undulations in the strata, which cause a repetition of the same beds to appear at the surface.

It seems probable that many of the small streams which flow into the river Wey from the elevated district of Hind Head and Woolmer Forest, have their origin in springs of the middle retentive division, similar to those observed at Folkestone and Shorncliffe. The river Wey itself in its course through the green sand district probably flows in a bed of drift, but the Rother, from Petersfield to its junction with the Arun, near Pulborough, seems to flow on the surface of this retentive middle division. Dr. Fitton observes, that most of the large ponds in the neighbourhood of Dorking, Godalming, Woolmer Forest, Frensham, and Pulborough, are situate on the same stratum.

The thickness of the lower green sand at Folkestone has been determined with tolerable accuracy, at about 400 feet, and Dr. Fitton thinks it does not much exceed this in Surrey, notwithstanding the greatly increased breadth it occupies. Mr. Prestwich, however, from a tolerably exact general measurement which he made with Mr. Austen found the thickness at Chilworth, between Guildford and Dorking, 680 feet. He also gives the following as approximations founded on general observation :—

		Thickness of Lower Green Sand. Feet.
Kent, Seven Oaks	. . . . .	500
Surrey, Farnham	. . . .	700
Wiltshire, Devizes, and Calne	. . . .	20
Oxfordshire, generally	. . . .	150
Buckinghamshire, Leighton Buzzard	. . . .	250
Bedfordshire { Woburn	. . . .	350
Biggleswade	. . . .	250

Mr. Prestwich describes the area occupied by the lower green sand surrounding the chalk basin of London as 650 square miles, but a great part of this area is covered by beds of drift. In Wiltshire, Oxfordshire, Kent, and Surrey, the covering of drift is altogether absent or of inconsiderable thickness. In Buckinghamshire the thick beds of drift by which the sand is covered, are generally permeable and sandy. In Bedfordshire, however, the drift gravel or sand is again overlaid by an impermeable formation, called the "boulder clay drift." In Cambridgeshire and Norfolk the covering of drift is more general and impermeable.*

Mr. Prestwich has evidently made many minute examinations of the lower green sand with a view to ascertain its capability to yield a supply of water for the metropolis, by means of wells sunk or bored around London, through the tertiary strata and the chalk. He has therefore investigated, with great minuteness and detail, the extent of permeable surface assignable to this formation, and also the proportionate thickness which the beds of sand bear to the argillaceous part of the lower green sand formation. The conclusion which he seems to arrive at on the latter point is, that the group may be considered as consisting of 117 feet in thickness of impermeable clays, and of 250 feet of permeable sands.

Although Mr. Prestwich assigns an area of 650 square miles to the lower green sand surrounding the London basin, he does not calculate on the whole of this as a contributing area,

* Prestwich on the Water-Bearing Strata of London, p. 88.

whose waters would drain or filter into wells sunk in the centre of the basin. Referring to the probable existence of argillaceous beds, similar to those described by Dr. Fitton in the neighbourhood of Folkestone, and which possibly form distinct water levels under London, he makes some large deductions from his effective surface on this account. He also makes some further deductions for parts covered by impermeable drift, and for interruptions caused by faults and flexures of the strata. As the data for these deductions are said by himself to be very general, it will be unnecessary to quote them here in detail. Suffice it to say, that the whole of the deductions being made, the result is, to reduce the effective drainage area of lower green sand to 230 square miles. He then assumes an annual rainfall of $26\frac{1}{2}$ inches over the whole of this area, and a probable absorption of 16 inches annually, which gives a quantity of nearly 146,000,000 gallons absorbed in 24 hours.

Mr. Prestwich gives a section across the green sand formation between the chalk and the weald, to represent generally the features of the country, for a breadth varying from three to six miles between the north downs and the Weald of Kent and Surrey.

In this section, he shows the lowest valley line at the level of the gault, where the green sand begins, and then shows the outcrop of the lower green sand and commencement of the weald at a considerably higher level.*

He further shows the weald dipping under the green sand, with an undulating outline, so that about the centre of the green sand district the weald is not many feet below the surface of the ground, whereas, if it dipped at the same angle as the chalk it would here have been several hundred feet in

---

* This form of section may be correct in the country about Dorking, where the green sand hills are very high; but it certainly does not apply to the neighbourhood of Merstham, where the Brighton railway intersects the green sand. On this section the gault is at a considerably higher level than the point at which the wealden commences.

depth. This bent, or undulating stratification of the weald has been confirmed by many observers, and by the evidence of shafts sunk through the green sand. In fact, in the neighbourhood of Dorking and Guildford the weald comes to the surface in the centre of the green sand district. Dr. Fitton says, the depth of green sand overlying the weald clay in Surrey is often very inconsiderable, and Mr. Hopkins, in his paper on the structure of the Wealden district,[*] has pointed out an almost continuous line of disturbance or flexure, extending from Farnham by Guildford and Dorking to Seven Oaks. This line of flexure everywhere brings up the weald clay very nearly to the surface.

Mr. Prestwich shows the presumed water level in the lower green sand by a line drawn from the outcrop of the gault to the commencement of the wealden. This line is not straight but somewhat arched or bent upwards. He remarks on the permanence of the springs, which break out where this line of saturation cuts the surface of the ground, and illustrates this by observing that a fall of the water level to the extent of only one foot probably sets free 50,000,000 gallons of water in each square mile. This calculation is made on the supposition that each cubic foot of the lower green sand is capable of absorbing two gallons of water, a result which his experiments lead him to anticipate.

Having arrived at the conclusion that the exposed area of the lower green sand which surrounds the London basin is capable of absorbing daily the enormous quantity of 146,000,000 gallons of water, Mr. Prestwich next draws attention to the immense volume of water which will be permanently held in the 4,600 square miles of subterranean area attributable to the lower green sand. He assumes the thickness at 200 feet, so that the whole capacity of the subterranean water-bearing mass will be equal to 920,000 square miles one foot thick. Now, according to the experiments of Mr. Prestwich, each of these 920,000 masses or square miles will hold more than a

* Transactions of the Geological Society, 2nd series, Vol. vii. p. 1.

day's supply for the whole metropolis ; so that we appear to have beneath our feet a subterranean reservoir holding at this moment a 25 years' supply—being a tolerably capacious reservoir.   Mr. Prestwich argues from this vast capacity a permanent and steady maintenance of the supply to be taken from these sands.

The conclusion which he seems to draw from all the researches he has made is, that the upper green sand would yield in artesian wells from 6,000,000 to 10,000,000 gallons a day, and that the lower green sand would yield 30,000,000 to 40,000,000.   Mr. Prestwich estimates the thickness and depths of the strata beneath London as follows :=

	Thickness. Feet.	Depth.
Tertiaries	200	
Chalk	650	
		850
Upper green sand	40	
Gault	150	
		1040 to the

top of the lower green sand.

He further estimates that the water from the lower green sand would probably rise in artesian wells to a height of about 120 feet above Trinity high water mark, and that from the upper green sand about 10 feet higher.

These views of Mr. Prestwich have lately been considerably modified, as there is now reason to suppose that the lower green sands are not everywhere continuous beneath the chalk.   For instance, at Calais, after boring through the chalk to the depth of nearly 1,300 feet, the strata of the carboniferous period were met with, and all the middle and lower secondary strata were wanting.   Again, at Harwich lower carboniferous strata were met with after passing through chalk, succeeded by about 60 feet of upper green sand and gault ; whilst at Kentish Town a series of red sandstones were found, after passing through 1,113 feet of chalk, succeeded by gault.   Mr. Prestwich * remarks that

* Report to Metropolitan Board of Works on the Boring at Crossness.

these facts, taken in conjunction with other phenomena observed in Belgium and the West of England, have led geologists to believe that a ridge of old rocks of unknown width ranges under the chalk from Belgium, passing under the valley of the Thames, and continuing to the West of England.

It is unfortunate that the boring at Crossness has been abandoned at a depth of 961 feet, at which point it had only penetrated through 147 feet of gault. This was not deep enough to determine the thickness of the gault, nor to solve the question of continuity in the lower green sand. So far as this boring is concerned, it is still unsettled whether the gault is succeeded by the lower green sand or by rocks of Palæozoic age, as at Calais, Harwich, and Kentish Town.

Whatever may be the case, however, as to the succession of strata below the chalk in or near the centre axis of the London basin, it is quite certain that all round the margin of the basin, from Folkestone, by Hungerford and Dunstable, and so on through Cambridgeshire into Norfolk, the chalk is everywhere underlaid by the gault and lower green sand.

SPRINGS OF THE LOWER GREEN SAND.

Where the weald clay dips with a uniform inclination under the green sand, it is probable that no springs are caused at the escarpment. The surface of the clay may be wet and marshy, but no perennial springs will appear. At places, however, where the dip of the wealden clay has an undulating outline like that of the gault or chalk marl as represented in fig. 16, p. 51, springs will probably break out from the escarpment if the crest at x be higher than the outcrop of the sand at y. However small the trough may be, it will have communication with a large area and mass of porous strata, so that the pressure of water to cause springs will be considerable. Such undulations do certainly exist, as the clay brought up in this way makes its appearance at the surface in the valley of Pease Marsh, near Guildford, also on the south-

east of Dorking, and at other places within the area of the lower
green sand. The same phenomena of springs will occur if we
suppose the strata fractured by a fault, as shown in fig. 15,
p. 49, representing a state of things in which the bending of the
weald clay has been so considerable as to produce actual dis-
ruption. Of course, in order that the spring may flow at e, we
must imagine the point h at the fracture, which corresponds
with the crest x in fig. 16, to be at a higher level than e.

It is probable that the springs of Leith Hill and Hind Head
are due either to fractures in the weald clay beneath the sand,
as represented in fig. 15, or to an undulation as in fig. 16.

We have already spoken of Artesian wells from which water
rises to or above the surface in the vicinity of Cambridge,
and of overflowing wells at Wrest Park, in Bedfordshire.

A similar phenomenon occurs in the depressed area of
gault near Biggleswade, where the ground is below the line of
saturation in the sand, and where the pressure of water from
the green sand ridges occasionally causes springs to break out
and discharge themselves into the adjacent streams.

THE WEALDEN AREA OF KENT AND SUSSEX.

This is a peculiar isolated tract entirely surrounded, except
at its eastern extremity, by the escarpment of the lower green
sand. Its shape resembles that of a horse shoe, the rim being
formed by the belt of green sand, while the open part faces the
sea from near Beachy Head to Hythe. Considered from a more
extended point of view, the whole wealden formation may be
described as an irregular ellipse with a curvilinear axis of
about 150 miles, and an extreme breadth of about 40 miles.
Measured in the line of the transverse axis, a breadth of about
40 miles has been broken through by the English Channel, and
as the French or eastern extremity of the ellipse is comparatively insignificant, the part of the ellipse which is left on the
English side is about 80 or 90 miles in length, from the coast
to Haslemere forest. The boundary of the English wealden
district may be roughly traced from near Beachy Head on the

Sussex coast, thence in an almost westerly direction towards Pulborough and Petworth. Then it passes on the east side of Alton, round by Petersfield, and thence in an easterly direction by Farnham, Guildford, Dorking, and Reigate, to the south of Seven Oaks and Maidstone, to Hythe on the Kentish coast.

The surface of this tract consists of two distinct parts, namely the Weald clay, and the Hastings sands. The former is an argillaceous deposit, which appears to dip everywhere beneath the lower green sand, but its continuity at distant points within the cretaceous area has not been satisfactorily established. The breadth of the surface occupied by the weald clay may be taken at about five miles, the breadth being somewhat greater in the northern part of the district than in the part which ranges parallel with the south downs. The Hastings sand occupies the whole tract inside the belt or zone of clay. The Wealden district is traversed by a winding anti-clinal axis, from which the strata dip in opposite directions, those on the north side dipping to the north, and those on the south in the opposite direction. The principal towns either situated within, or bordering the Wealden area, are Hastings, Battle, Pulborough, Petworth, Horsham, Tunbridge, Tunbridge Wells, Hythe, Rye, and Winchelsea.

The drainage of the Wealden district is effected in a remarkable manner, namely by rivers which rise in the elevated central parts and mostly break through the barrier of sand and chalk which forms the north and south downs. The principal rivers which rise in the Weald and pass in this way through the barrier, are the Ouse or Newhaven river, the Adur or Shoreham river, the Arun, the Wey, the Mole, and the Medway. Mr. Poulett Scrope, Mr. Martin, and other geologists, among the most recent of whom is Mr. Hopkins, all support the hypothesis that these rivers do not flow in simple channels of denudation, but in gorges produced by antecedent fissures cutting entirely across the Weald. The gorges of the Arun and the Wey are nearly opposite to each other, and are probably parts of the same fissure. In the same manner the gorges of the Adur and the Mole, and of the Ouse and the Medway, range nearly with each other in straight lines. The whole detail connected with the subject of these transverse fractures, and their connection with the great central axis of elevation, and with several other lines parallel to the latter, is very elegantly worked out in Mr. Hopkins' paper already quoted.

Not only does Mr. Hopkins agree with, and confirm the opinions of, former geologists, as to the transverse fractures through the boundary or encircling frame of the Wealden district, but he points out in addition many instances in which the Medway and other rivers break through the principal raised axis of the Weald itself. In gorges caused by transverse fractures of the main east and west ridges, several branches of the Medway flow through in a north and south direction, in the immediate neighbourhood of Penshurst and Lamberhurst.

THE JURASSIC OR OOLITIC SER. FS.

This consists of the following subdivisions :—

The Portland oolite or limestone.
Portland sand.
Kimmeridge clay.

Coral rag, calcareous grit or Headington oolite.
Oxford clay.

Cornbrash.
Bath or great oolite.
Fullers-earth, clay, and limestone.

Inferior oolite and sand.
Upper lias.
Lias marlstone.
Lower lias.

The tract of country which embraces these various subdivisions of the oolite formation, extends in a curved line from the coast of Yorkshire to Lyme Regis in Dorsetshire. A line passing through the centre of the oolitic district from Whitby to Lyme Regis, would be about 320 miles in length, while the breadth of the formation is extremely irregular, in some places not more than two or three miles, and in others as much as seventy. The average breadth may probably be about thirty-five miles. North of the Humber, the oolites of Yorkshire occupy an extreme breadth from Filey on the coast to near Thirsk, of rather more than forty miles, with about the same length in a north and south direction, from Redcar at the mouth of the Tees to beyond New Malton. South of this place they extend to the Humber, forming a narrow zone scarcely more than two miles in breadth.

The oolites of Yorkshire embrace the eastern Moorlands, the Hambledon, and the Howardian Hills, together with the fertile clay districts of Cleveland, the Vale of Esk, and the Vale of Pickering. The high country of the Moorlands, which rise to elevations varying from 1,000 to 1,500 feet above the sea, consists of rocks corresponding to the inferior and great oolite of the south, although the division of fullers-earth which separates them in the south appears to be wanting in Yorkshire. The clays of the south are in fact represented by shales in the Yorkshire oolites, while the limestones of Bath, Cheltenham, &c., are represented by sandstones and grits. The sandstone, shaly, and often ferruginous beds of the elevated districts rest on a platform of lias, which surrounds them in a semicircular form from Redcar on the coast, passing round near Northallerton and Thirsk to New Malton, and thence, still continuing in a narrow band, by Pocklington and Market Weighton to the Humber. The lias is also exposed entirely across the oolitic district from Stokesley on the east to Whitby on the coast. The river Esk flows in this valley of denudation, which is also traversed by a great basaltic ke, extending many miles inland as far as the Durham coal field. Resting in order on those divisions which correspond with the great and inferior oolites of the south, is a mass of the Oxford and Kimmeridge clay, which occupies nearly one third of the whole oolitic district of Yorkshire. These two divisions, the Oxford and Kimmeridge clay, are separated by about 200 feet in thickness of calcareous and coralline grits, which do not occupy much breadth on the surface, but which are remarkable for the occurrence of swallow holes, into which rivers and streams are absorbed and disappear for some miles of their course.

The drainage of the district is principally effected by the Esk and the Derwent, and by tributaries of the Ouse, which latter rise from the escarpment of the oolitic hills, and flow over the lias. The Esk flows chiefly through the lias, and is fed by numerous springs arising from the high Moorland district on each side of it. The Derwent lies chiefly in the Vale of Pickering, flowing through the Kimmeridge clay, but is fed by numerous tributaries from the Eastern Moorlands and the Howardian Hills, which flank the Vale of Pickering on three sides. These tributaries flow in succession over the great oolite, the Oxford clay, the coral rag, and the Kimmeridge clay, in the same direction as the dip of the strata; and it is under these circumstances that the water is so often engulphed in swallow holes, which correspond with fissures and cavities in the coral rag, and its representative grit-stone beds. Having received all its principal tributaries, however, in the Vale of Pickering, the Derwent, now a considerable river, breaks through the Howardian Hills in a direction opposite to the dip of the strata, and crosses in succession the coral rag, the Oxford clay, the great oolite, the lias, and the new red sandstone.

Few towns of much importance are situate in the oolitic district of Yorkshire. The principal are Whitby, Scarborough, Pickering, and New Malton. These would, probably, be able to derive abundant supplies of water from impounding reservoirs, constructed across the deep narrow gorges of the oolitic hills. The water of these hills not being so highly impregnated with calcareous ingredients, is very superior in softness to the ordinary water of oolitic districts in England. The Howardian Hills, the Hambledon Hills, and the Eastern Moorlands of Yorkshire will, probably, become of much importance in future years, as capable of yielding supplies of water to towns on the lias and new red sandstone districts lying to the eastward. It is probable that the rich and fertile valleys of the

Ouse and the Tees will so increase in population as to require supplies of water from these Yorkshire hills.

THE OOLITIC DISTRICT FROM THE HUMBER TO NEAR BATH.

This comprises the development of the formation in the counties of Lincoln, Leicester, Rutland, Northampton, Huntingdon, Bedford, Worcester, Gloucester, Oxford, Buckingham, and Wilts. The breadth at the Humber is nearly eight miles, from which it gradually increases in a southerly direction till, at Lynn, the breadth of the formation measured across the fens of Lincolnshire to its western boundary near Loughborough, is not less than seventy miles. In this breadth, however, about one half, consisting of the Oxford clay, is covered over by the fens of Lincolnshire. From this extreme development the breadth somewhat diminishes, as at Oxford a line measured across the district at right angles to its imaginary axis, would be about forty miles. From Stratford-on-Avon the outline of the lias may be traced all the way to Bath and Bristol, but it becomes very tortuous in Warwickshire, Worcester, and Gloucestershire, jutting out in many irregular hills and isolated prominences into the new red sandstone country.

The upper divisions of the oolite above the Oxford clay, are but slightly developed in all this extensive tract. The Kimmeridge clay is of inconsiderable extent in all the northern part, although it may be traced continuously under the green sand from the Humber into Buckinghamshire, where it attains a breadth of several miles in the forest of Brenwood, between Aylesbury and Oxford. The members of the oolitic series above the Kimmeridge clay are so little developed, as not to require notice in a hasty sketch of this kind.

The coral rag first appears a little north-east of Oxford, and extends in a narrow zone about three miles wide, by Abingdon, Farringdon, Highworth, and Wotton Basset, to Chippenham

F

and Calne, soon after which it disappears. The Oxford clay occupies a lenticular space between the Humber and Lincoln, where the fens commence. It may be traced, however, on the west side of the fens, by Sleaford and Bourne, almost down to Peterborough, and doubtless continues beneath the fens in all the eastern part of Lincolnshire.

South of Peterborough the Oxford clay expands to a considerable breadth in Huntingdon and Bedfordshire, except where it is denuded, and the great oolite exposed, in the valley of the Ouse around Bedford. Between Bedford and the neighbourhood of Bath, it occupies a very irregular zone, the eastern side of which is much encroached on, sometimes by the overlying Kimmeridge clay, and sometimes by the lower green sand. The breadth of the Oxford clay at Huntingdon is more than 35 miles, while at other places, as in the neighbourhood of Buckingham and Oxford, it is not more than two miles wide.

The great oolite has been already spoken of in the Eastern Moorlands of Yorkshire, and in the Howardian and Hambledon Hills. South of the Humber, the great oolite, with its subordinate beds of cornbrash and forest marble, everywhere accompanies and passes under the Oxford clay. North of Lincoln, it occupies a very narrow strip of country, but in the south of Lincolnshire it expands to a much greater width. This width, from Bourne in Lincolnshire to Saltby in Leicestershire, is not less than seventeen miles; three of which, however, are occupied by the denuded valley of the Witham, in which the lower oolite and the lias are exposed.

In Northamptonshire, Bucks, and Wiltshire, the great oolite is much developed, but has a very irregular boundary, frequently capping high grounds over considerable areas, and surrounded by zones of lower oolite and lias, which cut it off and isolate it in large irregular masses.

South of Bath the great oolite is only slightly developed. It extends in a narrow strip, by Frome as far as Bruton, where it begins to occupy a greater breadth, and spreads out in Gillingham Forest and Milborn Forest to a breadth of more than ten miles. South of Bradford Abbas the great oolite becomes extremely narrow, and can just be traced by Beamister and Abbotsbury as far as Weymouth Bay, where it appears as usual underlying the Oxford clay.

The fuller's-earth, or Frome clay, which separates the great and inferior oolites, is scarcely known in the north of England. It first appears in the neighbourhood of Winch-combe, in Gloucestershire, where it encircles several small irregularly-shaped conical hills, the top of each consisting of the great oolite, supported in a shallow basin of Frome clay, or fuller's earth, while the base of each consists of the lower oolite and lias.

From the neighbourhood of Northleach, under the Cottes-wold hills, to Bradford, Frome, and Bruton, the fuller's-earth may be traced continuously as a thin retentive motley-coloured stratum, everywhere separating the two great free-stone members of the oolite formation. This comparatively narrow strip of clay, which plays an important part in the hydrography of the oolitic country, continues by Bradford Abbas, Beamister, and near Bridport to the coast of Dorset-shire at Burton Cliff. Besides this continuous wavy line, the fuller's earth appears in all the outliers on the western side which are merely separated from the true oolitic country by valleys of denudation. Thus isolated hills in the neighbour-hood of Northleach, Stroud, and Bath, are all encircled by this narrow band of fuller's-earth, which commonly causes very copious springs to burst out from the lower beds of the superincumbent great oolite. There are no less than four hills in the neighbourhood of Bath which are thus encircled by the fuller's-earth. The most conspicuous of these are Lansdown and Claverton Down. The fuller's-earth formation may also be traced on both sides of the valleys around Bath, every-where occupying an intermediate position between the inferior and the great oolite, and presenting a broken irregular surface, showing abundant traces of landslips and watery action.

The inferior oolite, with its subordinate sands, is chiefly developed in Northamptonshire, Gloucestershire, and Dorsetshire. Its boundary is extremely irregular, and, like the great oolite, it caps numerous isolated hills where it rests on a base of lias, or other clay.

In this manner it caps the lias in the neighbourhood of Uppingham, in Rutlandshire, and also covers a considerable extent of lias country, extending from Market Harborough to Brackley, in Buckinghamshire. It appears at Northampton, Wellingborough, and Rothwell, underlying the great oolite which here forms the surface of the country.

The inferior oolite appears again in Campden and Bourton hills, near Evesham, and passes through Gloucestershire by Winchcombe, Northleach, and Painswick. It also occurs in Lansdown and other hills in the neighbourhood of Bath, forming the cap of Dundry and Stantonbury hills. South-west of Frome it appears in the platform of Doulting, extending by Bruton and Castle Cary to Bradford Abbas. From this point it increases considerably in width, and extends by Crewkerne and Bridport to the coast of Dorsetshire. The inferior oolite furnishes a famous building stone of a quality very superior to the ordinary Bath oolite. It is probable that the quarries in the neighbourhood of Bath first acquired their reputation from the employment of the lower oolites, which are now extensively replaced by beds of softer and more perishable quality from the great oolite. The inferior oolite of Dundry furnished the stone for building the beautiful church of St. Mary Redcliffe, at Bristol; while the Doulting stone, near Shepton Mallet, was used in Wells Cathedral and Glastonbury Abbey. In colour it is yellower than the great oolite, the oolitic grains are larger, and the cement which unites them stronger and more crystalline.

The lias formation is a great mass of clay with subordinate beds of limestone and marlstone which commonly underlies the inferior oolite, and is very extensively developed in England. This formation may be traced uninterruptedly from the coast of Yorkshire at Redcar and Whitby, to that of Dorsetshire at Lyme Regis and Axmouth. It ranges with a very variable width through all the counties lying between these extreme points. Its greatest breadth is probably in the neighbourhood of Towcester, where it occupies a tract nearly 60 miles in width, with some slight interruption by patches of the inferior oolite.

The oolitic system, like the cretaceous, has probably at one time been universal or nearly so. The members which compose it may be traced either in an entire or partial series around the edges of the three great chalk basins of France, namely, the Anglo-Parisian, the Pyrenean, and the Mediterranean. We find oolitic rocks in Spain, Portugal, Italy, Piedmont, Switzerland, Germany, Luxemburg, Swabia, Wurtemburg, Westphalia, Saxony, Bavaria, &c. They exist in Asia Minor, in the Crimea, covering a great extent of central Russia, and passing thence to the icy sea on both sides of the Ural Mountains. They appear in Indiana, North America; also in the Cordilleras of Coquimbo, in Chili. They have been recognised in the mountains of Himalaya, in the East Indies.

These distant points include an area which extends in the northern hemisphere over $60°$ of latitude, and in the southern hemisphere from the torrid zone to the 30th degree. The isolated outposts scattered at such immense distances over the surface of this planet prove that the oolitic formation is not a partial or local deposit, but that it has spread over by far the greater portion of the earth's surface.[*]

Taking a general stratigraphical view of the whole oolitic series, it may be said to consist of four great masses of calcareous or partially permeable strata separated by thick deposits of clay. Thus the Portland oolites, limestones, and sands repose on the Kimmeridge clay. 2ndly. The coral rag and calcareous grits of Headington rest on the Oxford clay.

* D'Orbigny.

3rdly. Beneath this come the cornbrash and Bath, or great oolite, resting on the fuller's-earth. Below this, again, is the inferior oolite, and its accompanying sands resting on the thick mass of the lias clay.

This remarkable alternate succession of porous and impermeable beds, gives rise to innumerable springs all over the oolitic country; and the repetition of similar beds gives rise to similar phenomena from the top to the bottom of the series. Thus, the Kimmeridge clay, which probably varies in thickness from 70 to 600 or 700 feet, throws out the water which sinks through the porous beds of Portland stone and sand. Hence in sinking wells through the Portland beds, which are usually less in thickness than the Kimmeridge clay, water will commonly be met with at no very considerable depth; whilst wells sunk in the clay may possibly have to penetrate 700 feet before deriving any water, which can only arise from the coral rag beneath. The Kimmeridge clay is usually of a bluish slate colour, and frequently contains beds of bituminous coal. Sometimes, however, the colour approaches more to that of grey, and is sometimes even yellowish or reddish brown. It has an unctuous, greasy feeling when rubbed, and has frequently a fissile, laminated appearance, owing to the presence of vegetable matter.

The French geologists estimate the upper division of the oolitic series, from the top of the Portland rock to the base of the Kimmeridge clay, at 700 feet in thickness, 200 of which they assign to the Portland rock and sand, and 500 feet to the Kimmeridge clay. In the neighbourhood of Weymouth, the Portland stone and sand are each about 80 feet in thickness; and the Kimmeridge clay, at the village of Kimmeridge itself, is about 600 feet, but it gradually becomes thinner, till at Oxford it is reduced to 70 feet.

Dr. Fitton gives the following thicknesses:—*Portland Stone*—In Portland Island, between 60 and 70 feet; at Swindon, 60 to 65; at Great Hazely, in Oxfordshire, 27 feet; at Brill, about 23 feet; near Quainton, and Whitchurch in

Buckinghamshire, from 4 to 20 feet. *Portland Sand*—Near St. Alban's Head, in the Isle of Purbeck, 120 to 140 feet; in the Isle of Portland, 80 feet; near Thame, in Oxfordshire, about 50 feet.

According to Sir Henry de la Beche and the Dean of Westminster, the thickness of the Kimmeridge clay in the bay from which it takes its name is 600 feet; but they state the thickness at Ringstead to be only 300 feet. Sir Henry de la Beche considers the general thickness in the south-west of England about 500 feet. Dr. Fitton says it is only about 20 feet thick in one of the Headington quarries near Oxford.

The Portland oolite and sand, as well as the Kimmeridge clay, are marine deposits, in which littoral shells appear to be mixed up with those inhabiting deeper water. Some parts of the Kimmeridge clay contain so much bitumen as to inflame spontaneously, and continue smouldering for long periods. Some of the bituminous beds are used for fuel in the neighbourhood of Kimmeridge.

Water issues abundantly round the edges of the Portland sand, where it is underlaid by the Kimmeridge clay. In sinking through the latter the small quantity of water which is met with in the partings is usually of inferior quality; and it is commonly necessary to sink through the whole thickness of the Kimmeridge clay before finding an abundant supply of good water. The supply for Boulogne-sur-Mer is obtained from springs at the top of the Kimmeridge clay, but the quantity is miserably deficient. An attempt is being made to procure an additional quantity by sinking in the clay. This attempt will probably fail, although the cliffs are extremely wet, owing to the percolation of water into the minute sandy partings of the clay.

THE CORAL RAG AND OXFORD CLAY.

The coral rag in Wiltshire, according to Mr. Lonsdale, is 230 feet in thickness. On the coast, near Weymouth, Sir H. de la Beche gives the thickness at 150 feet. The Oxford clay at Weymouth is about 300 feet in thickness, but Sir H. de la Beche considers the general thickness in England about 600 feet. The French geologists differ considerably from the English in estimating the thickness of these formations. They assign nearly 1,000 feet to the Oxford oolite and coral rag, and nearly 500 feet to the Oxford clay.*

The coral rag is so named because the central part of it consists of a loose rubbly limestone, which is almost entirely composed of an assemblage of branching corals belonging to the family of the madrepores. Above this central coralline mass are beds of calcareous freestone, tolerably close in texture, frequently very indistinctly oolitic, but sometimes consisting of such large oviform grains as to have occasioned the name of *Pisolite* (pea stone), which was applied to it by Mr. Smith, the father of English practical geologists. Below the coralline part are yellowish, sandy, calcareous beds, traversed by irregular strata, and concretions of indurated calcareo siliceous grit stone.

The water of the coral rag is frequently impregnated with iron, which is commonly abundant in the lower sands of the series. The coral rag of Westbury, in Wiltshire, has been worked for iron smelting ever since 1860. This formation is one of those which, owing to its loose incoherent nature, and to the fissures which traverse it, is found to engulph or swallow up the streams flowing over it, especially those whose course is opposed to the natural dip of the beds. Several examples of this may be seen near Headington in Oxfordshire, where the small streams which flow from the surface of the Kimmeridge clay towards the Cherwell are frequently lost in passing over the coral rag.

The Oxford clay is a thick mass of dark blue tenacious clay, frequently mixed with calcareous, and sometimes with bituminous matter, and abounding in places with septaria and argillaceous geodes, traversed by veins filled with carbonate of lime. Many of the beds contain sulphur in the form

* D'Orbigny's Cours Elémentaire.

of iron pyrites, besides sulphates of lime, gypsum, and magnesia. These mineral ingredients often impregnate the waters of this formation to such an extent as to render them valuable for medicinal purposes. Some of these waters are purgative, as at Stanfield, in Lincolnshire; Kingscliff, Northamptonshire; Cumner, Berkshire; Melksham and Holt, Wiltshire.

The water of Woodhall Spa, in Lincolnshire, which is drawn from a well in the Oxford clay, is strongly impregnated with iodine and bromine. Chalybeate and other mineral springs occur in the Oxford clay of Wiltshire. It is usually necessary to sink a considerable depth in order to procure water in the Oxford clay. A well, 478 feet in depth, has been sunk through it at Boston, in Lincolnshire, without obtaining an adequate supply. The well at Woodhall is said to be 840 feet in depth,* having been sunk as an attempt to procure coal. A brine spring, however, was met with at a depth of 500 feet, and this is the one which contains the iodine and bromine. From the account given by Dr. Granville of the strata passed through, the water appears to come from the cornbrash at the base of the Oxford clay.

THE GREAT OOLITE AND FULLER'S-EARTH.

The beds ranked by geologists under this name vary much in composition in different parts of the oolitic range. The cornbrash, however, generally appears to succeed the Oxford clay, and interpose between it and the oolitic beds, from which this part of the series takes its name. The cornbrash is a loose rubbly limestone, of a grey or bluish colour when broken, but on the outside usually brown and earthy. It rises when quarried, commonly in thin beds, and is seldom altogether more than 15 or 16 feet in thickness. The great oolite, in Yorkshire, consists of sandstones, shales, and limestones, about 230 feet in thickness, and is not succeeded here

* Granville on the Spas of England.

F 3

by fuller's-earth, but by a mass of sandstones and shales of 280 feet, which separate the great oolite from the inferior oolite. In Lincolnshire the cornbrash is about six feet thick. To this succeed beds of clays, shales, shelly, and marly oolites, and, in the lower part, sands which rest on the inferior oolite. The whole thickness of the great oolite, and of the sands which correspond with the fuller's-earth, is not more in Lincolnshire, according to Mr. Morris, than about 116 feet, while, in Yorkshire, the thickness is more than 500 feet.

In the south-west of England, as in the neighbourhood of Bath and Frome, the cornbrash is succeeded by sands, and by beds termed forest marble, which are sometimes 100 feet in thickness. Then comes the Bradford clay, 40 to 60 feet; then the great oolite, 40 to 120 feet, and below this the fuller's-earth, which at places is 130 feet thick, and at others, thins out almost to nothing. The great oolite and fuller's-earth together, in the south-west of England, vary from 80 to 410 feet.* The French geologists assign a much greater thickness to this part of the oolitic series. They estimate the *Callovien* formation, which corresponds with the lower part of our Oxford clay, and with the Kelloway rock, at 500 feet. To the *Bathonien* formation, which corresponds with our cornbrash and great oolite, they assign a thickness of 200 feet. The fuller's-earth (*terre à foulon*) ranks with them as a member of the *Bajocien* formation, which comprises also the inferior oolite, having a total thickness of 200 feet. The Bradford clay where it occurs overlies the great oolite, but is commonly wanting, except in the neighbourhood of Bradford, on both sides of which it thins out.

The great oolite, so well known from the large quarries at Box, Farley Down, Bradford, and Comb Down, consists at top and bottom of shelly, rubbly oolites, enclosing a central mass of fine-grained oolitic freestone, varying from 10 to 30 feet in thickness. It is this freestone which is so extensively

* Morris on the Lincolnshire Oolites. Geological Journal vol. ix. p. 317.

worked in the neighbourhood of Bath, but its quality is by no means equal to that of the lower oolite, which has a much tougher and more crystalline structure.

The same general rule which has been before alluded to, as to the occurrence of water on the surface of clays underlying permeable rocks, applies to this part of the oolitic series. Where clays support the cornbrash and separate it from the great oolite, springs of water are found and water is easily procured from wells. The fuller's-earth again throws out the water which sinks through the great oolite, in which the wells are commonly of considerable depth.

The cornbrash, like the coral rag, is remarkable for absorbing the water of streams which flow over it. Messrs. Conybeare and Phillips observe, that about 30 swallow holes may be noticed in the space of half a mile around Hinton, in the sandy cornbrash of Somersetshire. A similar phenomenon occurs to various branches of the Rye about Kirby Moorside, and Helmsby in Yorkshire.

The springs which break out from the fuller's-earth where this passes under the great oolite are usually very copious. These springs are very abundant in the town of Frome and the adjacent valleys It must not be supposed that the mass to which the name of fuller's-earth has been given consists entirely of this peculiar substance. At Bath, and in the neighbourhood of Frome, where the fuller's-earth formation is best developed, the upper part consists of 30 to 40 feet of blue and yellow clay, then 5 to 8 feet of true fuller's-earth, succeeded by 100 feet of brownish clay, frequently inclosing beds of tough rubbly limestone, called the fuller's-earth rock.

The town of Cirencester is supplied chiefly from wells sunk into a gravel bed resting partly on the great oolite and partly on fuller's-earth. The deeper wells sunk into the fuller's-earth yield very copious supplies of water. The engine for supplying the summit level of the Thames and Severn Canal pumps more than three million gallons per day from a well in the fuller's-earth.

In all the deep valleys in the oolitic range such as those of Cirencester, Stroud, and Bradford, immense volumes of water are thrown out by the fuller's-earth, many of the springs, like those at Boxwell and Ampney Crucis, near Cirencester, yielding several million gallons a day.

INFERIOR OOLITE AND LIAS.

In Yorkshire the inferior oolite is about 80 feet in thickness; in Lincolnshire, 20 to 50 feet; and in the West of England, in the neighbourhood of Bath and Cheltenham, 130 to 230 feet. Mr. Lonsdale gives 130 feet as the general thickness in the hills around Bath, and observes that the upper 60 feet consist of limestone, having a distinctly oolitic character, while the lower 70 feet consist chiefly of sand, lightly calcareous, with irregular concretions and courses of limestone. In the Cotteswold hills, near Cheltenham and Leckhampton, the upper portion consists of sandy beds, the middle of good freestone of a yellowish white colour, and the lower bed is a pisolitic rock composed of rounded or flattened grains about the size of a pea, cemented by a calcareous paste.

The lias consists usually of three principal divisions, namely, the upper marls and shales, the marl-stone or blue lias rock, and the lower lias marls. The thickness of the lias in England varies in different places, and according to different authorities from 400 to 600 feet, and in Leckhampton Hill, in Gloucestershire, the late Mr. Strickland estimated the thickness of the three members composing the lias at 750 feet. In the neighbourhood of Bath, according to Mr. Lonsdale, the whole thickness of the lias is under 300 feet. Among the French geologists the *Bajocien* formation, which includes the fuller's-earth and inferior oolite, is estimated at about 200 feet, while the united thickness of the three members of the lias, namely, the *Toarcien*, the *Liasien*, and the *Sinémurien* stages are estimated at nearly 2,000 feet.

The French geologists consider the whole thickness of the oolitic series from the top of the Portland rock to the base of the lower lias shales to be nearly 5,100 feet.

The lias throws out springs at its junction with the inferior oolite, just in the same manner as the fuller's-earth does where it underlies the great oolite. These springs are very copious in the Mendip district, where the inferior oolite rests sometimes on lias, but often overlaps the edges of the latter, and reposes on the highly inclined strata of mountain limestone, which flank the Mendip range. The towns of Gloucester and Cheltenham both derive their supply of water from springs which rise at the base of the inferior oolite, where they are held up by the lias. Shepton Mallet and other smaller towns in Somersetshire are supplied in the same way by streams of running water, which have their origin from springs arising in this manner.

The mineral waters of Bath are derived from springs thrown out by the lias, while those of Cheltenham and Gloucester are drawn from wells sunk in the lias. Cheltenham and Gloucester spas contain a large proportion of the muriates and sulphates of soda, while the Bath water contains a very small quantity of the salts of soda, but a notable proportion of sulphate of lime. The Gloucester and Cheltenham waters also contain a grain of bromine in 6 to 10 gallons of water, while the presence of this substance has not been noticed in any analysis of the Bath water. The thermal springs in Bath vary in temperature from 80° to 120° Fah.

The principal places situated on the oolitic formation in England are Whitby, Scarborough, Malton, and Market Weighton, in Yorkshire; Lincoln, Sleaford, Grantham, Bourne, and a few smaller towns in Lincolnshire. Proceeding southward, the following towns are also on this formation:—Oakham, Stamford, Peterborough, Uppingham, Market Harborough, Huntingdon, Higham Ferrers, Northampton, Bedford, Evesham, Banbury, Buckingham, Oxford, Witney, Stroud, Tewkesbury, Cheltenham, Gloucester, Cirencester,

Fairford, Lechlade, Faringdon, Abingdon, Highworth, Crick
lade, Calne, Malmesbury, Chippenham, Bath, Melksham,
Bradford, Trowbridge, Frome, Shepton Mallet, Bruton,
Glastonbury, Sherborne, Yeovil, Ilminster, Crewkerne, Axminster, Axmouth, Charmouth, Bridport, and Weymouth.

THE TRIAS AND PERMIAN GROUPS.

This is the nomenclature now commonly adopted by English geologists for the old names of new red sandstone and
the magnesian limestone on which it rests. The trias group
consists in the upper part of variegated red, white, and blue
marls and clays, frequently abounding in gypsum and selenite,
while the lower part consists usually of red clays and sandstones, being generally much more arenaceous than the upper
or gypseous beds. The Permian group consists of dolomitic,
or magnesian limestones, and, in its lower portion, of sandstones, conglomerates, and beds of indurated and cemented
pebbles or pudding stones.

The eastern boundary of the trias group may be traced in
something like a continuous line from the mouth of the Tees
in Durham to the English Channel in Teignmouth Bay, but
it is far otherwise with the western limits of this great formation. The trias and the Permian groups form the last or
lowest in a stratigraphical sense, of those formations which
appear to have been deposited in regular, concentric, and
parallel layers, one over the other. It is true that frequently
one member of the series above the new red sandstone overlaps the edges of another, or perhaps several of those beneath
it, so that the order of succession is found to have gaps in
which certain strata are wanting. Thus, the lower members
of the cretaceous series are frequently found to overlap
several of the upper members of the oolitic series, so that
the latter do not appear at the surface at all; and the formation No. 2, for instance, is found to be in contact with, and
immediately reposing on No. 5, the intermediate members

3 and 4 being wanting. Notwithstanding this peculiarity, which is chiefly due to what is called the thinning out of particular formations, there is a general order of super-position, and a general parallelism of dip, throughout all the formations we have been considering, and this parallelism extends to the base of the Permian series. This parallelism, termed by geologists the conformability of strata, indicates that during the vast intervals of time which the Permian and the superior groups have required for their deposit, no very striking elevatory movements of the earth's surface have taken place ; and, at all events, that no very important movement has taken place in the interval between the completion of one formation and the deposit of the next succeeding one.

But when we look beneath the Permian formation, we find a widely different state of things. We find that movements in the crust of the earth, on a scale almost too vast for com-prehension, must have taken place between the deposit of the palæozoic, or first-formed rocks, and the commencement of that great parallel series which commences with the Per-mian group. In the language of geology, we find the meso-zoic or middle-life series of rocks, resting unconformably on the palæozoic or early-life series. Sometimes the trias or the Permian group rests upon granite or syenite, as on the flanks of the Malvern Hills ; sometimes it rests on carboniferous limestone, or other members of the great coal series ; some-times on the old red sandstone ; and sometimes on the still more ancient rocks of the silurian series. Whatever forma-tion it rests on, however, the stratification of the older group is never conformable, or parallel with that of the newer. The beds of the trias and Permian group in England are usually horizontal, while the older rocks on which they rest are com-monly inclined at high angles, and are sometimes nearly vertical. This remarkable absence of parallelism shows that in the interval between the close of the palæozoic and the commencement of the mesozoic period, immense subterranean forces must have been at work to overthrow and break up the solid crust of the earth, and that these forces had generally ceased to act before the mesozoic deposits commenced.

The vast plains of new red sandstone and magnesian limestone which cover so great a portion of this island, have been compared to the figure of a sea composed of horizontal beds of red marls, sands, and sandstones, surrounding elevated islands of carboniferous rocks, old red sandstone, Silurian rocks and others, all variously and irregularly stratified.*

These plains commence north of Carlisle and Sunderland, on the opposite sides of England, and skirt the elevated mountain district of Durham, Westmoreland, Yorkshire, Lancashire and Derbyshire, as far south as Nottingham and Newcastle-under-Lyne, between which places they border the elevated district of Staffordshire and Derbyshire. The extreme horse-shoe boundary of the mountain district thus skirted by the Permian and trias groups from Tynemouth round by Nottingham and Newcastle-under-Lyne, and thence northward to Newcastle-upon-Tyne is not less than 350 miles. On the east side the breadth of the plains will, probably, not average less than 18 miles, while on the west side the breadth is seldom more than 3 miles, except in the plains of Cheshire, and in the valley of the Eden, near Carlisle.

Throughout the whole of this extensive boundary of the palæozoic rocks the magnesian limestone underlies the new red sandstone, except in the southern part from Newcastle-under-Lyne to Nottingham, and in a part of Lancashire between Ulverstone and Preston, where the magnesian limestone has not been observed. The principal development of the Permian group extends on the eastern side of the mountain district from the mouth of the Tyne to Nottingham, a length of about 150 miles. Throughout this extent it occupies a somewhat depressed range of hills resting on the coal measures of the Durham, Yorkshire, and Derbyshire coal fields. The average breadth of this tract of magnesian limestone on the surface does not exceed 3 miles, but it

* Conybeare and Phillips.

furnishes abundant supplies of excellent lime in the neighbourhood of Knaresborough, Knottingley, and other places in Yorkshire. In Derbyshire are the famous quarries of Bolsover Moor, Anstone, and other places, which have supplied the stone for the New Houses of Parliament, and for many ancient and celebrated edifices all over the north of England.

The magnesian limestone in its chemical composition consists of nearly equal proportions of carbonate of lime and carbonate of magnesia, and has commonly a more granular and sandy structure than common limestone. The varieties which make such excellent lime at Knottingley, Kinnersley, and Brotherston, are usually hard, bluish-white, thin bedded limestones, while the best building stones of the formation are usually buff-coloured or yellowish freestones, which dry to an almost white colour when they lose the quarry water. In the neighbourhood of Nottingham the Permian formation contains thick beds of quartz gravel. Throughout the western part of its range the magnesian limestone from Preston by Manchester, Stockport, and Cheadle, is seldom more than a mile in width, but in the valley of the Eden its breadth is about 3 miles.

The trias formation occupies the space between the magnesian limestone and the lias on the eastern side of the island, while on the western side, between Liverpool and Carlisle, it occupies the space between the sea and the magnesian limestone, or the older rocks where this is wanting.

In Cheshire and Salop, the trias or new red sandstone occupies an extensive and regular plain of about 50 miles from north to south, and about 25 miles in breadth, interrupted only by a patch of lias which exists in the neighbourhood of Wem.

South of the Trent, the great plain of the trias consists of a triangular area in the counties of Stafford, Leicester, Worcester, and Warwick. The base of the triangle from Leicester to the Shropshire coal field is about 55 miles, and its length, from the base to the apex, at Newnham, in Gloucestershire, about 75 miles. The eastern boundary of this triangular district is new red sandstone underlying the lias, and the western boundary is a narrow strip of the Permian rocks, averaging about two miles wide, either resting or abutting on the coal measures of the Shropshire and Colebrook Dale coal fields, or the syenitic rocks of Malvern, or the silurians of Micheldean and Newent. In the midst of this great triangular plain of new red sandstone are three islands, in which the older rocks are protruded through the surface and occupy a considerable area. One of these islands consists of the elevated syenitic district of Charnwood forest, and the coal field of Ashby-de-la-Zouch, near Leicester. Another is formed by the syenitic hills of Nuneaton and the small coal field which flanks them on the western side. The third island consists of the Staffordshire coal field, between Birmingham and Wolverhampton, with its remarkable accompanying trappean and silurian rocks in the neighbourhood of Dudley. A patch of magnesian limestone appears on the western side of the Tamworth coal field, and a similar zone entirely surrounds the Staffordshire coal field, being usually faulted against the coal measures, not resting on them as in the north of England and in the south-west.

The new red sandstone throughout this great triangular area has the usual typical characters of the formation; the upper beds being the most argillaceous, and containing in places large quantities of gypsum. The variegated colours are chiefly peculiar also to the upper beds. The more arenaceous beds, and those which become indurated into stone, as distinguished from the upper marls and clays, are usually the lower beds of the series, and these are generally either red, or, more rarely, buff coloured.

Occupying a narrow gorge in the valley of the Severn at Newnham, the new red sandstone passes through Gloucestershire, Somersetshire, and Devonshire, to the sea coast in Teignmouth Bay. It is usually accompanied by its underlying beds of magnesian limestone, which in the south-west is commonly a hard conglomerate or breccia, containing angular fragments of carboniferous limestone and old red sandstone. As in the northern parts of England the trias and Permian formations surround the older elevated coal fields in the neighbourhood of Bristol. In the same manner they mantle round the Mendip Hills and overlap the edges of the old red and carboniferous limestone of Devonshire and Somersetshire. The mode in which fragments and masses of magnesian conglomerate hang on the sides of the older rocks in the neighbourhood of Bristol, and in the deep picturesque valleys of the Mendip Hills, is highly suggestive of speculation. One sees in imagination that ancient sea filled with the huge detritus from disrupted mountains of limestone, millstone grit, and old red sandstone. One connects the broken truncated fragments which hang on the precipitous side of a valley with those which appear on the opposite side, and marvels at the mighty agencies which have thus in succession formed and broken up, and again and again repaired the shattered surface of this planet. Great and infinite are the wonders revealed by geology!

The trias formation is largely developed both on the continent of Europe and in those of America. It exhibits itself in the Pyrenees, in the Var, on both slopes of the Vosges, and in Normandy. In other parts of Europe the new red sandstone is represented in Germany, Belgium, Switzerland, Sardinia, Spain, Poland, Bohemia, Moravia, and Russia. Beds of it appear in the United States of America, and cover vast surfaces in the republic of Bolivia in South America. They have been traced in Columbia and in Mexico, and may be said to extend from 20° of south latitude to 48° north.

The Permian formation has a still more extensive range, and is said to have been identified by means of its fossils as far north as Spitzbergen, in the 80th degree of north latitude. It derives its name from the government of Perm in Russia, where the formation is largely developed.

Neither the trias nor the Permian groups rise in this country into escarpments, nor do they present that regular dip which distinguishes the more recent formations, and hence it is that measures of absolute thickness are difficult to obtain. Nor are such measures of much value in formations of this kind, as the depth of sinking required to penetrate through them will vary greatly in different localities, and the geologist's measure of thickness will throw very little light on this practical part of the question.

Messrs. Conybeare and Phillips quote several instances of pits sunk from 600 to 720 feet through the new red sandstone without reaching the bottom. Two of these sinkings were in the county of Durham, and one at Evesham, in Gloucestershire. They also observe that in shafts sunk from the lias down to the coal measures at Pucklechurch, in Gloucestershire, the trias and Permian groups together were not more than 153 feet in thickness. In fact, in the southwest of England where the new red sandstone is spread out over the upturned edges of the carboniferous rocks, there are many places where it thins out to only a few feet of thickness.

Mr. Binney* estimates the thickness of the red and variegated marls and sandstones of Cheshire at nearly 2,000 feet.

Messrs. Conybeare and Phillips state the average thickness of the Permian and trias groups together, in the south-west of England, at about 200 feet.

The Permian series is also exceedingly variable in thickness. In Derbyshire the thickness is given at about 300 feet by Messrs. Conybeare and Phillips, while they observe that in Glamorganshire the thickness varies in contiguous cliffs from 30 inches to as many feet.

M. D'Orbigny estimates the entire thickness of the trias group at about 2,360 feet; and that of the Permian at from 1,600 to 3,300 feet.

The surface of the new red sandstone is perhaps more

* Proceedings of Geological Society, vol. ii. p. 12.

covered than that of any other formation with gravel and other detritus, which is sometimes porous, and at other places impermeable to water. Professor Sedgwick describes the new red sandstone surface in the valley of the Eden, " as buried for miles together under great heaps of old alluvial detritus."

Where wells are sunk into the gravel lying on the stiff clays and marls of the upper new red sandstone, an abundant supply of water is commonly procured in domestic wells from the land springs which are sure to be met with in such situations. These supplies, however, are generally quite inadequate for a town of any considerable size. In deeper sinkings into the variegated clays and marls of this formation wells have to be sunk a great depth, namely, down to the sand or sandstone beds, in order to produce any large quantity of water. Nearly all the principal towns which have procured large supplies by sinking in the new red sandstone, as Manchester, Liverpool, Nottingham, &c., are situate on the arenaceous part of the formation, which is much more absorbent of water than the superior clayey and marly portions.

The new red sandstone of Cheshire and Lancashire may be separated into three principal divisions. According to Mr. Ormerod,[*] we have, in descending order—

The saliferous and gypseous beds of Church Lawton, near Congleton	649
The corresponding beds at Northwich and Middlewich	808
	1457
Take the mean of these, or	700
Waterstone beds	400
Sandstone beds corresponding with the Bunter Sandstein of continental geologists	600
Total	1700

[*] On the salt field of Cheshire.—Proceedings of the Geological Society, vol. iv. p. 262.

The depth of the beds has been well ascertained in Cheshire, in consequence of the numerous sinkings for rock salt and brine springs. Mr. Ormerod observes that all the above thicknesses are probably understated.

The saliferous marls of Cheshire, Staffordshire, and Worcestershire form a very interesting portion of the trias group. The whole of the new red sandstone district is so much intersected by faults, that the depths both of the rock salt and of the brine springs present great anomalies.

The thickness of unproductive gypseous marls, and clays overlying the salt measures, is frequently as much as 300, and, in some places, even 450 feet. The brine springs commonly rise in the shafts to a much greater height, often 100 feet higher than that at which they are first tapped.

At Middlewich, which appears to be the centre and focus of the salt-producing district, the first brine spring is met with below a layer of black gravel about 9 inches in thickness, between two horizontal beds of indurated clay or rock at a depth of about 78 feet from the surface. This brine spring rises to the height of 18 feet from the surface. Two other brine springs are met with at Middlewich at depths of 126 and 144 feet. The brine springs of Middlewich contain from 38 to 42 ounces of salt in each gallon. The Staffordshire brine springs in the neighbourhood of Ingestrie, near Stafford, contain about 34 ounces per gallon. The Staffordshire salt bed, although separated from the Cheshire by the North Staffordshire coal field, and by a tract of the lower sandstone, occurs like the Cheshire, beneath about 300 feet of red and gypseous marls. There are numerous mineral springs of medicinal water in the new red sandstone district, the most celebrated of which are those of Hartlepool, Croft, Harrogate, and Askern, in the North of England. In the southern and midland counties, mineral springs occur at Leamington, Tewkesbury, Ashby-de-la-Zouch, &c. Most of these contain large quantities of common salt, together with the muriate of magnesia and other purgative salts.

The Permian group, as described by Mr. Binney and Mr. Ormerod, in Lancashire, is about 330 feet thick, of which the upper 210 feet consist of red and variegated marls containing thin beds of limestone, and the lower 120 feet consist of red sandstone. The Permian group in the south-west of England consists usually of a conglomerate or breccia, enclosing large angular fragments of carboniferous limestone and old red sandstone, and passing upwards either into beds of red sandstone, or into strata of fine grained dolomitic or magnesian limestone of a yellowish colour. In the Mendip district the magnesian limestone is frequently cavernous, and often contains extraordinary subterranean reservoirs of water.

Professor Sedgwick, in a very valuable paper published in the Geological Transactions, vol. iii., Second Series, has thus described the subdivisions of the great belt of magnesian limestone which extends from Nottingham into Durham. It consists, in descending order, of the following :—

1. Grey thin bedded limestone.
2. Lower red marl and gypsum.
3. A thick deposit of yellow magnesian limestone, often cellular and earthy ; sometimes hard and crystalline.
4. Variously coloured marls, with thin beds of compact and shelly limestone.
5. Marl slate, associated with gray, thin bedded, and nearly compact limestone.
6. Lower red sandstone.

This part of the series in Durham is about 126 feet thick, and appears to yield an unusual quantity of water. Professor Sedgwick mentions several instances of the great volume of water met with in sinking through this formation. One of these is Eppleton coal pit, in which this bed was found to yield 48,000 gallons of water per hour, or considerably more than 1,000,000 gallons per day. Other cases are mentioned, as the new water-works at Bishop Wearmouth, and pits sunk by the proprietors of the Hetton coal works, in which very copious springs were encountered in this formation.

Mr. Stephenson, in his second report on the subject of a spring water supply from Watford, mentions the subject of the extraordinary yield from the lower bed of the Permian series in the county of Durham. He states that from two shafts sunk within a few yards of each other, the incredible quantity of 14,000,000 gallons a day has been pumped up from the stratum of sand which crosses them. This stratum of sand is described as lying between the magnesian limestone and the coal formation, and is, therefore, probably, identical with the lower red sandstone of Professor Sedgwick's subdivision.

The following are the principal towns in England situated on the trias and Permian formation :—

1. In the valley of the Eden—Carlisle, Penrith, and Appleby.

2. On the western side of the great Lancashire and Derbyshire coal fields are Lancaster, Preston, Liverpool, Manchester, Stockport, Chester, Northwich, Middlewich, Congleton, Wrexham, Market Drayton, and Shrewsbury.

3. On the eastern side of the Durham, Yorkshire, and Nottinghamshire coal fields are Sunderland, Darlington, Stockton, Thirsk, Ripon, Knaresborough, York, Pontefract, Doncaster, Gainsborough, Newark, and Nottingham.

4. In the midland district, south of the great Derbyshire coal field, are Stafford, Burton-on-Trent, Derby, Loughborough, Leicester, Lichfield, Penkridge, Shiffnall, Wolverhampton, Bridgenorth, Stourbridge, Kidderminster, Birmingham, Coventry, Warwick, Leamington, Droitwich, and Worcester.

In the west of England are Bristol, Wells, Bridgewater, Taunton, Exeter, and Teignmouth.

Although the capacity of the new red sandstone for absorbing water has been much insisted on of late years, it is remarkable how few of the towns situate on this formation are supplied from deep wells. The long and severe parliamentary contests with reference to the water-works of Liverpool, Manchester, and Birmingham are well known to most of the public. In all these cases the result has been that deep and expensive wells have been abandoned, and the corporation having obtained possession of the existing pumping establishments, is now, under the highest engineering advice, laying out very large sums in constructing storage reservoirs to collect the water from a drainage area, instead of pumping it from the sandstone.

The following particulars, as to the supply of a number of the principal new red sandstone towns, will serve to show the means at present commonly resorted to for large public supplies of water in new red sandstone districts.

Birkenhead.—Supplied from two wells, which are each about 95 feet in depth, with a boring of 300 feet in the bottom, thus making a total depth of 395 feet.

Each well, or shaft, is 9 feet in diameter, sunk in the red sandstone rock without lining or steining of any kind. The boring commences with a diameter of 26 inches, and diminishes gradually to 7 inches, which is the size of the last 110 feet. When not acted on by pumping, the water stands at 93 feet from the surface; and when reduced by pumping to the level of 134 feet below the surface, the well yields 2,000,000 gallons per day of 24 hours. (Latham.)

Birmingham.—Originally supplied from the river Tame. In 1855 the supply was increased by the addition of a large drainage area in the neighbourhood. In 1866 the Company was authorised to erect new works, and sink shafts into the new red sandstone, in order to meet the deficiency from the River Tame. The Company is bound to reduce annually the supply from this river until 1872, when the pumping from the Tame is to cease altogether.

The Birmingham Company was incorporated in 1826 for the purpose of supplying water to the town from the river Tame, their capital being £120,000. In 1854 they came to

Parliament, and increased their capital to £240,000, with a borrowing power of £30,000. In 1855 the Company again came to Parliament, and their share capital was fixed at £420,000, with borrowing powers not exceeding altogether £105,000.*

In the previous Acts the Company was confined to taking their supply from the river Tame; but the Act of 1855 authorised several new sources, viz., the Hawthorn Brook, Perry Stream, Upper and Lower Witton Pools, and the river Blythe.

The two first of these are small streams flowing into the Tame, the former at the village of Hampstead, and the latter at Perry, about 8¼ miles north of Birmingham.

The river Blythe flows through the gypseous part of the new red sandstone formation. Its general course is north and south, at a distance of about 8¼ miles east of Birmingham. It flows by Coleshill, where it joins the River Cole, and the united stream falls into the Tame at Ouston Grange, about 2 miles east of Water Orton.

In his evidence on the Bill of 1865 Mr. Hawkesley describes the very minute examination of all the sources of supply which the neighbourhood affords. The streams of the Blythe, the Rea, the Cole, and the Arrow were all examined, and rejected as unfit for use, because they were far too much polluted, and the Blythe on account of its hardness and the quantity of sulphate of lime which it contained, this being attributable to its flow through the gypseous marls and clays of the new red sandstone.

After most careful examination he recommended the water in the streams of Sutton Park, and that to be obtained from deep wells reaching the sandy part of the formation at Aston, Edgebaston, Witton, and Sutton Coldfield.

It appears that in 1865 the Company had already commenced sinking at Witton, which is on a small stream flow-

* Evidence on Birmingham Water Bill, 1865.

ing into the Tame at Aston, and although they failed to obtain power in 1865 to execute the other works recommended by Mr. Hawkesley, yet in the following year, 1866, they succeeded in passing their Bill.

The Act of 1865 authorised the Company to divert and appropriate both the streams flowing from Sutton Park, together with the water of certain pools formed by embanking those streams. They were also authorised to sink a shaft and drive adits in the neighbourhood of Sutton Coldfield.

Many persons interested in Bracebridge Pool, and other pieces of water on these streams, objected to the abstraction of the water. Bracebridge Pool is situate near the source of the northern branch of East Brook, about 2 miles northwest of Sutton Coldfield.

The inhabitants of Sutton Coldfield also petitioned against the Bill, on the ground that the operations of the Company would drain the wells in Sutton Coldfield and the springs in Sutton Park.

The referees, however, reported in favour of the undertaking, that there was a probability of good water being obtained both from the upper argillaceous beds of the new red sandstone, and also from the lower or Permian beds. That 2,000,000 gallons a day would probably be obtained from the streams flowing from Sutton Park, and an equal amount from the shaft and adit proposed to be made near Sutton Coldfield.

It was also proved before the referees that the present supply (1865) for the town of Birmingham is in the dry season 7,000,000 gallons a day, an amount barely sufficient for the then population of about 330,000.

About 4,000,000 of this quantity is derived from the Tame, which is becoming more impure every day, and must be soon abandoned. Hence the necessity for a further and better supply, especially with reference to new districts lying outside the present limits of supply.

New Forge Pool and Windley Pool, which were proposed

in 1865, are about a mile south-west of Sutton Coldfield, about 5½ miles north-east of Birmingham, and are situate on a stream called East Brook, which rises in the high grounds of Sutton Park, about 2 miles west of Sutton Coldfield, and was the line of the old Icknield Street at Thornhill. The East Brook flows south and east by Sutton Coldfield, where it is joined by another small stream which rises more to the north from the same elevated ridge, over which the Icknield Street is carried. The united stream then flows by Holland House, New Hall Mill, and Penn Mill, crosses the Birmingham and Fazeley Canal at Plants Brook Forge, and falls into the Tame at Berwood Hall, near Castle Bromwich.

Under the powers of their Act of 1866, the Company have formed a new reservoir at Aston, about 2½ miles north-east of Birmingham, adjoining and communicating with their old works on the river Tame. They have purchased about fifty acres of land at Plants Brook, near Minworth, about 5½ miles north-east of Birmingham, on the Birmingham and Fazeley Canal, and completed a large reservoir which impounds the water brought down from Sutton Park by the East Brook, which has been already mentioned. The supply of water from this source is said to be large and abundant. The Company are also sinking for water in the new red sandstone near Perry, which is a village on the Tame, about 3 miles north of Birmingham. This sinking promises to yield a satisfactory supply.

The Company are also sinking at King's Vale, on the south side of Barr Common, about 2½ miles south-west of Sutton Coldfield, and about 5 miles a little east of north from Birmingham. This sinking, however, does not realise the expectations of the Company. The sinking at Witton, which seems to have been in progress in 1865, is still incomplete; but the directors, in their report of September, 1869, still seem to entertain hopes of a considerable supply from this sinking.

The preamble in the Act of 1865 recites that the Company has raised a share capital of £420,000, in addition to borrowing £60,000 ; and by the same Act they were authorised to raise an additional share capital of £336,000, and to borrow an additional £84,000.   When this capital is all raised, it will amount, including borrowed money, to no less than £900,000, for which, on a moderate calculation a supply of 12,000,000 gallons a day should be given.   It remains to be seen whether this result will be attained.

There is an important provision in the Act of 1866, which limits the quantity to be taken from the river Tame after January, 1869, to 3,000,000 gallons a-day ; after January, 1870, to 2,000,000 gallons ; and after January, 1871, to 1,000,000 gallons a-day, and declares that after January, 1872, the Company shall cease to supply any water from the Tame for domestic purposes.

This is qualified by providing that, in certain emergencies, such as frost, drought, &c., or in case of the river Tame water becoming pure and fit for domestic purposes, that then it may be used under certificate from the Board of Trade.

Mr. Hassard, C.E., has lately prepared a scheme for supplying Birmingham from the Radnorshire hills, a distance of 52 miles from Birmingham.   He proposes an immediate supply equal to 22,500,000 gallons a day to be taken from the river Teme about 5 miles above Knighton.

The collecting ground has an area of 36 square miles, or 23,000 acres, and consists of mountain pasture on the clay slate formation.

Mr. Hassard estimates the cost of his scheme, including compensation, at £1,600,000, or about £71,000 per million gallons of daily supply.

Bridgenorth.—Water formerly pumped from a well in the new red sandstone ; but the supply was insufficient, and the town was often without water for days together.   Within the last eight years a 25-horse engine has been employed to pump the water from the river Severn.   There is also a duplicate engine, of 18-horse power. The pumping lift is about 279 feet. The consumption is about 210,000 gallons a day, and the supply has never been known to fail since the water has been taken from the river Severn.

Bristol.—Supplied by gravitation from the Mendip Hills.

Cardiff.—Supplied partly from the river Ely, and partly from storage reservoirs.

The Cardiff Waterworks Company obtained their first Act of Incorporation in 1850, and their powers were further extended under Acts of 1853 and 1860. The works were designed and carried out by the late James Simpson, Esq., who continued to be the consulting engineer to the Company to the date of his decease. The district originally comprised the town and port of Cardiff, and the parishes adjoining, and the supply was derived from the river Ely, at a point about three miles from the town. A pumping engine of 20-horse power was erected at this spot in 1851, and the water was delivered into a service reservoir (called the Penhill Reservoir) two miles and a half from the engine, and at about the same distance from the town, the reservoir having an elevation of about 60 feet above the general level of the district, with a capacity of 2,000,000 gallons. A second engine of 25-horse power was erected in 1855 near the first, and in 1859 a second service reservoir was constructed (called the Cogan Reservoir) of similar capacity and elevation to that at Penhill, but at a different part of the district.

The rapid increase in the town and population of Cardiff soon rendered a further extension of the works necessary, and the Act of 1860 was obtained, under which the district was enlarged by the addition of several parishes, and the authorized capital of the Company increased (from £15,000 in 1850, and £45,000 in 1853) to £145,000, exclusive of borrowing powers. Under this Act a third service reservoir was constructed (called the Landough Reservoir), having an elevation of 180 feet above the Cogan Reservoir, for the supply of Penarth, a somewhat elevated portion of the new district, and an engine of 15-horse power was erected at the Cogan Reservoir, for the purpose of pumping water therefrom into the new reservoir at Landough.

In 1862-3-4-5 the works were further extended, and an additional supply obtained from the drainage area to the north of Cardiff, and lying between the town and the southern ridge of the Welsh coal basin. The formation generally of this portion of the district is of the old red sandstone series, covered in places with a considerable depth of diluvial soil. A storage reservoir has been made here (called the Lisvane Reservoir), 1864-5, of a capacity of 70,000,000 gallons, together with filter-beds and other works, and a direct supply by gravitation afforded therefrom to the town, and the Penhill and Cogan service reservoirs. From this source the present supply is obtained during the greater part of the year.

The river Ely is pumped from during a portion of the summer, and its resources are also available for future extensions, or in the event of unusual drought or other emergency.

The supply is on the "constant" system; and it is gratifying to the Company that, whilst in many towns it has been found necessary to shorten the hours of supply during some part of the dry summers lately experienced, Cardiff has been afforded a constant and ample supply in the fullest sense of the word, and without an hour's exception.

The amount of capital expended by the Company up to the present time is £108,000, which, of course, includes land and all incidental expenses—the cost of works, with mains and services, &c., being about £85,000. Out of this sum, the Penarth Works (engine and house, Landough Reservoir, and mains for Penarth) cost £7,000, and the Lisvane works (including store reservoir, aqueduct, and intercepting tanks, filter beds, cottage, &c., four miles of 15-inch main into Cardiff) £25,000, exclusive of land, law and engineering, and compensation.

The service reservoirs contain 5,000,000 gallons. The average daily supply to Cardiff (including the docks and shipping) is 1,300,000 gallons, of which about 24,000 gallons are pumped into the Landough Reservoir for the supply to Penarth.

The engines employed for the Cardiff supply consist of one 20-horse high pressure condensing beam engine, one 25-horse high pressure non-condensing beam engine; total lift of pumps (exclusive of friction through mains), 80 feet; quantity, 2,000,000 gallons in twenty-four hours (both engines). Penarth supply (second lift) one 15-horse high pressure direct acting engine; total lift of pump, 190 feet; quantity, 340,000 gallons in twenty-four hours. (Information furnished by Mr. Henry Gooch, the resident engineer.)

Carlisle.—Supplied with about a million gallons a day, pumped from the river Eden.

Chester.—Water supplied by pumping from the river Dee, into an open subsiding reservoir. Formerly it was not filtered, and its quality was much complained of. It is now well filtered, and then stored in a covered reservoir and pumped up to a high level service tank, from which the city is supplied by gravitation with about 900,000 gallons a day. The supply both in quantity and quality now gives very general satisfaction. An unsuccessful attempt was made many years ago to procure water from the red sandstone by tunnelling under the river Dee.

Coventry.—Supplied partly from springs, and partly from artesian wells sunk into the red marl and new red sandstone rock. The principal artesian well is 320 feet deep, and has an iron pipe 18 inches diameter for the first 30 feet; then a 15-inch pipe for 130 feet; a 12-inch pipe for 20 feet; an open 12-inch bore-hole, without pipe, for 73 feet; then a 6-inch bore-hole without pipe for 67 feet: total, 320 feet.

A second artesian well is 250 feet deep, and the two others are each 75 feet deep. The supply is constant, and the water is pumped to a service reservoir 100 feet above the town, and over a stand-pipe 40 feet high, so that the water is delivered at a height of 140 feet above the town. The capacity of the service reservoirs is about 3,000,000 gallons, and the daily supply about 700,000 gallons in the summer. The quantity annually pumped, about 250,000,000 gallons. Two engines are employed, one of which is a 60-horse double cylinder beam engine, working a solid plunger, and raising 63 gallons at each stroke. The other is a 40-horse Cornish engine. The cost of coal used about £230 per annum. The whole cost of the works, which were erected in 1846, has been about £33,000.

Darlington.—Water pumped from the river Tees. The reservoir is a mile distant from the town, to which the water descends by gravitation, with a pressure in the town varying from 40 to 100 feet. Supply abundant, but water sometimes discoloured. The engine is of 30-horse power, and lifts 500 gallons per minute 100 feet high.

Derby.—Supplied by water which is partly collected in impounding reservoirs, and partly pumped from the Derwent. The service reservoir is at a height of 140 feet above the town, and contains 1,158,000 gallons. The impounding reservoir has an area of eleven acres and an average depth of 30 feet (? extreme depth). Works completed in 1851; since which time the supply has been adequate. Cost of works, £50,000. Daily supply 953,000 gallons.

Exeter.—Supplied by water pumped from the river Exe, at Pynes, about three miles above Exeter. The works were executed in accordance with an Act passed in 1833, and now supply about 960,000 gallons per day. Water power is used for pumping, but steam, as an auxiliary power, was added in 1856. The supply is intermittent, and is given to each district during about six hours each day, and on six days a week, Thursday being excepted on account of repairs, and especially in summer time, when the water is required for washing gutters and flushing sewers. The total daily consumption for all purposes is 23 to 25 gallons per head.

The Company are now making arrangements for taking the water from a point higher up the river Exe, namely, above the junction of the river Culm.

Lancaster.—An abundant supply of very soft water, obtained by gravitation, from a height of 1,148 feet above the town, and stored in four equilibrium reservoirs at various heights.

Leamington.—The supply is pumped from the river Avon into two summit reservoirs, 120 feet above the town. The works were executed in 1857, and the supply has been constant since 1867. The capacity of the two reservoirs is about 1,000,000 gallons, and the daily supply to the town about 350,000 gallons. Two 35-horse power engines are employed, and these are capable of pumping 500,000 gallons a day to a height of 145 feet.

The engines are also employed in driving a corn mill when not required for pumping. About two cwt. of coal are used per hour. (Information received from Mr. T. D. Barry, Engineer for the borough of Leamington.) The works belong to the Local Board.

Leicester.—Supplied by gravitation with water collected on the high lands bordering Charnwood Forest.

Liverpool.—Supply partly obtained from wells sunk in the new red sandstone, and partly from large impounding reservoirs at Rivington, distant about thirty miles from Liverpool. The average quantity supplied per day is now, in 1869,

From Rivington	10,500,000 gallons.
„ wells	6,000,000 „
Total . .	16,500,000 „

Population within area of supply about 620,000, so that the

supply for all purposes, trading and domestic, amounts to an average of nearly 27 gallons per head per diem.

The sum expended on the works for water supply up to this time amounts to about £2,000,000.

The drainage ground for collecting water has an area of about 9,000 acres, and the storage capacity 400,000,000 feet altogether. Mr. Hawkesley's estimate for the supply of Liverpool was 15,000,000 gallons a day; but the supply of late years has fallen to 10,000,000 and even 8,000,000 gallons a day.

Macclesfield is situate on the edge of the Cheshire new red sandstone, where this formation overlies the coal measures. The town is supplied from a drainage area of 2,000 acres, situate east of the town, and on the coal formation. The water is conveyed by covered channels to a reservoir capable of holding forty days' supply. The rainfall is about 40 inches, and about 53 per cent. is collected, or ·526 inches per inch of rainfall.

Manchester, although situate on the new red sandstone, is supplied from an extensive drainage area of millstone grit. It was formerly supplied by deep wells, one of which yielded more than a million gallons a day. The supply is now obtained by gravitation from large impounding reservoirs in the millstone grit district of Longdendale.

According to Mr. Baldwin Latham, the drainage area of the Manchester water supply is about 18,900 acres, which furnishes a supply of 12,000,000 gallons a day to a population of 550,000 souls, in addition to 55 cubic feet of water per second for twelve hours daily as compensation to mill-owners. Thus the total quantity of water collected is upwards of 26,000,000 gallons daily, although the actual supply to Manchester is said to be only 12,000,000.

Mr. Bateman,* however, who constructed the Manchester works, states the supply to Manchester at 25,000,000 gallons

---

* Pamphlet on Metropolis Water Supply, 1865.

a day, and quotes the cost of the works at £1,500,000, or £60,000 for each million gallons of daily supply.

Mr. Bateman also stated in his evidence before the Water Supply Commission of 1867 that the gross daily supply to Manchester was 21 or 22 gallons per head for 600,000 persons; and of this quantity about one-third was used for trade purposes, leaving 14 or 15 gallons per head for domestic use.

Mr. Bateman further states, in the same evidence, that 33 inches of rain have been actually collected, and measured out of a total rain-fall of 45¾. This proportion amounts to ·72 inches, or nearly three-fourths, for each inch of rain-fall. The whole deficiency or loss is stated at 16 to 9 inches, which would give respectively a depth collected from each inch of rain-fall = ·651 inches and ·800 inches. Taking the rain-fall on the Manchester gathering grounds to be 37 inches, the annual quantity of water which falls would be equal to 69,930,000 tons.

Mr. Bateman's quantity of 25,000,000 gallons a day probably includes the supply to millers as well as that for Manchester itself. The total supply of 25,000,000 gallons a day would amount to nearly 41,000,000 tons a year, showing a loss of about 42 per cent. out of the whole 37 inches, and leaving only 58 per cent. available for collection.

Middlewich.—Supplied by gravitation from high ground in the neighbourhood. Supply never known to fail.

Nantwich.—Supplied by gravitation from a source four miles distant.

Newark.—The water is pumped from the river Trent at a point two miles distant from the town, after going through a process of natural filtration. The water is pumped into a covered reservoir 100 feet above the highest part of the town. The daily supply is about 150,000 gallons, and the covered reservoir contains about 1½ days' consumption.

Northallerton.—Supplied from shallow wells in the drift, seldom more than 12 feet deep.

Nottingham.—This town has been supplied for many years partly from the river Trent and partly from wells sunk into the new red sandstone.

The original Trent Water Works Company were incorporated in 1827. In the following year the Nottingham Old Water Works Company were incorporated, and for some years afterwards obtained a supply of water from springs and streams in the parish of Basford. These two Companies were eventually amalgamated, and a new Company incorporated in the year 1845.

The Trent Water Works were begun in 1830, and brought into operation in the following year. They consisted of the Trent pumping station with a filtering reservoir on the banks of the river, and a service reservoir called the Park Road Reservoir, on high ground near the park. This is just under the Park pumping station. In 1834 the water of the Trent was so bad that the pipe had to be removed from the river, and the water was allowed to filter into the reservoir through a natural bed of gravel. Mr. Hawkesley said in his evidence of 1869 that even this naturally filtered water was gradually getting worse, and was now found to be contaminated with brine springs from the coal strata, to such an extent that about a ton of salt was daily pumped into the town.

The supply obtained from the stream of the Lene had also become much contaminated with sewage, and even the springs breaking out from the red sandstone cliffs at Scottholme and elsewhere were found to be unsatisfactory. Under these circumstances the Company were advised that a purer supply would probably be obtained by sinking a well into the new red sandstone, and accordingly in 1845-6 they sank a shaft within their own premises, and close to the river Trent. This shaft produced an abundance of water, but had not been sunk more than 80 feet before it was discovered that the quality was bad, and this water was accordingly never supplied to the town.

The next proceeding was to sink a well at Sion Hill, about half a mile from the market place of Nottingham. The red sandstone formation at Sion Hill is overlaid by marl which contains a large quantity of gypsum. This water contained a large proportion of mineral matter, and was found to be nearly 21 degrees of hardness by Dr. Clarke's scale. The water obtained from the Trent is still worse in quality, being of about the same hardness, and containing per gallon nearly 50 grains of solid matter, of which more than 20 grains is common salt.

Since 1854 the Company obtained from the Duke of Newcastle power to sink in Basford parish, and have there succeeded in obtaining about 2,000,000 gallons a day of very good water. This is pumped up by two steam engines, having an aggregate power of 120 horses.

A still more recent attempt to obtain water by sinking has been made a little north of Bestwood Park. The sinking passed through the new red sandstone, then through 83 feet of the Permian formation, and then reached the coal measures, at a depth of about 200 feet, beyond which a boring was put down. Less than half a million gallons a day were procured from these operations; but the water, unfortunately, was as salt as sea-water, and therefore the experiment was discontinued.

Notwithstanding all these numerous sources, Mr. Hawkesley states that in the dry summer of 1868 the Company had nearly exhausted all their powers of supplying water, and had very little to fall back upon. In 1869 Mr. Hawkesley calculated the whole daily supply as follows :—

The Trent Works supply	600,000 gallons.
„ Scottholme Works (springs) .	400,000 „
„ Park well, about	900,000 „
„ Basford well	1,750,000 „
	3,650,000 „

It was calculated also that by deeper sinking and more

engine power at the Basford well, an additional quantity of 750,000 gallons a day might be obtained, thus making a total of 4,400,000 gallons for a population of about 140,000 persons. Mr. Hawkesley thinks, however, the Trent water should be greatly disused, and this would take away 600,000 gallons. He also proposes to discontinue the supply from Scottholme springs, amounting to another 400,000 gallons. He calculates that in 1881 the Company will have to supply a population of more than 200,000, and in 1901 a population of 290,000 persons.

After all these harassing failures endured by the Company, it is not surprising to find them applying for largely increased Parliamentary powers in 1869. The following were the principal provisions contemplated in this Bill of 1869 :—

1. To take the water of the Dover Beck, which rises near Oxton, in the county of Nottingham, and of Grimes Moor Little Dyke, which rises near Calverton, in the same county.

2. To erect a pumping station between these brooks, at a distance of seven miles from the centre of Nottingham, and to pump the water a distance of three miles and a half to a reservoir about 270 feet in height above the pumping station.

3. To construct a new reservoir three miles and a half from centre of town, and convey the water by means of pipes into the town of Nottingham.

4. To supply a number of parishes and places adjacent to the town of Nottingham.

5. To increase the capital by a sum of £150,000, which is to be entitled to a dividend of 7 per cent.

6. To borrow, in addition to their present borrowing powers, a further sum of £37,500.

7. That new shares are to be disposed of, not by auction, but as directed or authorised by the Company.

The Corporation of Nottingham petitioned against this Bill on the grounds—

1. That the Company are monopolists, and at present obtained a filtered supply from the river Trent, also from

springs and wells in the parish of Basford, and from wells in Ropewalk-street, Nottingham, and can obtain further supplie from the sandstone rock of Nottingham and its immediate neighbourhood.

2. That the underground springs of the sandstone rock afford the purest and cheapest supply.

3. That the river Trent, on account of its great volume, rapid flow, and small population above Nottingham, is easily kept free from pollution, and that its purity will be much improved by impending legislation prohibiting the pollution of rivers.

4. That the Dover Beck and Little Dyke (from which the new supply is to be taken) have together not more than one four-hundredth part of the average volume of the Trent.

5. That Dover Beck and Little Dyke receive the drainage of 10,000 acres of cultivated land, and are liable to be much polluted by the manures employed.

6. That the outlying places proposed to be supplied contain 40,000 acres, and mostly contain a thin and scattered population, and do not present a remunerative field for water supply.

7. That the supply of the new district would probably cause a loss which would prevent reduction of price in Nottingham.

8. That the deposited estimate for the new scheme is £176,000, and would involve heavy compensation to water-mills below the proposed pumping station.

9. That the reduction of price contemplated by the Act of 1845 will be prevented or indefinitely deferred.

10. That the Company should be bound to raise and apply their already authorised capital before creating any new capital.

11. That the Company have a large amount of unraised capital, and that the new capital proposed is excessive.

12. That by the Act of 1854 the dividend on share capital is limited to 5 per cent., and shares are to be sold by auction, the premium not being entitled to dividend.

13. That the present shares bearing a dividend of 5 per cent. are at a premium of 20 per cent., so that the proposed dividend of 7 per cent. would correspond with a premium of 68 per cent., and as there is no auction clause, this premium would be pocketed by the shareholders, and would entail a charge on the consumers of about £100,000 on the new capital of £150,000.

14. That as water (*unlike gas*) is incapable of being superseded, there is no reason why new capital should bear higher rate than 5 per cent., nor why new shares should not be sold by auction.

15. The Corporation are willing to purchase the undertaking of the Company, or to establish other waterworks for the supply of the town, pursuant to the Public Health and Local Government Acts.

16. That the Bill should require the Company to supply by meter.

The above were the principal allegations made against the Bill, and the result shows the opposition was successful, as the Committee, after a very patient inquiry, declared the preamble of the Bill not proved.

Penrith.—Water pumped up from the river Eamont to two separate reservoirs, one of which is 130 feet above the river, and commands the greater portion of the town. The upper service reservoir is 370 feet above the river, and commands the higher portion. Water pumped by a wheel worked by the river, supplemented by a steam engine of 14-horse power, which is employed in times of low water and when the river is flooded. Daily supply about 300,000 gallons.

Preston.—Supplied by gravitation from a millstone grit district in the neighbourhood. Water collected in a series of reservoirs, afterwards filtered, and then stored in covered reservoirs. The drainage area on Longridge Fells has a capacity of about 3,000 acres. The supply is equal to 150 days' consumption, and is conveyed to the reservoirs partly in open channels and partly in stone culverts. The mean rainfall is said to be 43 inches per annum, and about 23 per cent. is collected in the reservoirs.

Rugby is situate on the lias formation, which here rests on the new red sandstone. Attempts have been made to procure water by wells sunk through the lias into the new red sandstone. The attempts to procure water for domestic consumption, however, have been unsuccessful, as the water has been found highly impregnated with common salt.

Stockton-on-Tees.—Supplied by water pumped from the river Tees.

St. Helen's, Lanc.—There are two wells sunk in the red sandstone here. Each well is 210 feet deep, beyond which there is a boring in the bottom, and the two wells supply daily about 572,000 gallons.

Selby, Yorkshire.—The supply of this town is chiefly obtained from a shallow well, succeeded by a boring of 320 feet in depth and 7 inches diameter. Supply about 120,000 gallons per day.

Stourbridge, Worcestershire, has a shallow well, about 80 feet deep and 6 feet in diameter, which yields about 150,000 gallons a day.

Sunderland.—The supply is from four deep wells sunk in the neighbourhood. The first well was sunk in 1846 at Humbledon, about a mile and a half west of Sunderland. The second was sunk in 1852 at Fulwell, about a mile and a half north of the town. The third was sunk in 1859, at Cleadon, about four miles north of Sunderland, or nearly midway between Sunderland and South Shields. The fourth well is just completed, and is sunk at Ryhope, about three miles and a half south of Sunderland.

The shaft at Humbledon Hill is sunk through magnesian limestone (Permian or lower new red sandstone) to a depth of 228 feet, and two 3-inch bore-holes are made in the bottom of the shaft to a further depth of 100 feet, making the total depth 328 feet.

The shaft at Fulwell is sunk through clay and limestone to a depth of 222 feet.

The shaft at Cleadon is sunk principally through limestone, to a depth of 270 feet, and the shaft at Ryhope is sunk also through limestone to a depth of 240 feet. The water from all these wells is pumped up to reservoirs, from which it gravitates to the town. The reservoirs at Humbledon Hill and Fulwell each contain 1,000,000 gallons. The reservoir at Cleadon Hill contains 2,000,000 gallons. All these three reservoirs are situate 220 feet above high-water mark. The reservoir at Ryhope is 230 feet above high-water mark, and contains 4,000,000 gallons. The supply is constant, being always laid on to the houses with the full pressure due to the height of the reservoirs; and the present daily supply to Sunderland and South Shields is from 3,500,000 to 4,000,000 gallons.

The following are particulars of the pumping engines :—

At Humbledon, a low-pressure single condensing beam-engine of 120-horse power, capable of lifting 1,000,000 gallons in twenty-four hours. At Fulwell, two double-powered rotative engines, with 36-inch cylinders, and 7-feet stroke, capable together of raising 1,000,000 gallons in twenty-four hours. At Cleadon, two single-powered expansive engines, with 60-inch cylinders, and 8-feet stroke, capable of raising 3,000,000 gallons in twenty-four hours. At Ryhope, two rotative double-cylinder expansive engines, with large cylinders 45-inches diameter, and 8-feet stroke; smaller cylinders 27-inches diameter, and 5-feet 4-inches stroke.

The quantity of water pumped last year, exclusive of the Ryhope supply, was 1,298,000,000 gallons.

The cost of the works, including mains, branches, &c., has been £268,120, and the present population supplied throughout the entire district is about 200,000. The cost of coal and cartage for the last year was £2,419. The oil, tallow, leather, &c., cost £222.

From the revenue account of 1868 it appears the receipts

For Water Rents amounted to . . .	£26,540
Less Expenditure	10,127
Profit . .	£16,413

on a paid-up share capital of £179,500, equal to more than 9 per cent.

Information furnished by Mr. William Dixon, who states that the works have been carried out from the designs and under the superintendence of Thomas Hawkesley, Esq., C.E.

Tranmere, Cheshire, has a well 9 feet diameter, and 120 feet deep. Supply about 150,000 gallons a day.

Wallasey, Cheshire, according to Mr. Latham, is supplied with 300,000 gallons daily, from a well sunk and bored 236 feet into the new red sandstone.

Warwick.—Supplied with water taken from the River Avon at Emscote. Water filtered and pumped into the town. The daily supply about 810,000 gallons. Supply constant during the day, namely, from 6 A.M. till 9 P.M. A well was sunk here about twelve years ago, but abandoned at a depth of more than 400 feet. The strata consisted chiefly of red marl (upper argillaceous and gypseous beds of the new red sandstone). Information from Mr. J. Fenna, Borough Surveyor.

Wellington.—Supplied from the Wrekin Brook.

Wells.—Supplied from wells and springs in the detrital gravel on which the town is built. The water is probably derived from the carboniferous limestone chain of the Mendip Hills.

Wolverhampton.—Formerly supplied in a very inadequate manner, partly by a Water Company, and partly from private wells sunk in the drift gravel, on which the town stands. The water of these wells was of very bad quality, containing a large quantity of organic matter, and from 11 to 24 grains per gallon of sulphate of lime.

In 1845 an Act was obtained by a Company for supplying

water from two wells sunk into the new red sandstone. One of these wells was sunk at Goldthorne Hill, and the other at Tettenhall, and their united yield in 1855 did not exceed 400,000 gallons, whereas the town then required at least a million gallons. In this year, 1855, a new Company, having purchased the old works, erected a pumping establishment on the River Worf, at a point about ten miles distant from the town ; and since that time there has been a plentiful supply of good water at constant pressure. The Worf water is collected in an impounding reservoir.

In 1867 the Corporation of Wolverhampton purchased the whole of the Water Company's property, and now supply Wolverhampton, as well as the neighbouring towns of Bilston, Willenhall, and Wednesfield. The total supply given by the Wolverhampton Water Works amounts to nearly 1,750,000 gallons per day. The water is of good quality, and is shown by analysis to rank high in purity amongst the waters from the new red sandstone. It is worthy of remark that while the district supplied by the Wolverhampton Water Works suffered very severely from the early visitations of cholera, the town has been remarkably free from that epidemic since the introduction of the present water supply. The daily supply is now about 1,250,000 gallons.

Worcester.—New works were constructed in 1857, at a cost of about £28,000. Two engines of 25-horse power each were then erected by the Haigh Foundry Company. The water is lifted from the Severn into depositing tanks, and flows thence into sand-filters, from which it passes to the pure water tank, and thence is pumped into the town against a head at reservoir of 160 feet. The water is of excellent quality, and the works have given greater satisfaction than any other of the public improvements in the city. The demand became so great that the works were extended in 1867 by the addition of two single-cylinder engines of 80-horse power each (erected by the Worcester Engine-Works Company), and the construction of large depositing tanks and filters of proportionate dimensions. The cost of this extension, including land, was about £14,000.

The four engines, working at 18 strokes per minute, deliver about 100,000 gallons of filtered water per hour. The supply is constant, and the highest houses in the city are supplied. Mr. Hawkesley, C.E., was the Engineer-in chief, and Mr. S. G. Purchas, C.E., is the Resident Engineer, and under his superintendence the whole of the works were carried out. Mr. Purchas has since laid down a new engine main of 18 inches diameter from the pumping station to the reservoir on Rainbow Hill, a distance of about 2,800 yards; the cost of the latter work is about £3,500.

York.—Supplied from the River Ouse. Water first pumped into subsiding tanks; afterwards filtered, and then pumped by two 40-horse power condensing engines into an elevated reservoir. The supply is constant, and amounts to 1,306,000 gallons daily.

Considerable change of opinion has necessarily taken place since the first edition of this book was published, as to the relative merit of wells and drainage areas for procuring large supplies of water for important towns. The estimates made by engineers as to the proportion of rain-fall capable of being collected in large impounding reservoirs have been found to be much exaggerated, especially in years of considerable drought, and although no general rule can be laid down as applicable to every locality, it is probable that in future the sinking of deep wells must extensively be resorted to for the purpose of procuring water where the drainage areas seem inadequate to yield the necessary quantity.

NEW RED SANDSTONE OF LIVERPOOL.

A great deal of information relative to the new red sandstone and Permian groups, has been elicited during the last few years, especially during the parliamentary inquiries into the Liverpool and Manchester corporation water works' schemes. The following are some particulars relative to the new red sandstone district of Liverpool, a subject which has been exceedingly well illustrated in Mr. Stephenson's valuable report to the Town Council.

Previous to the year 1850, the town had been supplied by several water companies, who obtained the water from wells, and the corporation having purchased the works of these companies, had their attention drawn about that time to other sources of supply. Accordingly, we find in the beginning of 1850 that Mr. Hawkesley was proposing his Rivington Pike scheme, while Mr. Newlands, the borough engineer, backed by the authority of Mr. Simpson, was recommending further works of sinking and boring in the red sandstone in the immediate vicinity of the town. Under these circumstances the Water Committee of the Town Council called for the advice and assistance of Mr. Robert Stephenson. At a meeting held on the 14th of January, 1850, the Committee passed the following resolutions. "That Mr. Stephenson having been unanimously appointed the engineer for the purposes of the resolution of the Council of the 9th of November, the desire of the Committee is that he should inform himself upon the subject in all its bearings by evidence, reports, or otherwise, so as to ensure that the views of all parties may be elicited before him to their satisfaction, and report his opinion to the Committee fully :—

" 1st. Whether a supply sufficient as regards quantity and quality for the present and prospective wants of the town and neighbourhood, including domestic, trading, and manufacturing purposes and shipping ; and for public purposes— viz., watering and cleansing streets, flushing sewers, extinguishing fires, and supplying public baths and washhouses—can be obtained by additional borings, or tunnels, or otherwise, at the present stations—viz., those purchased

from the companies respectively, and from the Green Lane
Works, now vested in the corporation; and the cost of
obtaining such sufficient supply.

"2ndly. Whether a sufficient addition to the present
supply can be obtained in the locality or neighbourhood of
Liverpool, as recommended by Messrs. Simpson and Newlands, or by borings or by any other course; and the cost of
obtaining and distributing the same?

"3rdly. Whether such supply can be obtained by means
of the Rivington Works; and the cost of obtaining and distributing the same, as recommended by Mr. Hawkesley?

"4thly. Under all the present circumstances of the cases,
what course is recommended to be pursued?

"WILLIAM SHUTTLEWORTH, Town Clerk.'

In pursuance of these resolutions, Mr. Stephenson held a
court at Liverpool, received a large body of evidence, caused
numerous experiments to be made, and on the 28th of March,
1850, furnished a very elaborate report to the Town Council, from which the following information has been ex-
tracted:—

On the Permeability of the Red Sandstone in the Neighbour-
hood of Liverpool.

The evidence adduced before Mr. Stephenson on this subject was very conflicting. The geological evidence had
chiefly reference to the faults and fissures with which the
new red sandstone is known to be intersected, some geologists contending that these fissures are filled with clay, and
that they completely cut up and divide the formation into a
series of boxes. Dr. Buckland is quoted as an authority
for this opinion. Others, again, are of opinion that the
fissures are not only sufficiently open to admit of the free
percolation of water through any particular bed, but that

they even act as channels to diffuse the water from one permeable bed to another; and thus they contend the whole mass of the new red sandstone, wherever it consists of other than argillaceous beds, is freely permeable by water. Mr. Stephenson himself appears to incline to this view, which, he says, is supported by numerous instances in which wells, at a considerable distance, were affected by pumping at Green Lane.

In dealing with this part of the subject, Mr. Stephenson makes a very shrewd comment on some evidence which had been given by advocates for sinking wells, to the effect that certain wells were not affected by pumping from those in the neighbourhood. "If," says Mr. Stephenson, "the sandstone were so impermeable as to prevent one well influencing another at a moderate distance, it would be exceedingly difficult, if not absolutely impossible, to obtain a very large supply of water from any one well. As regards, indeed, the main question of obtaining from the sandstone an adequate supply of water, it is of the utmost consequence to establish, indisputably, that the sandstone is extremely permeable." Mr. Stephenson here admirably exposes a vice which is far too common at the present day, that of overproving one's case—that unhappy mistake of cunning people, whose ingenuity is so extreme that it absolutely overreaches themselves. To return, however, to the subject of permeability, Mr. Stephenson sums up his conclusion in these words—"that the rock may be looked upon as almost equally permeable in every direction, and the whole mass regarded as a reservoir filled up to a certain level, to which, whenever wells are sunk, water will always be obtained, more or less abundantly; and a very careful consideration of all the facts that have come to my knowledge in the present investigation, leads me to consider this view as the simplest, and the only one capable of general application."

It must be fully borne in mind, that this strong opinion

H

as to the complete permeability of the new red sandstone at
Liverpool, is confined to this locality alone, and does not
apply to the new red sandstone generally. In that vast
aggregate thickness of marls and sands, clays, conglomerates, and sandstones, of which this formation is composed,
we may naturally expect to find very opposite and various
degrees of permeability in different districts. In the neighbourhood of Liverpool, the strata appear to be highly arenaceous with few interstratified beds or partings of clay ;
and it is remarkable that most of the cuttings on the railways
near Liverpool, as well as the exposed faces of rock in
quarries, &c., are remarkably dry, presenting seldom any
appearance of moisture trickling down the face and breaking
out above the partings of clay or marl. This absence of any
visible water, anomalous as it may seem at first sight, is
one of the best indications of complete permeability. It is
almost invariably found all over England in chalk and sand
cliffs, and cuttings where, from the absence of clay or marl,
the water is allowed freely to percolate. On the other hand,
there are many districts of new red sandstone in which the
sides of every excavation, both in quarries and in road, canal,
and railway cuttings are constantly trickling with moisture,
thrown out by the numerous clay partings, which are often
so thin as to escape observation, unless the attention be
directed to them by this unfailing symptom of the water
breaking out. In such a new red sandstone district as this
last, it is hopeless to expect water in abundance from sinking. We might as well expect to meet with water in a
thick mass of London or lias clay, in which the same appearance of water is presented, wherever they are exposed in
cuttings. The appearance is also due to the same cause
—namely, the alternation of beds porous in different
degrees, the water percolating through the one being
stopped and thrown out by the other. All engineers know
well the difficulties which are often met with in executing

works through clays containing interstratified beds, which allow of partial penetration by water; but although water is often met with in such quantity as to be very troublesome, affecting the stability of slopes, &c., yet no one would think of sinking in any such strata to procure a copious supply of water.

The following are some of the general conclusions which seem to have been established with reference to the phenomena of water and springs in the trias and Permian groups:—

1. That water abounds in the drift gravel covering the new red sandstone and the Permian rocks, but this is only sufficient for private domestic supply on a small scale, and cannot be depended on for the public supply of large towns.

2. That the water in the superficial drift is usually very impure, containing sulphates of lime and magnesia in large quantities, and being frequently in towns much contaminated with organic matter.

3. The gypseous beds of marl and clay forming the upper part of the trias group, are commonly destitute of water except in the thin sandy partings which intervene between the layers of marl. Therefore they cannot be depended on for any large supply of water. For single isolated houses they commonly yield a sufficient quantity, but at very uncertain depths. In sections taken across the district between the Staffordshire and Shropshire coal-fields, the water was found standing in the wells at extremely different levels, which seemed to follow no law except that the level was generally higher in the neighbourhood of faults, so that the water levels appeared to dip in each direction from the faults.

4. The whole of the new red sandstone is remarkably intersected by faults, which are sometimes pervious by the water, and sometimes oppose an impenetrable dam to its progress.

Faults of small extent which are not filled up with clay, appear to oppose no obstacle to the passage of water. The faults in the Shiffnal and Wolverhampton district, however, appeared, from actual observation and measurement, to exert a very decided influence on the height at which the water stood in wells.

The brick red sandstone beds, and also the conglomerate or pebble beds of the trias group, contain much more water than the marls, and appear to yield in the Liverpool wells very large quantities of water, some wells producing more than 2,500,000 gallons a day throughout the year. The corresponding beds at Wolverhampton, however, yield very little water, the Tettenhall well producing only 168,000 gallons a day. The point where this well is sunk is, however, 450 feet above the sea, whereas the Liverpool wells are sunk almost at the level of the sea. In addition to this, the beds are much more sandy at Liverpool, although probably having about the same geological horizon as the beds at Wolverhampton. The sandstone rock at Liverpool is usually remarkably dry in excavations, quarries, and railway cuttings —see those on the Liverpool and Manchester Railway at Edgehill, &c. On the contrary, the beds in which the Wolverhampton well is sunk are remarkably wet, and exhibit a great deal of moisture in all the partings and laminations. This is well seen in a road cutting at Tettenhall, which exposes a good section of the beds sunk through in the well. The dryness of the Liverpool rock shows that the water sinks freely into the sandstone, and affords a very probable reason for expecting to find an abundant supply in wells. The moisture in the Wolverhampton beds, on the other hand, shows that the water penetrates with extreme difficulty, and is thrown out by the alternations of marl and clay, of which the strata are composed.

5. Where the Permian rocks are faulted against the coal measures, as on the western side of the South Staffordshire coal-field, and the eastern side of the Coal Brook Dale field,

the water will usually be cut off by the fault, and a well, however deep, will yield scarcely any water. Thus, the Goldthorn well near Wolverhampton, which is sunk only about 400 yards west of the great fault which throws down the Permian rocks to the level of the coal measures, and even to that of the Silurian rocks, produces only 200,000 gallons of water per day.

6. Where the lower arenaceous beds and conglomerates of the Permian groups, however, overlie and repose on the coal measures, they will probably yield a very large supply of water, as in the instance mentioned at Monkwearmouth by Professor Sedgwick and Mr. Robert Stephenson.

PUBLIC WELLS OF LIVERPOOL.

The old works of Liverpool, from which the town was exclusively supplied for many years before the Rivington Pike Works were contemplated, furnish numerous examples of procuring water by means of borings in the bottom of wells or lodges.

For instance, the pumping-station at Bootle consists of an extensive lodgment about 45 feet in depth, the bottom of the lodgment being about the same level as high water of spring tides in the Mersey. In the bottom of this well are no less than sixteen bore-holes, some of which are 600 feet in depth, and these yield collectively rather more than a million gallons per day. From some experiments which Mr. Stephenson made on these bore-holes, it appeared that when all were plugged up except one, the yield was 921,192 gallons per day, and when the whole sixteen were opened the yield was only increased by 112,792 gallons, from which it would seem that a very unnecessary outlay of money must have been made in sinking so many bore-holes close together, since very nearly as much water was obtained from one as from the whole sixteen. When not acted on by pumping, the water stands in the Bootle lodgment 22 feet in depth above the bottom, but when reduced by pumping the general level is only about 6 feet from the bottom.

The Green Lane well is 185 feet in depth, the bottom being about 63 feet below high water of spring tides in the Mersey, and 44 feet below the old dock sill. When not acted on by pumping, the water stands 147 feet above the bottom of the well, but the general level during the pumping is only about 11 feet 7 inches above the bottom. When Mr. Stephenson reported in 1850, the yield of this well was 991,118 gallons per day on the average. Since that time a boring of 98 feet deep has been made in the bottom of the well, and the average yield has thereby been increased to 2,413,068 gallons per day, or more than double the produce of the well alone.

The Windsor well is 210 feet deep, the bottom being 87 feet below high water mark in the Mersey. The water when not acted on by pumping stands at a depth of 70 feet above the bottom, and the general level during the pumping is five feet; the average yield of this well in 1850 was 678,560 gallons per day. A boring 214 feet deep has since been made in the bottom, which has increased the yield to 1,020,423 gallons.

It has been asserted with some degree of probability that the large increase at these two wells, and especially at Green Lane, is due to the infiltration of water from the Mersey.

The public wells of Liverpool, from which the town has been supplied for many years, are seven in number. They are all situate within the area of a circle of three miles radius from the centre of the town, the most distant being the Bootle well, which is situate on the edge of such an imaginary circle. The following table contains the particulars of the wells as to depth, diameter, &c.

PARTICULARS OF THE PUBLIC WELLS AT LIVERPOOL.

Name of Well.	Size and shape of shaft.	Depth of well in feet.	Bottom above high water mark of spring tides.	Bottom below high water mark of spring tides.	Level at which water stands when not acted on by pumping.*	Level at which water stands when subject to pumping.*	Average daily yield at end of 1849.	Average daily yield in 1854.	Remarks.
							Gallons.	Gallons.	
Bootle ...		45	2		+24	+8	180,691	881,008	The sinking consists of a large lodge and 16 bore holes.
Beving-ton Bush ...	10 × 6 oval.	150		65	—19	—56	180,875	252,737	
Soho	8 × 6 oval.	123		39	—17	—34	497,869	509,732	
Hotham-street ...	7 × 6 oval.	110		26	—12	—21	216,381	229,201	
Water-street ...	9 × 6 oval.	150		52	—38	—45	419,264	402,344	
Windsor	12 × 10 oval.	210		37	+33	—32	678,560	1,020,493	Increase due to a boring 214 feet in depth.
Green-lane	10 feet circular.	185			+84	—51	991,118	2,413,068	Increase due to a boring 98 feet in depth.

* The + sign in these columns indicates that the height marked is *above* high water mark.

The — sign indicates that the depth is *below* high water mark.

Volume of water yielded by the seven public wells of Liverpool :—

At the time of Mr. Stephenson's report in 1850 the maximum yield of all the seven wells amounted to 5,170,486 gallons per day of 24 hours, the minimum yield to 3,320,990 gallons, and the yield at the ordinary working level was 4,216,784 gallons per day.

This being the yield of the wells at that time, Mr.

Stephenson observes in his report, that the expectation of much augmenting this quantity, either by sinking, boring, or tunnelling, cannot be entertained; and that any increase obtained by deepening the wells would only be temporary, and will only take place to the same extent as the private supply of water is diminished. He is further of opinion that the deepening of the public wells will render it necessary to deepen also the private wells; and then comes this most important observation, which has been before alluded to—"and it cannot be doubted that a large proportion of any increase would be derived from the river Mersey, as all the wells are now sunk to or below the level of low water, and many yield brackish water."

TABLE SHOWING THE COMPARATIVE YIELD OF THE WELLS IN 1850 AND 1854.

	Yield in 24 hours taken from the Dip Books for the last quarter of 1849.	Daily Yield for 1854, taken from Mr. Duncan's evidence in the Wolverhampton Water Works Bill, 1855.	Increase.	Decrease.
	Gallons.	Gallons.	Gallons.	Gallons.
Bootle . . .	850,691	881,008	30,317	..
Bush	180,875	252,737	71,862	..
Soho	497,869	509,732	11,863	..
Hotham Street.	216,381	229,201	12,820	..
Water Street .	419,264	402,344	..	16,920
Windsor. . .	678,560	1,020,493	341,933	..
Green Lane. .	991,118	2,413,068	1,421,950	..
	3,834,758	5,708,583	1,890,745	16,920

The large increase at Windsor and Green Lane has been occasioned by borings which have been made since the date of Mr. Stephenson's report. But independently of these two wells, there has been an increase in all the other five except Water Street. The increase in the first four is equal to 126,862 gallons per day, and if we deduct the falling off of

16,920 at Water Street, we have an increase in the whole group = 109,942 gallons per day.

This increase, however, is not confirmed by the observations which were made in 1850 by Mr. Stephenson's assistants, of which a table will now be given:—

Name of Well and Date of Observation.	Yield in 24 hours at practical working level.	Yield in 24 hours according to Mr. Duncan's table for 1854.	Increase.	Decrease.
Bootle, Jan. 3, 1850	979,944	881,008	..	98,936
Bush, March 11, 1850	224,688	252,737	28,049	..
Soho, Jan. 16, 1850	547,715	509,732	..	37,983
Hotham St., March, 1850	354,552	229,201	..	125,351
Water Street, Jan. 22, 1850	511,488	402,344	..	109,144
	2,618,387	2,275,022	28,049	371,414

So that according to this table the falling off in this group of five wells is 343,365 gallons per day, which may be spread over a period of 4½ years = 76,303 gallons a year, or nearly 2·9 per cent. per annum.

It is true the records kept at the stations show rather an increase than a decrease between 1850 and 1854, but this is inconsistent with what has been observed elsewhere, and what has always been admitted in the case of new red sandstone wells. As Mr. Stephenson's experiments in the beginning of 1850 gave a higher result than the dip books, it is probable that the delivery of the pumps had been underestimated at that time, but were corrected for the future. Hence I think it most fair to compare Mr. Stephenson's quantities, as given in the last table with Mr. Duncan's for 1854.

Table of maximum yield of wells, at lowest level, in the beginning of 1850, and the same for 1854:—

	Maximum yield in 24 hours in the beginning of 1850.	Maximum yield in 24 hours in 1854.	Increase.	Decrease.
Bootle . . .	1,102,065	1,102,000	..	65
Bush	395,983	333,984	..	61,999
Soho	664,385	609,528	..	54,857
Hotham Street.	436,692	364,000	..	72,692
Water Street .	715,550	578,907	..	136,643
Windsor . . .	660,864	1,028 000	367,136	..
Green Lane . .	1,248,816	2,605,812	1,356,996	..

Showing a considerable falling off in all except the two last, in which additional borings have been made to obtain an increased supply of water.

Table showing the falling off in the wells, comparing the whole quantities raised in 1849 and 1854, respectively:—

	Total water raised according to Mr. Stephenson in 1849.		Total raised according to Mr. Duncan in 1854.
Bootle . . .	329,486,250	.. ..	321,567,770
Bush	95,433,850	.. ..	92,250,018
Soho	168,812,589	.. ..	186,052,194
Hotham Street.	80,783,436	.. ..	83,658,450
Water Street .	150,038,675	.. ..	146,855,645
Windsor . .	252,922,650	.. ..	372,480,000
Green Lane .	367,378,629	.. ..	880,769,922

Mr. Duncan makes the following observation on the Liverpool public wells, in his evidence before the Commission on Metropolis Water Supply in 1868:—

"The original supply from the wells has been partly discontinued, the Hotham Street, Bush, and Soho Street wells being abandoned; but the present supply from Green Lane, Windsor, Bootle, Water Street, and Dudlow Lane is between 38,000,000 and 39,000,000 gallons per week. The quantity originally raised from the well sources was about 35,000,000 a week, and the greatest amount ever raised was between 42,000,000 and 43,000,000 gallons. In the month of November, 1866, the amount drawn from the present well sources averaged between 39,000,000 and 40,000,000 gallons a week. There appears, with one exception, to have been a gradual falling off in the supply from the well sources; in fact, the continual pumping has occasioned a great depression in the water level. In the case of Bootle, which is here referred to, the original supply was 900,000 gallons per diem, but upon sinking the well a further depth of 70 feet, making a total of 100 feet, the supply has been increased to 1,000,000. At Green Lane, in 1855, the supply per diem was 3,300,000, but in 1866 it was down to 3,000,000. Dudlow Lane is a new station, and the yield per diem is equivalent to 400,000 gallons. In 1867, three more wells were in course of sinking, but it is not known to us with what results. The water from the red sandstone is very hard, and as a fact it was found that at Green Lane the original degree of hardness was 4½, whilst in 1867 it had increased to 7 degrees."

Some representations were made in 1866 by Dr. French, the medical officer of health, against the practice of supplying part of the town with hard water, and other districts with soft, or Rivington water. The consequence is that the practice has been discontinued, and a perfectly mixed supply is now distributed to all parts of the town.

COST OF PUMPING FROM WELLS AT LIVERPOOL.

TABLE SHOWING THE QUANTITY OF WATER RAISED IN 1849, AND THE COST OF PUMPING AT THE SEVERAL STATIONS.

Name of Station.	Total water raised.	Water used for condensing at Windsor.	Total cost per annum of raising water.			Cost per annum of raising 1 million gallons.			Water raised per day.
			£	s.	d.	£	s.	d.	
Bootle . . .	329,486,250	..	1445	3	3	4	7	8	902,702
Bush . . .	95,433,850	..	716	3	5	7	10	1	261,463
Soho . . .	168,812,589	..	833	17	1	4	18	9	462,500
Hotham St. .	80,783,436	..	603	4	8	7	9	4	221,334
Water Street	150,338,675	..	874	7	10	5	16	6	411,065
Windsor . .	252,922,650	20,233,812	949	0	3	4	1	6	637,504
Green Lane .	367,378,629	..	920	2	7	2	10	1	1,006,517
									3,903,085

The following table shows the comparative cost of raising the water in 1849 and 1854. As the water at each well is pumped into a reservoir, the second column of the table shows the lift in feet at each well, and enables us to derive the cost per 100 feet of lift.

Name of Station.	Lift in feet.	Cost per million galls. in 1849.			Cost per million galls. in 1854.			Cost per million gallons raised 100 feet h'gh.					
								In 1849.			In 1854.		
		£	s.	d.	£	s.	d.	£	s.	d.	£	s.	d.
Bootle . . .	170	4	7	8	3	2	11½	2	11	7	1	17	0
Bush . . .	228	7	10	1	5	10	11½	3	5	9	2	8	8
Soho . . .	247	4	18	9	4	6	3	2	0	0	1	14	11
Hotham St. .	205	7	9	4	8	3	4	3	12	10	3	19	8
Water Street .	257	5	16	6	4	4	5¾	2	5	4	1	12	10½
Windsor . .	287	4	1	6	2	12	0	1	8	5	0	18	1½
Green Lane .	270	2	10	1	2	2	5¾	0	18	6	0	15	9

One or two facts are worthy of notice in this table, in which the cost includes every expense of labour, coals, oil, tallow, &c., but no allowance in either case for wear and tear, or depreciation of engines, pumps, &c. In the first place, the only two stations which afford any proper guide to pumping expenses, are Windsor and Green Lane. The engines at all the other stations are old, and are only employed until the new supply is obtained by means of the Rivington Pike Works. It will be observed that in 1849 the cost of raising water 100 feet high was four times greater at Hotham Street station than at Green Lane; while, in 1854, the disproportion is still greater, the cost at Hotham Street being now five times as much as at Green Lane. So much for bad engines and machinery.

The absolute cost of raising the water seems to have been reduced at all the stations since 1849, except at Hotham Street, where the cost is greater now than at the time of Mr. Stephenson's report. The diminution at all the other stations is due to the watchful care and supervision exercised by Mr. Duncan, the able engineer of the corporation, whose system of tabulating the accounts, and keeping the chief results constantly under his own eye, cannot be too highly commended.

The cost of raising 1,000,000 gallons, 100 feet high, at the East London works by various kinds of engines, is given below on the authority of Mr. Wicksteed.

	£	s.	d.
By a single-acting engine by Boulton and Watt (average of 2 years' working)	2	5	3
Two single pumping engines by Boulton and Watt (average of 10 years' working)	1	9	10
Two other single pumping engines, also by Boulton and Watt (average of 10 years' working)	1	7	9
By a single-acting Cornish engine, erected by Harvey and Co. (average of 4 years' working)	0	12	6

COST OF PUMPING STATIONS AT LIVERPOOL.

Mr. Stephenson says the cost of the Windsor station was nearly £30,000, but there is a valuable piece of land attached to it. The Green Lane station, at the time of his report, was still incomplete, but had then cost upwards of £19,000. The following are the details of the entire cost of this station, as obtained from Mr. Duncan :—

Cost of the well	£6,600
Cost of the John Holmes engine and two pumps. (The engine has a 50-inch cylinder, and worked on the average in 1854 up to 64-horse power)	5,782
Building engine house, boiler house, and stand-pipe tower	4,278
Cost of the George Holt engine of 52-inch cylinder, (worked to an average of 76 horse-power during 1854) including buildings, boilers, pumps, and fixing	6,500
Cost of Green Lane Station	£23,160

At the time of Mr. Stephenson's report (March, 1850) the population of Liverpool to be supplied with water was about 400,000, for which Mr. Stephenson estimates a quantity of

8,000,000 gallons per day, being at the rate of 20 gallons per head. The increase of inhabitants between 1831 and 1841 appears to have been 39·6 per cent., so that, according to this rate of increase, Mr. Stephenson assumes the population in 1861 will be 557,500, and the volume of water then required will be increased to 11,150,000 gallons per day.

Supposing a supply of only 8,000,000 gallons a day to be required, he estimates that as Green Lane and Windsor wells yielded together 2,000,000 gallons, or one-fourth of the required quantity, six new stations should be erected equally complete in every respect with these two large ones. He estimates that each of these new stations would cost £20,000, because, although the land for some of them would cost less than at Green Lane and Windsor, a greater amount of engine-power would be required.

ESTIMATE OF THE COST OF A SYSTEM OF WELLS FOR LIVERPOOL.

Mr. Stephenson then assumes six new stations, including shafts and steam engines	£120,000
Mains to connect these stations with proposed reservoir, at Kensington	48,000
	168,000
Contingencies	17,000
Total	£185,000

to supply 8,000,000 gallons, which would have to be increased to £277,000 for the supply of 11,000,000 gallons, the quantity which will be required at the end of about ten years.

ARGUMENT IN FAVOUR OF HAVING SEVERAL STATIONS FOR THE SUPPLY OF WATER.

The value of concentration, according to Mr. Stephenson, is not so important in water works as in the case of some manufactures. On the other hand, the number of stations is supposed to afford an opportunity for dispensing with duplicate engines. It is obvious that when several stations are at work, the failure of one will produce slight inconvenience, and a surplus may always be provided in the shape of an additional station, to be worked in case of need.

There are, probably, no large water works with so small an amount of surplus engine power as the present pumping stations at Liverpool, where the engines work, for the most part, night and day without intermission, and nearly up to their full power. No inconvenience has, hitherto, been experienced from this, because if the engine at one station is out of order, there are six other stations from which the supply still goes on. Another argument which Mr. Stephenson gives in favour of detached pumping stations, is that all the water need not be pumped to the highest reservoir, but that the different levels of the town may be supplied from the most suitable stations.

ANNUAL EXPENSE OF PUMPING STATIONS.

Mr. Stephenson thus estimates the annual cost of raising 1,000,000 gallons per day at Green Lane and Windsor stations.

For current expenses, including superintendence . .	£1,100
Depreciation upon engines and machinery, engine houses and cooling ponds, £11,200 at 2 per cent. .	224
	£1,324

At each new station he estimates the expense as follows :—

Current expenses, including superintendence . . .	£1,100
Depreciation upon engines and machinery, engine houses and cooling ponds, £12,000 at 2 per cent. .	240
Depreciation of mains, £8,000 at ¼ per cent. . . .	20
Interest on capital, namely, £30,800 at 4½ per cent. .	1,386
Compensation to landowners	250
Annual cost of each new station . .	£2,996

Mr. Stephenson then gives the following table, showing the

annual cost of obtaining from 8 to 14,000,000 gallons per day, estimating each of the old stations at an expense of £1,324, and each of the new ones at £2,996.

To obtain	Old Stations.	New Stations.	Cost a year.
			£
8 million gallons	2	6	20,624
9 „	2	7	23,620
10 „	2	8	26,616
11 „	2	9	29,612
12 „	2	10	32,608
13 „	2	11	35,604
14 „	2	12	38,600

A well or system of wells being already in existence, whether it is advisable to seek for an increase in the yield by driving adits from the bottom, or by sinking new wells ?

Mr. Stephenson then discusses this question with much clearness and ingenuity. He first explains what is called the cone theory in well sinking, namely, that the mass of earth drained by a well may be viewed as an inverted cone, of which the apex is the bottom of the well, and the base is a circle at the surface of the ground. Also, in this cone, the sloping sides represent the inclined surface of the water, flowing in all directions towards the well.

As the pumping proceeds, the angle formed by the sides of the cone becomes more and more obtuse, or in other words, the sides of the cone become less vertical, and more nearly horizontal, " until," in the words of Mr. Stephenson, " an inclination is established, when the friction of water in moving through the pores and fissures of the rock, is in equilibrium with the gravity upon the plane." Mr. Stephenson observes that when once this condition of equilibrium is established, all further pumping is useless, as the water will simply be gradually lowered, until the well is exhausted ; and, of course, no addition of pumping power will increase the yield of the well.

When an existing well is thus exhausted, three courses present themselves. 1st, to deepen the well by sinking bore holes at the bottom; 2ndly, to drive adits from the bottom, and thus increase the surface drained by the well; and 3rdly, to sink a new well in another spot. Mr. Stephenson does not recommend the first expedient in the case of Liverpool, because all the public wells, except the Bootle, have already been sunk to the level of low water mark, and he is of opinion, if the sandstone be as pervious as it has been proved to be, any addition of the depth, either by sinking or boring, would have the effect of admitting water from the Mersey, and consequently very much impairing the quality of the water.

Mr. Stephenson next shows that the second expedient, that of driving tunnels, so as to make what is technically called a *lodge* or *lodgment* for the water, is principally useful where the pumping is periodical, but that it does not much increase the drainage area of the well. In illustration of this view he gives a diagram (fig. 17), which he thus describes: " Let us

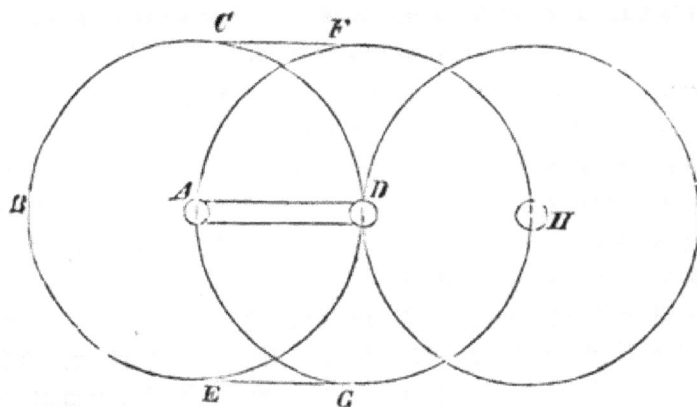

Fig. 17.

suppose that a well is sunk at *A*, and that it drains an area represented by the circle *B C D E*, and that a tunnel is driven from *A* towards *D*, say one mile in length, and that another well is sunk at *D* upon the extremity, or upon the terminus of this tunnel. The only effect of this would be to increase the drainage area of the well A by the area $F H G$, together with the small triangular pieces shown on the figure ; whereas, instead of the tunnel being driven from A to D, if the well at D had been sunk at H the area drained would have been double that which was originally drained by A." After viewing the subject in all its bearings, Mr. Stephenson is of opinion that increasing the number of wells is likely to be a more permanent source of supply than extensive tunnelling. At the same time he observes that the latter admits of an easy mode of connecting the various sources of supply, and, consequently, of concentrating the whole of the pumping establishments.

It appears from the evidence given before Mr. Stephenson, and from the evidence in 1843 on the Liverpool Watering Bill, that the Windsor well, which is 210 feet in depth, affected the surrounding wells to a maximum distance of a mile and three-quarters.

ON THE FLUCTUATION OF LEVEL IN THE WATER OF THE LIVERPOOL WELLS.

There are some interesting notes and observations made on this subject. A series of dip books, kept in the year 1844, show that in the Windsor well the water usually stood on Monday morning at about 75 feet above the bottom, and on Saturday night at an average of about 40 feet. This effect of the pumping during the week appears to have continued without much variation until June, 1845, when the engine began to work 24 hours a day instead of 12. The effect of this increased amount of pumping is, unfortunately, not recorded.

In October, 1845, the engine commenced its previous mode of working during the day only. The levels of the water at this time were 75 feet above the bottom on Monday mornings, and on Saturday nights about 32 feet. In June, 1846, when accurate records were again kept, the depths on Monday morning and Saturday evening were respectively 66 feet and 24 feet, and these depths seem to have continued nearly the same till April, 1847, at or about which period a larger pump was applied. At the end of May in this year the levels of the water on Mondays and Saturdays were respectively 39 feet and 14 feet. These levels diminished from this time till February, 1848, when they were 17 feet and 9 feet, and still further diminished till December, 1849, when they became 14 feet 5 inches and 8½ inches, respectively. This state of the water on Saturday night clearly showed that as much water was being pumped from the well as it was capable of yielding.

Bearing on this part of the subject, the appendix to the report contains some very valuable tables showing the remarkable falling off in the yield of certain wells.

It appears from the evidence of Mr. Thompson on the Liverpool Watering Bill in 1848, that the Windsor well in April of that year yielded, per day, 1,152,000 ;

In May, 1848, after an interval of 5 years, the yield was 807,061 gallons ;

In January, 1850, a further interval of 21 months, the yield was 705,667 gallons ;

Another observation in January, 1850, made the yield 634,752 gallons ;

and this latter agrees nearly with the records of the Duty book.

It appears, then, that in the first 5 years the yield diminished at the rate of nearly 6 per cent. per annum.

In the next 21 months, the diminution was at the rate of 7·2 per cent., or, taking the smaller yield to be correct, at the rate of 11 per cent.

Similar experiments were made on the Green Lane well, from which it appeared that the falling off in 22 months, from May, 1848, to March, 1850, was at the rate of 6 per cent. per annum, according to one set of observations, and at the rate of 4·7 according to the other.

Several observations are given to show the falling off in the yield of two railway wells at Edge Hill, but as this appears to be due to the sinking of bore-holes in the vicinity, they are not suitable cases to show the gradual and general falling off in the yield of new red sandstone wells.

The following are the conclusions which Mr. Stephenson draws from a careful consideration of the facts which he col-lected :—

" That an abundance of water is stored up in the new red sandstone, and may be obtained by sinking shafts and driving tunnels about the level of low-water.

" That the sandstone is generally very pervious, admitting of deep wells drawing their supply from distances exceeding one mile.

" That the permeability of the sandstone is occasionally interfered with by faults or fissures filled with argillaceous matter, sometimes rendering them partially or wholly water-tight.

" That neither by sinking, tunnelling, nor boring, can the yield of any well be very materially and permanently in-creased, except so far as the contributing area may be thereby enlarged.

" That the contributing area to any given well is limited by the amount of friction experienced by the movement of the water through the fissures and pores of the sandstone ; and

" That there is little or no probability of obtaining per-manently more than about 1,000,000 or 1,200,000 gallons a day from each well, and this only when not interfered with by other deep wells."

THE PALÆOZOIC SERIES.

It is proposed to glance at the great formations composing this series in a more rapid and general manner than that which has been adopted for the mesozoic and tertiary groups. In this country especially, although highly important in a geological point of view, the Palæozoic rocks are much less within the range of hydraulic investigations than those which have hitherto been considered. With the exception of a few large towns, situate on the coal measures, and a still smaller number on the old red sandstone and Silurian rocks, most of these old formations are without any large centres of population, and therefore do not possess the same interest for the hydraulic engineer as the new red sandstone and other more recent formations. It is true, many of the older rock districts, especially the carboniferous limestone, the millstone grit, and even the slate rocks of the Silurian and Cambrian series, furnish the gathering or drainage ground, from which some of the largest and most important towns in the kingdom are in future to be furnished with their supply of water. Liverpool and Manchester are both now supplied from the elevated region of millstone grit which surrounds them. Bristol is already supplied from the carboniferous hills of the Mendip district, and Plymouth has been supplied for two hundred years from the granitic district of Dartmoor. Most of the large Scottish towns, as Edinburgh, Glasgow, and Aberdeen, already draw their water from drainage areas on the surface of the oldest rocks in the country.

Principal subdivisions of the Palæozoic rocks below the Permian group :—

Coal formation. Consisting of a series of indefinite alternations of shales and grits, or sandstones of various kinds, with layers of coal and ironstone, and occasionally but rarely thin beds of limestone, thickness about 3,000 feet.

Millstone grit. Consisting sometimes of pebbly, quartzoze gritstones, and sometimes of compact felspathic sandstones and shales, with occasionally thin beds of coal and ironstone, several hundred feet in thickness; but in Yorkshire attaining a thickness of 1,000 feet.

Carboniferous limestone. A mass of calcareous rocks, consisting in the upper and lower divisions of thin, laminated, shaly beds, sometimes with argillaceous partings, and in the centre portion of a thick mass of nearly pure thick bedded limestones without clay partings. In Yorkshire containing, in the lower part, beds of coal. Thickness varying from 500 to nearly 2,000 feet.

Old red sandstone. Consisting of arenaceous conglomerates, sandstones, and argillaceous rocks, sometimes containing impure limestones and flagstones. Exceedingly variable in thickness, but probably not less than 3,000 feet in Herefordshire, and still more in Devonshire.

Silurian group. Consisting of sandstones, limestones, shales, conglomerates, and calcareous flagstones, many thousand feet in thickness.

Cambrian group or lower Silurian. Consisting of sandstones, indurated argillaceous rocks, dark laminated limestones, fine and coarse grained slates, &c., also several thousand feet in thickness. Below these are micaceous and chloritic slates, quartz, and other rocks in Anglesea and Carnarvonshire.

The coal formation, owing to the alternation of porous grit and sandstone beds with retentive strata of shales and clay, commonly yields a moderate supply of water wherever wells are sunk into it. The water, however, is usually of inferior quality, being often much impregnated with iron. It is doubtful, also, whether any shaft sunk in the coal measures would yield a sufficient quantity of water for the supply of a large town; whether, in fact, any such shaft would yield even a million gallons a day. Water is indeed often troublesome in coal mines, and we frequently hear of large quantities having to be pumped out in order to keep the workings dry. These quantities are, however, commonly quite insignificant when compared with those required for consumption in large towns. It is probable that the numerous faults which occur in all coal measures, and which are frequently filled up with impermeable matter, operate to cut off and isolate the water-bearing strata into sections of limited area; and this provision, so beneficial to the miner, is evidently one which interferes with the supply of water to be obtained from wells or shafts sunk for the purpose of water works.

The following are the principal towns in England situate on the coal formation :—

Newcastle and Durham in the coal field of Durham.

Leeds, Huddersfield, Barnsley, Rotherham, Sheffield, Bradford, Wakefield, in the Yorkshire coal field.

Cardiff, Merthyr, Swansea, Newport, and Pontypool, in the South Wales coal field.

The millstone grit has scarcely any towns situate upon it, but is very important, especially in Lancashire and Yorkshire, as furnishing the drainage area which supplies water to several of the principal towns in the kingdom.

Lancaster, Preston, Liverpool, Manchester, Stockport, Bradford, Leeds, and many others, are all supplied from the united springs and surface water of the great millstone grit formation which constitutes the axis of what has been distinguished as the *Penine chain of England.*

The deep, precipitous valleys of the millstone grit, the porous strata resting on the impervious limestone shales which throw out the springs, and the extraordinary rainfall in the bleak and elevated district which it occupies, all contribute to make this a highly desirable water-collecting formation. A comparatively small surface suffices for a drainage area; the volume of drainage water is further swollen by springs which usually break out at the heads and sides of valleys where the limestone shale begins, and the deeply grooved form of the valleys gives peculiar facilities for the construction of embankments to hold up the water in storage reservoirs.

The carboniferous limestone, like the millstone grit, is equally destitute of large towns situate within its limits, although there are several large towns which derive their chief supply of water from the springs and streams of carboniferous limestone districts.

The great central mass of the carboniferous limestone admits water very freely to pass through by means of the numerous fissures by which it is fractured. The upper and lower shales of this formation are probably not so freely permeable by water.

The phenomena of springs caused by faults filled up with argillaceous matter in limestone districts have been already alluded to. These springs are very common in Derbyshire, in the Gower district of South Wales, and in the district of the Mendip hills. Copious streams of beautifully clear limpid water are constantly flowing through Leigh-upon-Mendip, Downhead, and other villages on the Mendip hills. These springs, which break out on very high ground, are usually caused by faults damming up the water. Nearly every range of carboniferous limestone in England presents examples of streams, and sometimes large rivers, being engulphed for a time and pursuing a subterranean course through the fissures and cavities of the limestone. The river Manifold and its tributary, the Hamps, are well-known examples of subterranean rivers in Derbyshire. Both these rivers flow from the millstone grit, and after passing over the lower limestone shales they become lost in caverns of the great limestone; and after pursuing a subterranean course of about four miles in length, the two rivers, having united under ground, appear again near Ilam as one stream, which flows in a deep cleft of the rocks and joins the river Dove, near

Thorpe.* The Ribble, the Nid, and others of those rivers which rise in the elevated mountain district between the counties of York and Westmoreland, are also subterranean in the upper part of their course.

The sources of the Ayr and the Wharfe which rise near the base of Pennygent mountain in the western moorlands of Yorkshire, are also subterranean for several miles during dry seasons. The small river Greta which rises in Yorkshire, but flows westward to join the Lune in Lancashire, is also subterranean for some part of its course.

The same phenomena of swallet holes and subterranean rivers present themselves in the district of Gower and in the Mendip hills. Both in the Mendip and in Derbyshire the business of stopping the swallet holes in order to prevent the loss of the water is a regular profession, pursued by some of the most cunning and artful of the native inhabitants. Their services are highly valued, and are in great request in dry seasons when every drop of water becomes valuable. Many of the subterranean courses in limestone valleys are of comparatively small capacity; at least there are certain points in them at which no unusual volume of water can pass. In consequence of this the surplus water which cannot pass off by the subterranean channel appears at the surface, generally at the bottom of a deep romantic valley, cleft through the solid rocks which hem it in on both sides. In wet seasons, therefore, it is not unusual to find a brook flowing at the surface in the bottom of a valley, while, a few weeks afterwards during a drought, the brook may have disappeared from the surface and be confined to its subterranean channel.

Many of the springs from the carboniferous limestone are extremely copious. The one named after St. Winifrid, at

* In the Peak district of Derbyshire, many small tributaries of the Derwent are engulphed in the middle limestone rocks, and re-appear in the neighbourhood of Castleton.

I

Holywell, in Flintshire, mentioned by Messrs. Conybeare and Phillips, yields nearly seven million gallons a day. This stream in its short course to the sea, of about one mile, turns no less than eleven mills, three of which are placed abreast. The springs dedicated to St. Mary and St. Osward, in the same neighbourhood, are almost equally copious. Other celebrated springs occur at Giggleswick scar, on the road from Settle to Kirby Lonsdale, and at other places in the western moorlands of Yorkshire.

The hot springs of Buxton and Matlock, in Derbyshire, are derived from mountain limestone, and so are the hot waters of Clifton, near Bristol.

The old red sandstone. This formation is chiefly developed in Devonshire, and in the counties of Hereford, Monmouth, and Brecon. The principal towns on it are Leominster, Hay, Hereford, Ledbury, Ross, Brecon, Abergavenny, Monmouth, and Newport. In North Devon, Barnstaple, Ilfracombe, and one or two smaller towns are situate on the old red sandstone; while, in South Devon and Cornwall, there are Falmouth, Bodmin, Tavistock, Plymouth, and many others of less consequence.

The scenery of the old red sandstone is characterised by large and massive features, as extensive plains, broad valleys, and occasionally mountain masses of considerable height. There have been few embankments made across old red sandstone valleys in this country for the storage of water, nor does the physical conformation of the valleys afford the same facilities for this as in the millstone grit and slate districts.

The alternation of argillaceous and sandstone beds throughout this formation occasions the water to be held up at various levels, and generally throughout the districts of Herefordshire and Monmouthshire the wells of private houses meet with abundant supplies of water at no great depths. There is no instance of any well being sunk in the old red sandstone for the purpose of procuring any large supply of water.

The towns situate on the Silurian and Cambrian strata of this country, and on the granitic and syenitic formations, are so unimportant that it will not be necessary to enter into particulars respecting the hydrography of these groups.

WELLS AND BORINGS FOR PROCURING SUPPLIES OF WATER.

The commonest form of well is that sunk into a bed of sand or gravel resting on an impervious substratum of clay. Such are the shallow surface wells found in abundance all over the metropolis, and such afforded, probably, at one time the best supply which its inhabitants were able to procure. Wells of this kind are still existing, and are still used in many towns which are built on deposits of sand or gravel resting on a bed of clay. In digging such wells it is not necessary to sink down to the clay, but only to the depth at which the gravel or other porous stratum is saturated; and this is frequently found to be by no means a uniform level even in wells which are situate very near to each other.

The superincumbent stratum of sand or gravel being usually situated in a trough or depression of the clay, it may be presumed, theoretically, that the basin is permanently saturated below the level of the lowest natural point at which the water can escape or flow over the edge of the basin, and that at any given point inside the basin, the water will stand at a certain fixed height above this level of overflow or saturation. This height, again, it has been sought to determine theoretically from considerations connected with the permeability of the overlying deposit. There are many circumstances, however, which conspire to render theoretical deductions erroneous even in the most simple cases. For instance, the surface of the clay basin may be very irregular, as indeed it commonly is. It may have been traversed by ancient water courses which have left ridges and hollows, while, beside all this, there may have been subterranean movements either before, at, or after the deposit of the gravel which may have produced rolls or ridges in the clay. Hence it happens that from natural causes alone, apart from the effect of other wells, it can seldom be predicted with certainty at what height the water will be found or will permanently stand in such surface wells. In all basins of this kind the water must be conceived as gradually percolating from the surface downwards towards the level at which it can escape, sometimes flowing with comparative freedom through materials tolerably porous, and at other times obstructed, now by changes in the nature of those materials, and now by dams and other obstacles interposed by nature.

I have said that such wells probably furnished at one time the best supply obtainable in the metropolis; but all this is widely altered at the present day, for since the immense increase of buildings, the soil has become saturated with impurities of every description, which can scarcely fail to have frightfully contaminated the water of all such wells. The churchyards of the modern Babylon, her sewers reckoned by hundreds of miles, her gas works, her chemical and other manufactures whose proceeds are injurious to human life, have all contributed their share to the poisoning of wells which might once have been wholesome, although the water could never boast of the freshness and life of deep spring water.

The other class of wells is widely different from the first, and commonly yields a supply of good spring water. These are wells sunk through an impervious material which is itself destitute of water, but which serves to keep down the water in a porous bed lying beneath it. For instance, after sinking through the thick mass of the London clay in the neighbourhood of London, we reach a series of beds containing sand and gravel in such proportions as to make them readily permeable by water; so that it is no uncommon thing, on sinking down to, and disturbing the remarkable pebble bed which usually forms the basis of the London clay, to find the water burst up with some violence and stand permanently in the well many feet higher than the level from which it rose.

The same thing occurs in sinking through other hard beds before reaching the chalk, and, again, at the top of the chalk, the bed of green-coated flints, which usually separates the chalk from the overlying sands, gives rise to the same phenomena of water bursting up. On sinking through the chalk also, the chalk marl is usually found to keep down the water of the upper green sand in a similar manner, and again on a much greater scale the gault performs the same office with respect to the water of the lower green sand.

Now in all such cases as these where a superior overlying impervious stratum has to be sunk through in order to reach a water-bearing stratum beneath, it is not necessary to dig a well to the entire depth, but after sinking down a certain distance and forming a well into which the water may rise as into a reservoir, the rest of the process is simply completed by boring down to the water-bearing stratum by means of a small perforation varying from three or four inches up to eighteen inches. Through this small perforation the water will find its way up to the larger well above, from whence it may be pumped up as required.

In certain favourable situations where a boring is made through an impervious stratum at a point considerably lower in level than the porous district, at the line where the water filters into the earth, the water rises above the surface, giving rise to the condition termed an artesian well, from their first general establishment in the French province of Artois. Such wells were once common in the valley of the Thames about Brentford, but are not now overflowing in consequence of the great number of them which have been sunk into the same stratum. Overflowing wells are still common, however, on the flat lands of Essex, about Foulness

Island, &c., and also on the coast of Lincolnshire, from Spalding to the Humber, in which part of the country they are called blow wells.

There is some ambiguity in the name artesian as applied to these wells, as it is uncertain whether it should be confined to those which overflow, and rise above the surface, or should be applied to all wells in which the water is procured by boring instead of sinking. I think myself that a general name should be applied to all those wells in which the water is procured by boring through a retentive soil down to a water-bearing stratum, and the name artesian would do for these as well as any other. Then those wells, in which the water rises above the surface, should be distinguished as overflowing artesian wells.

Some such distinction as this is obviously necessary, because different persons understand different things by the term artesian well as commonly applied; in addition to which many wells which were once overflowing are not so now, as already mentioned in the case of those bored in the valley of the Thames. Such wells are of course still artesian wells, but are no longer overflowing artesian wells. It need scarcely be observed that ordinary borings are quite distinct from artesian wells. These are exceedingly common in such districts as the chalk and new red sandstone, where they are sunk at the bottom of shallow wells or lodgments, as they are sometimes called, in order to increase the supply.

For instance, in procuring the supply for Watford, which is situate on the chalk, a well or lodgment was first made 15 feet in diameter and 33 feet in depth. A boring of 12 inches clear diameter was then directed to be made to the depth of 70 feet below the bottom of the well, and this it was calculated would produce 300,000 gallons per day of twenty-four hours.

CONSTRUCTION OF WELLS.

In hard and homogeneous rocks such as chalk, the oolites, and new red sandstone, wells frequently do not require a lining of any kind, and have merely to be excavated to the required size. The chief difficulty in hard rocks is, that blasting by gunpowder is frequently necessary, so that the process of sinking becomes tedious and expensive. All shafts sunk in argillaceous, marly, and sandy strata require, however, to be lined or steined, as it is technically called. Even in sinking through the harder rocks, also, partings and beds of clay and sand are frequently met with, and sometimes require a lining as much as if the shaft were entirely in clay. Various materials are used for lining wells, and preventing the earth through which they are sunk from falling in. The principal of these are brick, stone, and iron, while wood is only employed for temporary purposes, as in the shafts of mines.

For large wells sunk in the tertiary strata around London a lining of one brick in thickness is frequently employed, and for smaller wells a lining of only $4\frac{1}{2}$ inches, or half a brick, may be used. When the lining is half a brick in thickness, there is only one mode of laying the bricks, namely, in courses breaking joint with each other. Where the work is a whole brick thick, it is sometimes built in two half-brick rings, and sometimes the bricks are laid all as headers. Wherever the common statute brick is used, it is evident that all the horizontal joints are lenticular, or wedge-shaped, so that the sides are never parallel. This is a great inconvenience and can never make good work, no matter in what way the gaping space at the back of the joint may be filled up. Formerly there was some difficulty connected with the excise duties, in procuring bricks of any other shape than the regular statutory rectangular form. Since the repeal of these duties, however, bricks may without inconvenience be procured of any shape and size, and ought certainly for steining wells to have ends radiating to the centre of the well. It would not be necessary of course to have bricks moulded specially for every particular well, but there might be forms to suit wells, increasing in diameter by steps of two feet up to ten or twelve feet in diameter. Radiated bricks of this description are now extensively made for sewer work, and might be employed with equal advantage for wells. They would at once effect a great superiority over the ordinary steining of wells.

The material to be employed for steining must, of course, be decided by the judgment of the engineer, according to the particular circumstances of each case, as in many localities and in certain peculiar cases, stone or iron may possess advantages over brickwork.

In sinking through the tertiary clays and sands in the neighbourhood of London, it was formerly very common to employ a wooden ring or curb on which a cylindrical mass of brickwork was built. Then the ground beneath the curb was excavated, and the whole mass of brickwork gradually sunk down as additional courses were placed on the top of it. This method of building wells is not now so common as that of building underneath in the fashion of underpinning. This method of building underneath the already executed steining is well described by Mr. Swindell in his treatise on wells and well-digging. Suppose eight or ten feet of the well dug out, the steining of this first part may be built from the bottom, and then a further depth being excavated, it becomes necessary to complete the steining beneath the part already executed. This is effected by laying at the bottom of the recent excavation three courses of brickwork in cement, and building the other courses of ordinary brickwork upwards until the space is filled up. The well, when completed, will then consist of a cylinder of brickwork, with occasional nine-inch rings set in cement, and the interval at which these rings occur will depend on the nature of the strata, being usually in the tertiary beds from five to twelve feet. In working through sandy strata, which alternate with clay and are frequently saturated with water and converted into veritable quicksands, many precautions are necessary. Frequently the whole brickwork in such cases has to be set entirely in cement or in the best hydraulic mortar, and very often the steining already executed has to be suspended or held up in its place by iron rods attached to an iron plate underneath the brickwork. Even these precautions are sometimes not sufficient, and resort must be had to iron cylinders in lieu of brickwork. These iron cylinders are sometimes of wrought, and sometimes of cast iron, bolted together by means of flanches. They frequently require considerable force to drive them down through quicksands and other incoherent beds. Mr. Swindell says:—" In some wells that have been executed in sandy soils, cast-iron curbs have been inserted at intervals, each curb slung to the one above it by tie-rods. The gravel or sand can then be excavated under the curb, as the clay can under the brickwork rings set in cement; the curbs, in fact, bearing the same relation to the cemented brickwork in the case of sandy soils, as the cemented rings do to the dry brickwork in clayey ground."

Although Mr. Stephenson, in his able report on the well supply of Liverpool, prefers the sinking of separate wells to the employment of adits or drift-ways, yet the latter are extremely useful as lodges or reservoirs to store a quantity of water in cases when the pumps are not at work. The drift-ways at the bottom of the well in Trafalgar Square form a reservoir capable of holding 122,000 gallons of water, and the storage rooms at the bottom of some of the Liverpool wells is much greater than this.

In many wells extraordinary precautions are necessary to exclude land-springs and water of inferior quality, and sometimes the interests of adjacent canals and of the neighbouring well proprietors call for such precautions. For this purpose the brickwork for a considerable depth is sometimes set in cement or hydraulic mortar, besides being puddled or con-

creted at the back.   In the well sunk by the London and
North-Western Railway Company at Camden Town, there
was an inner and an outer steining of brickwork, the inner
being 4½ inches thick, and the outer 9 inches.   The space
between the two cylinders of brickwork contained a seg-
mental cylinder of iron, backed with 9 inches of concrete.
The whole thickness of the lining in this case, therefore, was
about 2 feet.   This mode of construction was adopted to a
depth of 28 feet from the surface, below which was a brick
steining of 9 inches, with bonding curbs of iron at intervals.*
The diameter of this well is 9½ feet in the clear, and the
whole of the steining is executed in cement.

COST OF WELL-SINKING.

There are so many contingencies connected with this kind
of work that we must be prepared to find great anomalies in
the cost which has been incurred.   The comparative hardness of the strata to be sunk through, the frequent occurrence
of quicksands occasioning peculiar precautions and much
expenditure in the sinking, and the necessity for frequently
keeping out land springs, are all subjects which very materially influence the expense.   Most cases of well-sinking
require a special estimate from experienced practical persons
well acquainted with the strata to be sunk through, and with
similar works.   Such estimates, although probably more to
be depended on than any other, are frequently very erroneous,
and lead to great disappointment.   The following attempt to
show the cost of wells in the neighbourhood of London is
founded on some statements by Mr. Prestwich; but I have
endeavoured, as will be seen, to separate the well-sinking
from the boring, having in certain cases deducted from the
total cost a sum equal to £2 per foot, which may be taken as

* Swindell on " Wells and Well Digging ;" Weale's Rudimentary
Series.

a fair allowance for boring and tubing at the bottom of a deep well.

Situation of Well.	Depth of Shaft.		Total depth of shaft.	Cost of shaft.	Cost per foot in depth.
	In tertiary strata.	In chalk.			
				£	£
Truman's brewery .	196		196	4056	21
Reid and Co.'s ditto	136	123	259	7454	29
Zoological Gardens, Regent's Park	220		220	1900	9
Model Prison, Pentonville . . .	220		220	1300	6
Lunatic Asylum, Colney Hatch .	188		188	991	5

The price for the well at Reid's brewery includes the hire and repair of temporary pumps during the execution of the work, and also the cost of two new sets of permanent pumps. It will be seen from the following account of this well, taken from the description by Mr. Braithwaite, in the Proceedings of the Institution of Civil Engineers for 1843, that it comprises considerably more than a mere shaft, that the sinking in the chalk was enlarged into a capacious reservoir, and that a considerable extent of adit was driven at different levels. The yield of this well is 277,200 gallons per day, the water being used for refrigerating purposes, although it costs the brewers more to raise it than they would have to pay the water companies for a similar quantity. The depth down to the chalk at this well is 138 feet. At the depth of 87 feet a cast-iron cylinder, 5 feet 3 inches by 3 feet 2 inches, is inserted and continued down to 135 feet, to within 18 inches of the first bed of flints in the chalk. The sinking in the chalk was gradually increased in size as it proceeded downwards, till, at the depth of 178 feet from the surface, it was 16½ feet diameter. The excavation was continued at this diameter as far as 202 feet below the surface. Water was found under the second, sixth, eighth, and tenth beds of flints, and the total supply at 202 feet was 72,000 gallons per day. At the depth of 196 feet from the surface the first tunnel was driven 91 feet north-west, in the direction of another well, but the tunnel only increased the yield 14,400 gallons a day. The eighth bed of flints, at 154 feet from the surface, yielded the largest quantity of water, namely, 10,800 gallons per day. Mr. Braithwaite, therefore, drove a second tunnel at this level, 6 feet high by 5 feet wide, distance of 16 feet from east to west, and then north and south for 108 feet, by which he obtained an increase of 54,000 gallons per day. The shaft being now sunk to the depth of 202 feet from the surface, it was ascertained by boring that a further supply could be obtained at a depth of 20 feet lower. The shaft was continued 22 feet deeper, with a diameter of 7 feet, when the water was found flowing from two horizontal fissures in the chalk without flints. At this depth two tunnels were driven, one to the north-west, being connected with the first tunnel of 91 feet, by which an increase of 121,600 gallons a day was obtained. The other tunnel was driven 24 feet in a south-easterly direction, and produced a further increase of 28,800 gallons a day. The total quantity obtained from the shaft and all the tunnels is about 277,200 gallons a day; the well being formed into a reservoir at the bottom, which is capable of holding 100,800 gallons.

It appears from this account that, in addition to the great enlargement of the shaft in the chalk, about 110 yards of headings were driven, and the cost of these could not have been less than £550, making the cost of the shaft alone about £27 per foot vertical.

It appears from the instances which have been quoted, that small shafts may be sunk in the London clay and other tertiary beds to the depth of about 200 feet, at the rate of from £5 to £10 a foot vertical, including the steining and temporary pumping when necessary. Also, that large shafts of 8 or 10 feet in diameter, or an equivalent area, may be sunk from 200 to 300 feet deep for the cost of £21 to £27 per foot vertical. True it is that many estimates are made, and even contracts taken at much lower prices than this. For instance, the well at the Model Prison, which is 6 feet diameter in the clear within the brickwork, is said to have been let to Mr. Clarke, of Tottenham, at the following prices :—

£	s.		
67	10	for the first 30 feet.	
57	0	,,	second ditto.
58	10	,,	third ditto.
60	0	,,	fourth ditto.
61	10	,,	fifth ditto.
67	10	,,	sixth ditto.
£372	0		

So that Mr. Clarke's price for 180 feet was only £372, whereas, the well which was sunk 220 feet, cost in reality £1,300.

The Green Lane well at Liverpool, which is 185 feet deep and 10 feet diameter, cost £6,600; but this includes a large lodgment in the bottom of the well, and nearly 100 feet of boring in the bottom.

For wells in the lower new red sandstone in the neighbourhood of Wolverhampton, without steining, a price of about £5 per yard has been paid for a shaft 5 feet diameter down to 150 feet in depth. Headings driven in the sandstone rock, 5 feet high and 3 feet 6 inches wide, have cost £10 per yard forward.

In the Wolverhampton Water Works Bill of 1855, the existing company sought for powers to sink additional wells, with the view of increasing their supply of water, and the following was the estimate given in evidence by Mr. Marten, their engineer, and Mr. Hawkesley, their consulting engineer.

Sinking two shafts at Oaken in the red sandstone rock, each 150 feet deep, 100 yards at £5 . . .	£500
Standage at bottom forming the enlargement necessary for the lodgment of the water	1,500
Driftways, 200 yards at £4	800
Pumping and sundries . . . . ⸴	1,200
	£4,000

The estimate for these *two* shafts, each 150 feet deep, is somewhat inconsistent with the fact that the Green Lane single shaft at Liverpool, 185 feet deep, cost £6,600.

Some very valuable information as to the construction of wells, especially in the neighbourhood of London, will be found in Mr. Swindell's "Rudimentary Treatise on Well Digging and Boring." The method of constructing wells with sinking curbs, and of steining in brickwork, and lining with iron cylinders, are there described in a very practical manner.

There are also some very useful observations on the most improved methods of boring.

ARTESIAN WELLS.

The chief localities in which these wells have hitherto been sunk in this country have been already described. The expense of boring them is of course much less than that of shafts.

Mr. Prestwich gives several examples of artesian well borings in the neighbourhood of London, from which their cost may be judged of with tolerable accuracy. Thus, a boring of 252 feet in depth through tertiary strata in Lombard Street, cost £200. One at Water Lane, Edmonton, 66 feet deep in tertiary strata, cost £13; one at Waltham Abbey, through 90 feet of tertiary strata, cost £16; one at Wigborough, in Essex, through 300 feet of tertiary strata, cost about £120; and one at Mitcham, through 190 feet of tertiary strata and 21 feet of chalk, cost £100. From these examples, it appears that the cost of boring up to 100 feet in tertiary strata, does not exceed 4s. per foot. For depths between 100 and 200 feet, the price seems to vary from 6s. to 10s. per foot; and between 200 and 300, from 10s. to 16s. per foot.

A deep boring sunk at Loughton, in Essex, through 535 feet, of which 324 were in tertiary strata, and the remaining 211 in chalk, cost £750, or nearly 30s. a foot.

Mr. Clarke's price for boring in tertiary strata, at the bottom of a well 180 feet deep at the Model Prison, was 45s. per foot; the boring to be made with a 10½ inch auger, and to have cast-iron pipes inserted in it 8 inches diameter ⅝ thick, fitted together with turned joints and wrought-iron collars, and fitted with screws; the whole to be flush inside and out. His price for boring in chalk with a 7½ inch auger, at the bottom of the 10½ inch bore, was 27s. a foot; no pipe to be inserted in this boring. He, however, offered for 10s. 2d. per foot extra to insert a perforated copper pipe, weighing 6 lbs. per foot in the chalk boring.

The artesian borings through the gault at Cambridge are usually 130 to 150 feet deep, and cost from £15 to £20.

The price paid at Liverpool for a 6 inch bore in the bottom of Green Lane well, which is 185 feet deep, was :—

£	s.				
2	10	per yard for the first	20 yards.		
3	0	,,	,,	second	,,
3	10	,,	,,	third	,,
4	0	,,	,,	fourth	,,
4	10	,,	,,	fifth	,,

The following are some tenders recently made for boring through chalk at Guildford, where the chalk is very near the surface :—

Mr. Thomas Collins, Kensington.	Mr. Thomas Clarke, Gray's-Inn Sq., London.	Messrs. Baker & Sons, London.	Mr. Robert Peyton, of St. Alban's.	Mr. Isaac Denning, Mile End Road.	Mr. Tiller, Bond Court, Walbrook.
Price of boring per foot without pipes.	Boring in chalk without pipes.	For a 17-inch boring, with pipe, at a depth from surface not exceeding 100 feet.	For a 12-inch bore including pipes.	For a 14-inch boring without pipe.	For a 12-inch boring without pipes.
per ft. s. d. 6 inch .. 11 0 9 ,, :: 14 0 12 ,, :: 17 0	per ft. s. d. 1st 20 feet 20 0 2nd ,, 22 6 3rd ,, 25 0 4th ,, 27 6 5th ,, 30 0 6th ,, 32 6 7th ,, 35 0 8th ,, 37 6 9th ,, 40 0 10th ,, 42 6 11th ,, 45 0 12th ,, 47 6 13th ,, 50 0 14th ,, 52 6 15th ,, 55 0	25s. 0d. per foot. Depth from 100 to 150 feet. 32s. 0d. per foot. 150 to 200 feet. 37s. 0d. per foot. 200 to 300 feet. 46s. 0d. per foot.	per ft. s. d. 1st 30 feet 30 0 2nd ,, 40 0 3rd ,, 50 0 4th ,, 60 0 180 feet 80 0	Per ft. s. d. 1st 50 feet 16 0 2nd ,, 18 6 3rd ,, 20 6 And so on, increasing 2s. 6d. per foot for each 50 feet of depth.	per ft. s. d. To 100 feet 20 0 200 ,, 25 6 300 ,, 29 0 400 ,, 35 0 500 ,, 40 0 600 ,, 43 6 With pipes. 100 feet 43 6 200 ,, 49 0 300 ,, 53 6 400 ,, 58 6 500 ,, 63 6 600 ,, 70 0
With pipes. 6 inch .. 16 6 9 ,, :: 20 6 12 ,, :: 24 0 To extend boring through chalk till water met with.					

In the case of the Crossness well, the following are tenders for a boring of 17 inches diameter, or as nearly of this diameter as the strata will admit of, the boring to commence at the end of an existing 18 inch bore, at the depth of 720 feet below surface :—Mr. Clarke, of Gray's Inn Square, London, offered to bore the first 25 feet below 720 feet at £7 per foot, the second 25 feet at £7 5s. per foot, and so on increasing 5s. per foot at each 25 feet of depth. Messrs. Docwra and Son, of Balls Pond Road, Islington, tendered for boring from the depth of 720 feet below surface at £6 a foot for the first 20 feet, £6 5s. per foot for the second 20 feet, and £6 10s. for the third, and so on increasing at the rate of 5s. per foot for each 20 feet.

The following were the tenders of the same two firms for the best wrought-iron brazed and collared pipes with steel shoes :—

Clear diameter in inches.	Tenders per Foot.	
	By Mr. Thomas Clarke.	By Messrs. Docwra & Son.
	s. d.	£ s. d.
16	..	2 2 0
15	2 0 0	..
14	..	1 14 0
13	1 12 0	..
12	..	1 10 0
11	1 8 0	..
10	..	1 6 0
9	1 5 0	..
8	..	1 2 0

The deepest borings in the world are probably, that made in the Abattoir of Grenelle in Paris, 1,800 feet, and one made 1,878 feet deep through the new red sandstone at Kissengen, in Bavaria.

The boring at Grenelle passed through 148 feet of tertiary strata, 1,394 feet of chalk, and 256 feet of green sand and gault. Its cost, according to Mr. Prestwich, was £14,500; but this includes some extraordinary expenses, such as double sets of tubes and the constructions over the well. The contract for the first 1,312 feet was £4,000; and M. Mulot, the engineer, states that the whole work could now be executed for £10,000.

This boring was commenced in 1835, to supply with water the Abattoir of Grenelle. The auger penetrated the water-bearing stratum in February, 1841, when the water rushed up with great force, and overflowed the surface of the ground. The boring was commenced with a diameter of 20 inches, and gradually diminished, till at the depth of 576 feet it was only 12 inches. Wrought iron tubes were first used to prevent the bore choking up with sand, but these have been replaced by copper tubes. The tubes commence with a diameter of 12¼ inches, which gradually diminishes till the fifth tube, which is 10 inches diameter at a depth of 1,148 feet. A sixth tube goes down to 1,345 feet, with a diameter of 8¼ inches; a seventh to 1,771 feet, with a diameter of 6¾ inches. The remaining part of the bore is not tubed. According to an article in the *Constitutionnel*, 4th of March, 1841, the yield of the well was 880,387 gallons in 24 hours. The water is said .to have risen at first to a height of 65 feet above the surface, but this statement appears doubtful. The temperature of the water, according to Sir John Robison, in 1843, was 82½° Fahr.; that of the water in the United Mines, in Cornwall, which are 1,770 feet deep, being 92° Fahr.; and the highest recorded temperature of water in these mines being 96°.

The boring at Kissengen, in Bavaria, 1,878 feet deep in new red sandstone, is said to have cost £6,666. The expense of several borings in chalk of about 1,000 feet deep seems to have been about £3,000; but there are several instances of much cheaper work. For instance, Mr. Prestwich states that M. Degouzeé has lately much reduced the expense of boring by the use of steam power, and by the introduction of new machinery. He has lately contracted to bore an artesian well at Rouen to the depth of 1,080 feet, through the lower cretaceous and oolitic rocks, including expenses of every description. M. Degouzeé has also constructed three artesian wells in different parts of France, to an average depth of about 825 feet, at a cost, including tubes and all expenses, of £600 to £1,000. Shallower borings down to 600 feet have been made in France, at a rate varying from 5s. to 20s. per foot, while one of 666 feet, chiefly in green sand and gault, cost £1,216. Messrs. Degouzeé and Laurent, of Paris, undertake to bore to the depth of 200 English feet for about £120, and to go 1,000 feet deep for £600.

BORING MACHINERY.

The method sometimes used on the Continent of boring by means of a rope in place of a boring-rod, was well described by M. Jobard in his report on the Paris Exhibition of 1840.

For perforating hard rock, a boring head, termed the *mouton* or ram, is used. This is a cylinder of cast-iron, commonly about 8 inches in diameter and 39 inches in length, weighing from one to three cwt. The upper part of the cylinder is hollowed in a conical form, and the sides are fluted so as to allow the broken *débris* of the rock to pass up and lodge in the hollow conical top. The *mouton* is attached to the rope by means of wrought-iron handles, which are double, in case one should break. The lower part of the *mouton* is prepared to receive a number of blunt-pointed steel chisels, which are firmly secured to it. There are several ways of giving motion to the rope which works the *mouton*. One of these is by means of a long plank, placed obliquely about 12 or 15 feet above the bore hole. The *mouton* is suspended about 15 or 20 inches from the bottom, and is made to fall through a height of 2 to 3 feet about 25 to 30 times in a minute. The dust or powder resulting from the cutting action of the steel chisels would soon, if not moistened, impede the action of the tool by deadening the blows. If, therefore, no water be naturally present, it must be supplied, and then the sort of liquid mud thus formed rises or spouts up through the fluted openings in the *mouton*, and enters the hollow cone at the top. When it is observed by the workmen at the surface that the tool has gone down sufficiently to fill the cone the *mouton* is withdrawn, and is then found to be filled commonly with a hard conical mass, the result of several hours' working.

The Chinese are said to have bored successfully with the *mouton* alone to the depth of 1,800 feet. This is quite possible, if the strata be hard and solid throughout; but if they consist of sand, gravel, wet clay, or of such material as requires tubing, the *mouton* is not applicable alone, but requires to be replaced by what the French call the *emporte-pièce*, or shell pump, as it has since been termed in England. The *emporte-pièce* is a cylinder fitted at its base with two common D valves, turning on the diameter as a hinge, and opening upwards. This cylinder is lowered down to the bottom, and caused to penetrate by a blow from the *mouton* or ram before described, and which when thus employed is of course without its cutting chisels, and is merely a hollow cast-iron cylinder, weighing about three cwt. The ram having an opening through it longitudinally, slides on a metallic rod attached to the *emporte-pièce*. When it has gone up a few feet, it is stopped by a projection, and then falls down on the *emporte-pièce;* which latter sinks down at every blow, and at the same time allows the semi-fluid mud to rise up through its valves. When drawn up, the space above the valves is found to be filled with a cylindrical lump of mud, more or less indurated.

In the strata which require tubing, it is of course necessary that the boring tool which passes through the tube should be capable of making a hole larger than the external diameter of the tube itself. The form of *mouton* employed in France for this purpose is called the *allezoir* or instrument for enlarging, and the contrivance is sufficiently ingenious. Instead of the *mouton* being suspended from the rope in a truly central axis, it is suspended somewhat on one side of the centre, so that on being lowered the steel cutters or chisels strike obliquely, and thus excavate a hole of larger diameter than the *mouton* itself. This *mouton*, when employed to act through tubing, has a conical receptacle for the mud, like the one first described. There are several minute points to be attended to in working with these boring tools, such as the degree of torsion to be given to the rope, and the proper time for drawing up the *mouton*, so that its working be not too much impeded by friction. These are points which the workmen are soon enabled to master after a little practice. This method of boring is much used in France, Saxony, and other parts of the Continent. The *moutons* are not expensive, one of three cwt. costing only about £2. They can be made and repaired by the ordinary village blacksmith.

MESSRS. MATHER AND PLATT'S EARTH-BORING MACHINERY.

The arrangement which has just been described has been somewhat modified by Messrs. Mather and Platt, of Salford Iron Works, Manchester. The boring head employed by them consists of a wrought-iron bar, about 8 feet long, on the lower part of which is fitted a block of cast-iron in which the chisels or cutters are firmly fixed. There is no conical cup or other contrivance in this boring-head for containing any of the loosened matter, as in the French machines; but, when the boring-head has been in action for some time, and the chisels have loosened and broken up a certain quantity of *débris*, the boring-head is raised, and another contrivance, called the shell pump, is lowered, in order to raise the *débris*. The invention of Messrs. Mather and Platt, therefore, consists essentially of two parts, the boring-head and the shell pump, sometimes called the sludge pump, both of these being

put in action by the power of steam. One great advantage
in Messrs. Mather and Platt's machinery is the rapidity with
which it executes the work. The makers state that the
machine is capable of boring in chalk at the rate of 18 inches
per hour. At this rate the machine will execute a boring of
15 inches diameter at the depth of 1,000 feet. They also
state that in the red sandstone rock of Cheshire, Lancashire,
and the adjacent counties, they can bore at the rate o. 12
inches an hour. This result has, in fact, been considerably
exceeded, by an experimental trial of the boring machine,
which has recently been made at the Salford Iron Works,
Manchester. The machine there executed a bore hole of 15
inches to the depth of 212 feet in 141 hours, principally
through hard sandstones and grit stones, such as are quarried
at Chester for grindstones. This was at the rate of 18 inches
per hour, under the obvious disadvantage of working only
an hour or two each day. Had the machine worked continuously, its performance would doubtless have been much
greater. Messrs. Mather and Platt announce the capacity of
the machine to form artesian and other wells up to 3 .eet in
diameter, through strata of the hardest description, and to
a depth of 3,000 feet if required.

TEMPERATURE OF WELLS.

Land springs, according to Mr. Braithwaite, have usually
a temperature of 52°, and water in wells 600 feet deep is
usually 52° or 53°. He has understood that the temperature increases 1° for every 65 feet after a certain depth.

The water in the artesian well at Grenelle, 1,794 feet deep,
has a temperature of 82° or 82½° Fahr. (Sir John Robinson
and Sir William Cubitt).

The highest temperature recorded in the United Mines of
Cornwall, which are 1,770 feet deep, was 96° Fahr. (Mr.
Taylor).

The temperature is not invariable in mines even at the

same depth; for instance, Mr. Fox found in his experiments (Report of 7th meeting of British Association) that at the depth of 1,740 feet, where the lode was first reached in the cross cut, the temperature of the water was 92°; but on proceeding along the same cross cut at 10 fathoms from the lode the temperature decreased to 86°·3, and at 24 fathoms distant it was 85°·3. Mr. Fox's experiments show an increase of 1° of temperature in 48 feet calculated from the surface.

The heat of the water is undoubtedly influenced by the strata, as the mines in North Wales, although about the same depth as those of Cornwall, are much colder.

In making observations on the temperature of wells it is important to observe the depth at which the water really issues from the spring, because it will of course be affected by mixing with other water in the well or boring; and if standing some time in the well, the sides will exercise a cooling influence. Mr. Clarke, the experienced well-sinker of Tottenham, says he found water taken from the bottom of a well 540 feet deep, at St. Alban's, was 4° warmer than that which was commonly pumped from the same well. He also said the water from the bottom of Barclay's brewery well, 367 feet deep, is 3° warmer than at the usual water-level in the same well.

Local causes sometimes influence the temperature of springs, which are occasionally found warmer at the surface than at a greater depth.

DESCRIPTION OF SOME REMARKABLE WELLS IN AND AROUND LONDON.

Well at Messrs. Meux's Brewery.—This well is situated at the Horseshoe Brewery, at the corner of Tottenham Court Road and Oxford Street, where the ground is about 70 feet above Ordnance datum. The sinking of this well commences at a point 22½ feet below the surface of the ground, which level is called on the plans of the brewery the "floor-line of regions."

The following is a tabular statement showing the structural work of which this well consists :—

Depth in
feet.

0	Floor-line of regions 22½ feet below surface. Top of 7 feet brick shaft.
40	Bottom of brick shaft lined with brickwork built in two rings of half a brick each.
51	Bottom of shaft widening out from 7 feet to 9½ feet in the clear, also lined with 9 inch brickwork.
96	Top of iron cylinder 8 feet in diameter, which commences within the brick shaft 5 feet above the base of the latter.
96	Top of inner iron cylinder 6 feet in diameter.
101	Bottom of cylindrical shaft 9½ feet diameter in the clear, also lined with 9 inch brickwork.
135	Bottom of 8 feet cylinder.
144	Bottom of 6 feet inner cylinder (the top part of this down to about 129 feet has been removed so that only the lower 15 feet are left).
164½	Base of conical excavation in the chalk spreading out to 14 feet diameter at bottom.
164½	Excavation contracts to 9 feet diameter.
165	10 inch bore commences.
173½	Bottom of 9 feet excavation, which is then contracted to 8 feet diameter.
176	8 inch bore commences—top of 8 inch pipe.
188	Bottom of excavation 8 feet in diameter.
343	Bottom of 8 inch bore in hard grey chalk.

The following are particulars of the strata passed through in sinking this well, the depth to the bottom of the stratum being given in each case :—

Depth in
feet.

0	Floor-line of regions 22½ feet below surface. This depth above the blue clay consists of made ground, diluvial gravel, &c.
44	Blue clay.
46½	Black sand, rather wet.
54	Jointy blue clay.
54½	Black sand.
59½	Jointy clay with sandy joints, clay stone, and shells (? septaria).

Depth in
feet.

61	Hard jointy clay.
63	Hard blue clay.
68½	Brown red clay with light blue joints.
72	Light blue clay with stones.
73	Blue sand.
73½	Hard red clay.
74½	Dark brown clay.
76	Ditto mixed with blue.
77½	Dark yellow clay, rather soft.
79½	Dark red clay with sandy joints or partings.
80	Blue sand.
81½	Yellow clay with blue joints
84¾	Dark brown clay with blue joints.
85½	Dark brown shelly clay.
88	Dark blue clay.
90	Light blue clay.
91½	Black clay with sulphur, very shelly.
92	Black shelly clay.
95	Mixed yellow, white, and red clay.
96½	Clay mottled with large patches of red and white.
98	Hard pink clay mottled red and white.
99½	Clay mottled with red and brown, and with blue joints.
101½	Mottled red and white clay.
104½	Dark brown clay.
109	Dark brown clay mixed with blue.
111	Dark blue jointy clay
113	Hard black sand.
115	Red clay and sand.
135	Sand.
136	Gravel.
137	Black flints on top of chalk.
292	White chalk and flints.
300½	Hard sand rock.
323	Hard grey chalk.
331	Hard sand rock.
338	Hard grey chalk.
338	Hard sand rock parting.
343	Hard grey chalk with a sand parting at 340½ feet.

Water-levels.—The original water-level some years previous
to 1843 is marked on the section about 93 feet below surface

K

of ground. The water-level in 1843, the date of the section, is marked at 115 feet below surface.

The present ordinary level of the water is about 198 feet below surface of ground. When the pumps were stopped for some days five years ago the water rose 6 feet above this level, but has never since been so high.

At the level of 209 feet from the surface several adits, about 5½ feet high by 4½ feet wide, have been driven in various directions. The entire length of these adits is about 500 feet, and as the water stands above the roof of the adits, these are consequently always full of water, and form a reservoir of considerable capacity. A large quantity of water was met with in driving the adits.

The occurrence of water in the bore holes below the shaft sinking is rather uncertain; nevertheless, water was noticed at the following depths in sinking the bore hole, namely, at 215 feet from floor-line, or 237½ feet below surface of ground :—

Also at 255 feet below floor-line.

300½ ,,	,,	,,	in hard sand parting.
328½ ,,	,,	,,	
331 ,,	,,	,,	
338 ,,	,,	,,	in sand rock parting.
340½ ,,	,,	,,	in hard sand rock.

On one occasion, seven years ago, when the pumps both at this brewery and at that of Combe, Delafield, and Co. had been idle for some days, the water stood at 173 feet below the surface of ground.

The well has a duplicate set of pumps, each set consisting of three similar pumps. These are each 6¼ inches in diameter, with an 18 inch stroke. During the hours of working only one set of pumps is employed, the other being only used in case of accident to the first set.

The barrels of the pumps usually worked are about 165 feet below the floor-level, or 187 feet below surface of ground. The other set of three pumps have their barrels about 3 feet below those of the first set.

The suction-pipe from each set of pumps is 5 inches in diameter, and reaches to about 188 feet below the floor-line, or 210 feet below surface of ground. Thus they are 10 or 12 feet below the ordinary surface of water, even when the pumping is going on.

The three working pumps are capable of raising per hour about 250 barrels of 36 gallons, or about 9,000 gallons an hour. It must not be supposed, however, that this rate of working is continued during the whole 24 hours. In point of fact, during the four days of the week, from Monday to Thursday inclusive, the pumps only work about 7 hours out of the 24. On Friday and Sunday they do not work at all, and on Saturday only about 5 hours. Hence the whole work of the week is :— .

	Hours.
From Monday to Thursday	28
On Saturday	5
	——
	33

and $33 \times 9000 = 297,000$ gallons per week; so that the average for each day throughout the year may be taken at $\frac{297000}{7}$ = 42,430 gallons per day.

Messrs. Courage and Donaldson's Well, at the Anchor Brewery, Horsleydown.

This well consists of a shaft 6 feet in diameter and 100 feet deep, with a boring in the bottom 350 feet deep below the bottom of the shaft. The shaft consists, at the top, of an iron cylinder 35 feet in length, resting on a brick shaft 65 feet deep and one brick thick.

The following strata were passed through :—

Depth from surface. Feet.		Thickness. Feet.
32	Gravel and sand and made ground	32
82	Blue clay	50
97	Coloured clay	15
112	Rock sand	15
126½	Fossil sand	14½

Depth from surface. Feet.		Thickness. Feet.
140¾	Green sand and pebbles	14½
155	Sand with springs	14⅓
259	Chalk with flints	104
350	Chalk	91

The bore hole is lined with an iron tube 18 inches diameter at top, then diminishing to 12, and afterwards to 9
inches. The top of the 18 inch pipe stands 8 feet above
the bottom of the shaft, or 92 feet from the surface of the
ground. The 18 inch pipe is about 50 feet in length. The
12 inch pipe is let down inside the 18 inch, and goes from
the top of this. The total length of the 12 inch pipe is
about 65 feet. Then begins the smallest, or 9 inch pipe.
The water now overflows the top of the 18 inch pipe, and
stands 9 or 10 feet above the bottom of the shaft, or 90 feet
from the surface of ground.

The manager cannot speak to any decrease in the level of
the water, but believes that in the absence of pumping it
would stand about 15 feet higher.

The well has two sets of pumps, placed between 60 and 70
feet in depth. The suction-pipe draws from the bore hole.
Each set of pumps will throw about 120 gallons a minute,
or 7,200 gallons an hour.

One set of pumps will work at this rate for 48 hours without materially decreasing the water-level, and the pumps do
habitually work 8 or 10 hours a day for 6 days in the week.

This well was sunk in 1859 or 1860 by Messrs. Easton
and Amos, of the Grove, Southwark; and they employed
Mr. Tilley, of Edmonton, as their sub-contractor.

Well at Bow Brewery.

The site of the brewery is close to the Lee navigation,
about 12 feet below level of Bow church.

The shaft here is about 120 feet, with a boring in the

bottom 180 feet below this, making the total depth about 800 feet. The top of the shaft is formed by an iron cylinder about 12 feet diameter and 20 feet deep. To this succeeds a brick shaft 8 feet diameter. The boring in the bottom is 7½ inches diameter. The depth to the pump-barrels about 50 feet; level of water, 60 or 70 feet below surface. The pumps work 14 or 15 hours a day.

The present proprietors of the brewery had only been in possession a few months, and were not in possession of any reliable information as to the strata passed through, nor could I learn with any accuracy the quantity of water daily pumped. The water is slightly tinged with a milky colour when first pumped, but becomes perfectly clear after standing in the backs. It is used for all purposes, both for brewing and refrigerating; and is said to be well adapted for pale or bitter ale, which was made at this brewery some years ago.

Well at Messrs. Webb's Mineral Water Works, Islington Green.

Level of ground, 105 feet above ordnance datum.

This is a boring 225 feet deep into the chalk. The shaft or well is only about 20 feet deep, and the boring in the bottom is 9 inches diameter for the first 100 feet. This is lined with a cast-iron tube, and then succeeds a 6 inch boring lined with a wrought-iron tube.

The following are the strata passed through :—

Depth from surface. Feet.		Thickness. Feet.
12	Sand and gravel	12
60	Blue clay	48
130	Mottled clay	70
140	Light sand with shells	10
147	Dark blue clay	7
165	Dark sand and pebbles	18
175	Green sand and oxide of iron	10
178	Black sand and shells	3
181	Flints	3
225	Chalk and flints	44
	Total	225

There being no shaft in this case, the pump-barrel, 4 inches diameter, is simply placed in the 6 inch bore pipe, at a depth of about 180 feet from the surface of ground. The water stands, however, at a much higher level than this ; namely, at about 155 feet ; and formerly rose to within 95 feet.

On the sinking of the neighbouring deep well at the Pentonville Model Prison some years ago, the level of water in this bore hole sunk about 15 feet ; but the spirited proprietor then sunk his bore hole 100 feet deeper, and obtained a plentiful supply, which now never sinks below 155 feet. The pump-barrel contains about 1 gallon, and will deliver at the rate of 80 gallons a minute, but the usual rate of pumping is from 480 to 600 gallons an hour ; and if this were continuous during the 24 hours, it is said no diminution of level would ensue.

Independently of the interest presented by this boring into the chalk, the mineral works of Mr. Webb are well worthy of a visit. The processes of decomposing the chalk by sulphuric acid, storing the carbonic acid which is driven off, saturating the spring water with carbonate of soda in crystals, and then impregnating the water with the carbonic acid so as to form a real solution of bicarbonate of soda, are all extremely well worthy of observation. Not less so are the means taken to avoid all contact both of the gas and water with metallic surfaces, and the consequent use of slate and earthenware pipes, and even in some cases of silver cylinders and pipes. Mr. Webb deserves all the prosperity he has achieved for the care and ingenuity bestowed on every department of his works.

The Well at Trafalgar Square.

The site of this well is 37½ feet above sea-level. The well is 384 feet deep, and consists of an open shaft sunk to a depth of 148 feet, with a bore hole in the bottom 236 feet ın depth, making a total of 384 feet below the ground surface.

The following strata were penetrated :—

Depth from surface. Feet.		Thickness. Feet.	
15	Made earth	15	
31	Sand and gravel	16	
169	London clay	138	Reading and Thanet series, 72ft.
197	Mottled clay	28 ⎞	
207	Sand and gravel	10 ⎟	
241	Green sand	34 ⎠	
384	Chalk	143	
	Total . . .	384	

The water now stands about 108 feet below the surface of the ground, and according to Mr. Beardmore, yields 65 cubic feet per minute, or more than 580,000 gallons in 24 hours.

Well at the Bank of England.

The bore hole here reaches to 307 feet below the surface of the ground, the depth to the chalk being 207 feet.

The water on the 1st of January, 1852, stood at 61 feet below Trinity high-water mark, and on the 1st of January, 1869, at 74 feet, showing a diminution of the water-level equal to 13 feet during the last 17 years. In January of each year the water usually stands 10 feet higher than in August, when the level is lower than at any other part of the year.

Well at the Royal Horticultural Gardens, South Kensington.

This well is 401 feet deep, and consists of an open shaft 226 feet deep, with a boring carried down to a further depth of 175 feet.

The following strata were passed through :—

Depth from surface. Feet.		Thickness. Feet.
18	Made earth	18
40	Gravel and loam, containing a little water	22
238	London clay	198
280	Mottled clay	42
292	Pebbles and sand	12
312	Green sand	20
316	Running grey sand	4
317	Flint	1
401	Chalk	84

Reading and Thanet series, 79ft.

Total . . . 401

The water now stands about 109 feet below the surface, or 129 feet above the base of the London clay.

This well was executed by Messrs. Easton and Amos, who undertook to sink it to a depth of 400 feet for a stipulated price, and also guaranteed a supply of water equal to 75 gallons a minute, or about 108,000 in 24 hours. Not only is this quantity available, but it is said the well is capable of yielding a million gallons a day if larger pumps and a more powerful engine were employed.

Well at Guy's Hospital.

This well is 298½ feet deep. It consists of a shaft 9 feet in depth and 8 feet in diameter, lined with 9 inch brickwork. The shaft is then reduced to 4½ feet in diameter, the first 25 feet in depth being lined with 5 cast-iron cylinders, each 5 feet in length. Below the iron cylinders the shaft is the same diameter, namely, 4½ feet in the clear, and is lined with 4½ inch brickwork. A 12 inch bore-pipe extends from the bottom of the well to a depth of 298½ feet below the surface, and is continued upwards to within 60 feet of the surface.

The following are the strata passed through :—

Depth from surface. Feet.		Thickness. Feet.
Made ground and drift.	8 Superficial or made earth	8
	10 Yellow clay	2
	11 Black loam	1
	14 Peat	3
	33 Gravel	19
London Clay.	96 Blue clay	63
	118 Mottled clay	22
	122 Dark blue clay	4
	Total	122

Well at Messrs. Whitbread's Brewery in Chiswell Street, near Finsbury Square.

Depth from surface. Feet.

0 Engine-house floor.

18 Bottom of brick shaft 12 feet diameter in the clear, and about 3 feet in thickness.

12 Top of iron cylinder 9 feet diameter.

34 Bottom of ditto, and top of brick shaft 8 feet in the clear, lined with iron.

123 Bottom of brick shaft 8 feet in diameter, and one brick in thickness.

120 Top of iron cylinder 7 feet diameter, standing 3 feet above bottom of brick shaft.

178 Bottom of iron cylinder.

143 Top of inner iron cylinder 5 feet diameter, this being 25 feet above bottom of outer iron cylinder.

183 Bottom of 5 feet cylinder at top of chalk.

163 Top of 12 inch pipe, being 20 feet above bottom of 5 feet iron cylinder.

408 Bed of 12 inch bore-hole in the chalk.

Strata passed through :—

Depth from surface. Feet.		Thickness. Feet.
18	Made ground and clay	18
27	Gravel and little water	9
62	Blue clay	35
82	Coloured clay, very sandy	20
92	Black and green sand and shells	10

Depth from surface. Feet.		Thickness Feet.
111	Grey sand	19
129	Sandy, coloured clay	18
139	Green sand	10
151	Green sand, stones and pebbles	12
183	Sand	32
283	Chalk	100
	Chalk continues.	283

The water in this well, from observations made in 1866, stood at about 132 feet below the surface. This was on Monday morning, when the pumps had been out of action for 18 hours. The usual water-level now is from 170 to 174 feet below surface, but the water would rise to 162 if the pumping were discontinued.

There are two sets of pumps, both fixed in the 5 feet iron cylinder. One set consists of 3 barrels at a depth of about 160 feet, with a long suction-pipe which reaches to 173 feet in the bore-hole; the other is a single barrel-pump, fixed at 171 feet, with a shorter suction-pipe.

Messrs. Whitbread have another well situate very close to this, and very similar in all respects.

The pumps in one well throw 3,780 gallons an hour, and in the other 3,456, making a total of 7,236 gallons. The pumps work on the average 20 hours a day during 6 days a week, so that the quantity pumped is about 868,320 gallons, or 144,720 on each working day.

An analysis of the water was made in 1866 by Mr. Houghton Gill, of University College. In this analysis only 1·53 grains of lime are reported in a gallon of water. The gallon also contained ·77 grains of silica, 1·04 grains of magnesia, and 17·88 grains of soda. The acids are given separately, namely—chlorine, 6·25 grains; sulphuric acid, 7·91; and carbonic acid, 6.90.

Combining these acids with the alkaline earths in the usual manner, the analysis will thus stand :—

	Grains per gallon.
Carbonate of lime	2·73
„ magnesia	2·17
„ soda	11·01
Sulphate of soda	13·91
Chloride of sodium	11·69
Silica	·77
Organic matter	·98
	43·26

| Hardness of the water by Dr. Clarke's test | 2·26 |
| „   „   „   after boiling two hours | ·15 |

It appears from the comparative softness of this water, as well as from the analysis, that this can by no means be considered a chalk water. It is evidently derived, in a great measure, from the sands above the chalk. The quantity of chloride of sodium (common salt) is remarkable ; but in this respect, as well as in the large quantity of sodium, the analysis closely resembles that of numerous deep wells in the chalk.

Messrs. Whitbread very kindly allowed me access to an excellent section of this well, made in 1866 by Messrs. R. Moreland and Son, of Old Street Road.

Well at the Lion Brewery, Belvedere Road, Lambeth.

This was formerly the brewery of Messrs. Goding, but now belongs to a company.

The following statement shows the construction of this well :—

Depth from
 surface.
 Feet.

0   Surface of ground, being about the level of Trinity high water mark, or 12½ feet above Ordnance datum.   Top of iron cylinder 6 feet diameter.

42   Bottom of iron cylinder and top of brick shaft 6 feet diameter and one brick thick.

Depth from
surface.
Feet.

150	Bottom of 6 feet brick shaft.
115	Top of old 6 inch bore pipe.
132	Top of new 12 inch pipe.
308	Bottom of 6 inch pipe.
408	Bottom of 12 inch pipe.

The following are the strata passed through :—

Depth below
surface.
Feet.

		Thickness. Feet.
14	Made ground	14
24	Sand	10
33½	Shingle	9½
131½	Blue clay	98
172½	Various-coloured clay	41
182½	Pebbles, with water	10
194½	Green sand, no water	12
214½	Sand, main springs	20
408	Chalk	193½
		————
		408

This well was sunk in 1837, when the water-level was about 40 feet below the surface. It is now about 86 feet below the surface.

There are 2 sets of pumps in this well, 3 in each set. They are of the same capacity as in several of the other large breweries, each set throwing about 200 barrels an hour, and as each barrel contains 36 gallons, this is equal to 7,200 gallons an hour. No deficiency of water complained of.

The average quantity of water pumped from this well is about 72,000 gallons a day. After 12 hours' pumping the water-level is reduced about 10 or 12 feet. The well was erected in 1837 at a cost of about £1,200. The contractors were Messrs. Baker of Southwark Bridge Road. (Information furnished by Mr. T. J. Thompson, secretary to the Lion Brewery Company, Limited).

Messrs. Reid's Well at Liquorpond Street.

This well is 222 feet 5 inches deep from surface of ground to bottom of well.

The following strata were passed through :—

Depth from surface. Feet.		Thickness. Feet.
7	Made ground	7
16	Gravel	9
18	Yellow clay	2
58	London clay	40
102	Clay and sand (Woolwich)	44
156	Sand (Thanet)	54
222·5	Chalk	66·5
		222·5

Two trials were recently made of the height at which the water stands with the following results :—Sept. 23, 1869. After an entire cessation of pumping during 84 hours the water stood 33 feet deep in the well, or 189·5 feet below surface of ground.   Sept. 24, 1869. After cessation of pumping during 24 hours the water stood 17½ feet deep in the well, or 205½ feet below surface of ground.   The latter depth is very constant, and the one usually allowed to accumulate.

Well at Kensington Gardens.

Sunk for supplying the Serpentine when the Bayswater Brook, owing to the admixture of sewage, became too offensive for the purpose.

This well is 321 feet in depth, and the water rises to within 105 feet of the surface.   For 203 feet in depth the well is 6 feet in diameter, lined with the following thicknesses of brickwork :—

25 feet with 9 inch
67 „ „ 4½ „
5 „ „ 9 „
10 „ „ 14 „
5 „ „ 9 „
91 „ „ 4¼ „

203 feet.

The remainder of the well is lined with iron cylinders 4 feet 6 inches in diameter. These cylinders are continued inside the steining to within 173 feet of the surface.

Depth from surface Feet.		Thickness. Feet.	
122	Made ground and London clay . . .	122	Reading series.
127	Shells and sand	5	
137	Mottled clay	10	
141	Sand and pebbles.	4	Thanet sand series.
145	Mottled clay, green-coloured sand, and pebbles	4	
149	Green-coloured sand and pebbles . .	4	
152	Green-coloured sand	3	
196	Grey sand	44	
196½	Layer of flints	½	
298½	Chalk	102	
	Total . . .	298½	

Plastic clay. (brace spanning rows 122–196)

Amwell Hill Well, near the source of the New River, in Hertfordshire.

This well is sunk entirely in the chalk, the entire depth of well and boring being 161 feet. The shaft is lined with 9-inch brickwork for a depth of 84 feet from the surface; then succeeds a shaft 10 feet diameter, without any lining. From this shaft headings are driven 6 feet high, and 4 feet 6 inches wide. In the centre of the shaft is a 2 feet bore, which at some depth is reduced to 9 inches. This well is said by Mr. Mylne, in his evidence on the Metropolis Water Supply, 1852, to yield 2,466,000 gallons a day.

Cheshunt Well of the New River Company.—London clay formation.

The entire depth of well and boring is 171 feet.
The following is the construction :—

	Depth. Feet.
Shaft 11½ feet diameter, lined with 14 inch brickwork .	12
Shaft 9 feet diameter, lined with 9 inch brickwork . .	44
Cast-iron cylinders 8 feet in diameter, carried up to within 15 feet of the surface, making depth of cylinders 90 feet	49
Cast-iron cylinders 6 feet 10 inches in diameter, rising 35 feet within 8 feet cylinders, equal in depth to 36 feet.	1
Bottom of cast-iron cylinders 6 feet in diameter and 13 feet below that of 8 feet	12
Brick steining, forming foundation for 6 feet cylinders .	7
Bottom of cone 12¼ feet diameter at base.	19
Headings are driven at this level 7 feet high and 4½ feet wide.	
Bore-hole 3 inches diameter	27
Total . . .	171

The following strata were passed through :—

	Depth below surface. Feet.		Thickness. Feet.	
	1½	Superficial earth.	1½	
	9½	Gravel	8	
London Clay.	54½	Blue clay	45	
	56½	Yellow clay	2	Reading and Thanet Series.
	68½	White sand	12	
	107½	Dark-coloured sand	39	
	171	Chalk	63½	
		Total . . .	171	

This well is said to yield 702,000 gallons a day.

Sir Henry Meux's Well at Cheshunt.

This well is 71 feet deep, with borings in the bottom extending to a depth of 202½ feet from the surface.

The actual shaft of 71 feet in depth is lined with 4½ inch

brickwork. The remaining 131½ feet consist of a bore-hole
commencing at 7 and then reduced to 4 inches.

The following strata were passed through:—

Depth below surface. Feet.		Thickness. Feet.
5	Gravel	5
64	Blue clay	59
76	Coloured clay	12
77	Dark-coloured sand	1
82½	Sand and pebbles	5½
85½	Bright sand	3
120½	Dark sand	35
124½	Flints and chalk	4
202¼	Chalk	78
	Total	202½

Well at Crossness.

This well has been sunk to endeavour, if possible, to obtain
a large supply of water for condensation of the steam produced by the large pumping engines employed by the Metropolitan Board at Crossness. The average quantity of water
required here for condensation alone is about 600 gallons
per minute, or 864,000 gallons per day of 24 hours; but
the maximum required for that period would be double this
quantity. In addition, about 2,500 gallons per day would
be required for domestic purposes.

In the construction of this well a shaft, lined with iron
cylinders, has been sunk to a depth of 81½ feet. In this an
18 inch bore-pipe has been inserted, commencing at 50 feet
below surface, and carried down to a depth of 166 feet below
surface. To this succeeds an 18 inch bore, without any
lining, carried down through the upper and white chalk into
the chalk marl, about 552 feet, or to a depth below surface
equal to 718½ feet. The boring then diminishes to 5 inches,
a pipe of this size commencing at 785½ feet, and extending to
884½ feet below surface. At 849 feet a 4 inch pipe commences, namely, 35½ feet above bottom of 5 inch pipe, and

is carried down 81 feet, namely, to 930 feet below surface. A 3 inch pipe commences at 909 feet below surface, and is carried down 48 feet, or to 957 feet below surface. Below this is a $2\frac{3}{4}$ inch pipe, which only extends about 4 feet, or to a total depth below surface of 961 feet, when the further prosecution of the work was stopped.

The following strata have been encountered :—

Depth from surface. Feet.		Thickness. Feet.	
12	Made ground	12	
$13\frac{1}{2}$	Alluvial deposit	$1\frac{1}{2}$	
17	Light brown clay	$3\frac{1}{2}$	
$20\frac{3}{4}$	Blue silty clay, with vegetable matter .	$3\frac{3}{4}$	
27	Peat, with remains of forest trees . .	$6\frac{1}{4}$	Alluvium.
$28\frac{1}{2}$	Dark grey silty clay	$1\frac{1}{2}$	
30	Peat and clay in thin layers with decayed wood	$1\frac{1}{2}$	
$32\frac{1}{2}$	Dark grey silty clay	$2\frac{1}{2}$	
$34\frac{1}{2}$	Silty sand	2	
$83\frac{1}{2}$	Grey rectangular flint gravel (sometimes partaking of the character of running sand), intermixed with iron pyrites and blue clay	49	Diluvium.
95	Sand with flint and shells, very hard .	$11\frac{1}{2}$	
$99\frac{1}{2}$	Fine sand, with flints, pebbles, and small shells	$4\frac{1}{2}$	Thanet and Woolwich sand, lower beds of the tertiary series.
101	Fine green sand	$1\frac{1}{2}$	
$103\frac{3}{4}$	Fine grey sand with small flints . . .	$2\frac{3}{4}$	
$113\frac{1}{2}$	Fine dark sand and flints	$9\frac{3}{4}$	
149	Fine light Woolwich sand	$35\frac{1}{2}$	
154	Sand strongly cemented by iron pyrites	5	
$155\frac{1}{2}$	Loam and pebbles	$1\frac{1}{2}$	
156	Layer of flints	$\frac{1}{2}$	Upper and lower chalk.
602	Chalk with layers of flints from 2ft. 6in. to 6ft. apart	446	
802	Chalk marl with few flints	200	
814	Sandy green marl	12	Upper green sand.
961	Gault clay	147	Gault.

961

The bore-holes of this well should evidently have been of much larger capacity, and should not have been reduced so rapidly. At the depth now reached the bore-hole should have been 10 or 12 inches diameter, instead of $2\frac{3}{4}$.

The present bore-hole has not penetrated the gault by 60 or 70 feet. Below this depth the lower green sand would probably be met with, so that, if the boring were continued another 100 feet, an abundant supply of water would probably have risen in the bore-hole.

Well at the Crystal Palace.

This well is $8\frac{1}{2}$ feet in diameter, and is sunk to a depth of 245 feet.

At this depth a boring was commenced, and fitted with pipes 15 inches diameter.

At 150 feet from the surface several headings were driven from 30 to 50 feet in length. These were commenced at 4 feet in height, but considerably increased as the work proceeded.

At a depth of 259 feet the boring passed through a bed of sand between two beds of plastic clay, and from this sand the water rose in the well to a height of 142 feet, and filled the headings in eight hours.

The boring was continued through the lower tertiary sands till it reached the chalk at 360 feet in depth, and had penetrated the chalk 190 feet, thus reaching a depth of 550 feet below surface.

It must be observed that the main supply of water in this well is derived from the sand spring at 259 feet, and very little addition to this was gained by the deep boring into the chalk.

The well was sunk in 1853-5, and no accumulation of sand took place until this year, when it was necessary to take out about 25 cubic yards, leaving still some accumulation which will require early removal.

Well at Cold Bath Fields.

This well was sunk in 1866-7 by Messrs. Baker and Sons for the Visiting Justices of Middlesex. The well commences with a diameter of 6 feet 10 inches, being lined through the gravel and diluvial clay with five iron cylinders of this diameter. Below this extends a shaft lined with 9 inch brickwork to a depth of 102 feet from surface. Iron cylinders 5 feet 2 inches in diameter were then inserted down to the chalk, which was reached at 132 feet from surface. The well was then excavated to a diameter of 10 feet, and by the time 20 feet of this size had been sunk, a supply of water was obtained equal to 150 gallons a minute, or about 216,000 gallons in twenty-four hours. The water issued with great force through the horizontal partings of the chalk. This supply was deemed by the Justices sufficient for their purpose, otherwise a much larger quantity could doubtless have been obtained by sinking deeper.

Wells of the Kent Waterworks Company.

These wells afford fine examples of the supply derived purely from the chalk. One of these wells, sunk by Messrs. Baker and Sons, at Shortlands, passed through about 6 feet of gravel, and then nearly 60 feet of Thanet sand into the chalk. All the gravel and sand springs were cut off and prevented from entering, and after this a very copious supply was obtained from the chalk.

Well at Walker's Brewery, Limehouse.

This is a well just completed by Messrs. Baker and Sons, in which a large volume of water was found in the gravel overlying the chalk—about 180 gallons per minute. The well is sunk through very variable strata down to the chalk, in which a boring is made to the depth of 160 feet.

The water from the chalk rises in this well upwards of 100 feet above the spring, and it is calculated the supply

of chalk water alone will be over 25,000 gallons an hour.
The brick lining of this well is formed of the best bricks,
burnt from the gault clay, and set in Roman cement; but
the whole of the well below the water-line is cased with iron
cylinders.

Well at the North Surrey Schools, Anerley.

A well was sunk here by Messrs. Baker in 1867, to a
depth of 220 feet, in London and coloured clays. A boring
of 43 feet was then made through pebble beds and mottled
clays. At a depth of 243 feet from the surface a bed of
sand 9 feet in thickness was reached; this bed was very
fully charged with water, which rose rapidly till it reached
a height of 118 feet in the well, or 102 feet from the surface.

This spring has ever since entirely supplied the schools.
Three pumps, each 5½ inches diameter, were put to work,
and have been daily in action ever since, always with an
ample supply.

The level of water in this well is considerably affected
when extensive pumping is going on at the Crystal Palace;
showing that the supply is probably derived in both cases
from the same sand springs, and that no impervious fault
exists between them.

Well at the Lunatic Asylum, Colney Hatch.

Was executed by Messrs. Baker and Sons. The chalk was
found in this well at a depth of 189 feet from the surface,
and a large supply of water was obtained both from the
chalk and from the tertiary sands. The depth sunk in the
chalk amounted to 141 feet.

Well at the Victoria Terminus of the London, Brighton, and South Coast Railway at Pimlico.

A well was sunk here by the Messrs. Baker in 1861 of
the following construction :—Cast-iron cylinders 6 feet in

diameter were driven through the gravel into the London clay, after which the shaft was lined with 9 inch brickwork to a depth of 140 feet below the surface. This was entirely through London and mottled clay. A boring was then commenced, and this, at 177 feet below surface, came to sand and water. About 9 feet, or to a depth of 186 feet, the sand was so much indurated as to resemble stone, and the colour very white. Sand continued to 190 feet below surface, then came white and mottled clay to about 206 feet. Then Thanet sand for about 58 feet down to the chalk, which was reached at 264 feet below the surface. This well yields a good supply of water entirely from the sand springs, and when the well was first sunk this water rose to within 7 feet of the surface.

Well at Old Malden.

This is a well executed by Messrs. Baker in 1859, for the London and South Western Railway Company. The sinking passed through 281 feet of London and mottled clay into a bed of sand about 4 feet in thickness, and very full of water. The water from this bed rose very rapidly, and so high as to overflow the surface. It is said this overflow has continued ever since.

Well at Messrs. Waltham Brothers' Brewery, Stockwell.

This well is sunk and bored to a depth of 380 feet. The first 100 feet is an open shaft lined with brickwork, and the remaining 280 feet are bored. The pipe rises about 40 feet above the bottom of the bore-hole, and the water overflows the pipe within 60 feet of the surface.

The first 21 feet consist of made ground and reddish gravel, then 79 feet of London clay, making the 100 feet depth to which the well is sunk. The bore-hole then passes through the lower tertiary beds, which are chiefly sand, with a few strata of clay. The depth of these is 92 feet, when the bore-hole reaches the chalk, into which it penetrates nearly 190 feet.

Well at Greenwich Hospital.

This well is sunk to a depth of 155 feet, and continued by boring to a further depth of 150 feet, making a total of 305 feet.

The following strata were passed through :—

	Ft.	in.	
Surface soil and alluvial	11	0	
Gravel	33	0	
Black sand	4	10	Lower tertiary.
Blue clay and shelly rock	4	10	
Red clay	6	0	
White sand and water	4	0	
Green sand and pebbles	4	0	
Dark sand and water	55	10	
Bed of flint	1	0	
Chalk	180	6	
	305	0	

This well produces 120 gallons per minute, the water rising to within 19 feet of the surface.

Chalk Wells at Watford.

The water in this neighbourhood is met with so near the surface of the ground, that it is usual to sink a shallow well of large diameter, and put down a bore-hole at the bottom. The large size of the well affords a lodgment for the water, and serves the purpose of a small underground store reservoir. The well recently sunk at Watford for the supply of the town is 15 feet in diameter, 33 feet deep, and lined with 9 inch brickwork, all laid as headers. The upper 15 feet is laid in cement. The well is domed over with 9 inch brickwork, laid in Portland cement, a circular man-hole,

3½ feet diameter being left in the centre, and covered over with a York landing-slab.

A bore-hole, sufficient to admit a pipe of 12 inches clear internal diameter, has been made to the depth of about 70 feet in the bottom of the well. A pipe of this diameter, 10 feet long, is inserted in the bore-hole. The pipe is of galvanized cast-iron, ¾ of an inch in thickness. It is furnished with a flange at the top, to prevent it sinking into the bore-hole, and is so fixed as to have its upper part or flange 8 inches above the floor of the well.

ADDITIONAL WELLS MENTIONED BY MR. BEARDMORE.

Well in the Hampstead Road, sunk by the New River Company, on a site 90 feet above mean sea-level.

Well at Woolwich, on a site 22½ feet above mean sea-level. Depth in chalk, 580 feet; yield, 160 cubic feet per minute, or 1,400,000 gallons per day.

Well at Brompton, on a site 152 feet above mean sea-level. Depth in chalk, 160 feet; yield, 33 cubic feet per minute, or nearly 297,000 gallons per day.

RECENT WELLS SUNK BY MR. PATEN.

Well at Harrow, for the supply of the town, sunk on a spot where the surface of the ground is 266 feet above Ordnance Datum, or mean level of the sea.

This well is 6 feet in diameter, and is sunk to a depth of 193½ feet, partly in London clay, partly in the underlying coloured clay and sands, and partly in chalk. A boring of 15 inches diameter then passes through 210 feet of chalk, making the entire depth of 403½ feet below the surface.

The following are the strata passed through:—

	Ft.	in.	
Brown clay	32	0	London clay.
Blue clay	79	0	
Mottled clays	26	0	Plastic clay.
Sand and pebbles	23	0	
Chalk with flints	243	6	
	403	6	

The usual pebble bed was met with at the base of the London clay.

The water rose in this well to within 125 feet of the surface, or 141 feet above Ordnance Datum. When the water is pumped at the rate of 200 gallons per minute, the water stands permanently at 12 feet below this level, or 129 feet above Ordnance Datum.

It is unnecessary at present to pump more than 200 gallons a minute, as the supply required for the town is only about 150,000 gallons a day, in addition to which the Harrow Station of the London and North-Western Railway has to be supplied. The well would doubtless yield considerably more if it were required.

Edgeware Public Well.

Site of well 188 feet above mean sea-level, and the water rises to within 90 feet of the surface, or 98 feet above mean sea-level.

The well is 4 feet 6 inches diameter, and 90 feet in depth, and below this is a 7 inch boring 201 feet deep, making a total depth from surface of 291 feet.

The following strata were passed through :—

	Feet.	
Brown clay and gravel	20	London clay.
Blue clay	56	
Coloured clay, sand, and pebbles, through which 50 feet in length of 8 inch pipes were driven into the chalk	64	Lower tertiary clays and sands.
Chalk and flints	151	
	291	

The quantity pumped is uncertain, and is considerably more in summer than in winter. If the present pumps, however, were worked regularly, the well would yield at least 50,000 gallons per day.

Well at the London Orphan Asylum, Watford.

This well is sunk on a site about 190 feet above mean sea-level, and the water rises to within 52 feet of the surface, or 138 feet above mean sea-level. The well is 6 feet in diameter, and 51½ feet deep. In the bottom is an 8 inch boring, 255½ feet deep, making a total depth of 307 feet.

The following are the strata :—

	Feet.
Brown clay	12½
Gravel	19½
Sand	18
Chalk and flints	257
	307

Yield of well about 80,000 gallons a day.

Well at Alperton, near Ealing.

This well is sunk on a level with the rails at Sudbury Station, on the London and North-Western Railway.

The well is 5 feet in diameter, and 195 feet deep, with a 10 inch boring in the bottom 205 feet deep, making total depth 400 feet.

The water stands 35 feet below top of well.

The following strata were passed through :—

	Feet.	
Yellow clay	25	London clay.
Blue clay	140	
Coloured clay	36	Lower tertiary.
Sand and pebbles	17	
Chalk and flints	182	
	400	

The permanent pumps have not yet been fixed; but it is probable the well will yield nearly 300,000 gallons per day.

Well at Berkhampstead, for supply of the Town.

The site of this well is 340 feet above Ordnance Datum, and the water stands within 8 feet from top of well, or 332 feet above Ordnance Datum.

The actual well is 4 feet 6 inches in diameter, and is only sunk 10 feet deep, and is then succeeded by an 8 inch boring 200 feet in depth, making a total of 210 feet.

Strata passed through :—

	Feet.
Clay	6
Gravel	6
Chalk and flints	198
	210

The quantity pumped from this well and boring is about 50,000 gallons a day.

Wells sunk at Dancers' End, near Tring, for the Chiltern Hills Water Company, to supply the towns of Aylesbury and Tring.

These works are of considerable magnitude, and consist of three wells, each 236 feet deep, with adits 541 feet in length, and five borings of 7 inches diameter, each about 55 feet deep. The surface of the wells is 562 feet above sea-level, and the water line is 178 feet below this, or 384 feet above mean sea-level.

The three wells are respectively 4½, 5½, and 6 feet in diameter, and are wholly sunk in chalk. The adits are 5½ feet wide and 7 feet high, and are driven at 226 feet below top of wells.

The average quantity of water pumped per day is about 400,000 gallons.

It seems clear from the analyses of water from most of the brewers' wells in London that the supply is chiefly from the mass of grey sands overlying the chalk, and not from the chalk itself. The waters are never hard enough for true chalk waters, nor do they contain any large proportion of carbonate of lime. The alkaline salts, in fact, are chiefly sulphates and chlorides.

It appears that the most certain method of obtaining water in the chalk is by driving adits, headings, or driftways, and not by boring.

ON SUPPLIES OF WATER FROM THE LOWER GREEN SAND IN THE NEIGHBOURHOOD OF LONDON.

The success of the artesian wells at Passy and Grenelle, near Paris, has been sufficient to induce several attempts to procure water by sinking or boring into the lower green sand near London.

The Paris wells and borings have successively penetrated a great mass of tertiaries overlying the chalk, then the chalk itself, then the upper green sand and the gault, and finally have reached an abundant supply of water in the lower green sand.

The succession of strata in the London basin underlying the chalk has usually been considered the same as that in the Paris basin, and hence considerable works have been undertaken with the view of procuring water by sinking through the chalk down to the lower green sand.

The borings at Paris, which penetrate the lower green sand, are 1,800 feet in depth; but the configuration of the London basin induced geologists to suppose that beneath London the lower green sands would be reached at a depth considerably less than this. Numerous sinkings which have recently been made through the chalk between Calais and London have proved that the same succession does not everywhere prevail, and that much older rocks seem in places to succeed the chalk than those which are found beneath the chalk of Paris. Thus at Calais, it was found, after the chalk had been penetrated to a depth of nearly 1,800 feet, that all the secondary rocks were then absent, that there was no green sand, no oolites, nor even red sandstone, but that rocks of the carboniferous series actually succeeded and underlaid the chalk. At Harwich, again, it was found that the chalk, which was passed through at a depth of 1,000 feet below the surface, was immediately followed by slates which probably belonged to the carboniferous series.

At a well sunk by the New River Company at Kentish Town, the gault was found beneath the chalk; but after a depth of 1,113 feet had been reached a series of red sandstone beds appeared, in place of the lower green sand.

All these facts seem to lend great authority to the view that some great axis of elevation connects the coal districts of Belgium and the Bas Boulonnais with the granitic country of Cornwall, the Channel Islands, and Normandy.

Coal is now worked beneath the chalk in the neighbourhood of Guines and Marquise, between Boulogne and Calais; and it is quite possible that the same great axis of elevation which has there brought up the coal formation has caused a vast upthrow of Palæozoic rocks between Dover and the West of England before the deposit of the chalk.

At the same time it is quite certain that the northern part of the London chalk basin contains the strata in regular succession. At Reigate and Dorking, on the south side, and at Dunstable and Ampthill, on the north, the gault is seen regularly succeeding the chalk, and the lower green sand is seen regularly passing under the gault. This regularity is further evidenced by numerous borings and wells, which are sunk for water all along the outcrop of the chalk from Leighton Buzzard to Cambridge, and by the coprolite workings, which frequently pass through the gault and penetrate the lower green sand, for the purpose of procuring water.

The tube-well sunk on Lord Brownlow's property at Eddlesborough, Bucks, under the direction of Mr. T. McDougall Smith, shows the regular succession of gault and green sand, as follows:—

The boring was 6 inches in diameter, and passed through—

	Feet.
Lower chalk	⎫
Chalk marl	⎬ 63
Upper green sand	⎭
Gault 205
Lower green sand 33
	301

The bore-hole is lined for the entire depth with iron tubing.

72 feet from surface, with 4½ inch tube.
216 „ deep, with 3 „ „
13 „ „ „ 2 and 1¼ „ „
———
301

The water in the old wells probably arises from chalk springs, and is obtained at a depth of 4 to 5 feet from the surface; but the boring, which passes through the gault, has tapped and given access to the water of the lower green sand. This water rose to the height of 79 feet from the surface, and during the next six months it rose 9 feet higher.

The working barrel in the tube is 2¾ inches in diameter. This well affords no criterion of the quantity which might be obtained from the lower green sand, as the boring is excessively small, and the quantity required is only for the supply of four cottages—probably about 100 gallons a day.

The well recently sunk at Arlesey for the Three Counties' Asylum is on a much more extensive scale, and affords results which are much more striking. The works at Arlesey consist of two wells, each 6 feet in diameter and

120 feet deep, and in one of these a boring 10 inches diameter has been made to a further depth of 360 feet.

The following were the strata passed through :—

	Ft.	in.
Loam and sand	7	0
Lower chalk, chalk marl and upper green sand	113	0
Gault	204	6
Lower green sand	155	6
	480	0

The supply of water is equal to 60,000 gallons a day, being equal to the requirements of a very large asylum built for three counties, and in which the use of water is encouraged on a very liberal scale for sanitary and other purposes.

The water from chalk springs in the Arlesey wells stands about 60 feet below surface, and the green sand water from below the gault rises to a height of 110 feet below surface, or a little above the level of the gault. The well is sunk on a site about 233 feet above mean sea-level.

The artesian wells of Cambridge have been already alluded to. These wells are commonly dug through the gault down to the lower green sand, and furnish usually a very copious supply of water. The numerous coprolite works in the gault district north of Hertford, such as those of Stonden, Shillington, and Arlesey, have numerous wells and borings sunk through the gault into the lower green sand. The gault in these workings is usually about 200 feet thick, and the supply of water about 40,000 or 50,000 gallons from each well.

The numerous overflowing artesian wells in Wrest Park and the neighbourhood of Silsoe, south of Ampthill, are very interesting. These are mostly bore-holes, which penetrate the gault down to the lower green sand. In most of these the water rises 8 or 10 feet above the surface, and flows perpetually through a bent pipe into a reservoir formed to receive it.

WELLS IN THE NEW RED SANDSTONE OF BIRMINGHAM AND WOLVERHAMPTON.

Mr. Thomas Clarke, who was called to report on the water supply of Birmingham, made in the autumn of 1852 an experimental boring 137 feet deep, from which he obtained 115,200 gallons per day of twenty-four hours. From this result, and from surveys which he says he has made, he reports that a supply of 700 gallons per minute, or upwards of a million gallons per day, may be relied on from wells. He recommends that four shafts should be sunk, each 8 feet diameter in the clear; and that besides these a pumping-shaft, 10 feet diameter in the clear, should be sunk to the depth of 90 feet, and all the other shafts connected with it by means of oval-shaped driftways or adits. He recommends that all the shafts be lined with 14 inch brickwork, and that in the bottom of each shaft a boring of 15 inches diameter should be made to the depth of 140 feet below the surface of the ground. In each boring is to be fixed a cast-iron pipe 12 feet in length, which is to be securely driven to the depth of 8 feet below the bottom of each shaft. The tops of the shafts to be covered with wrought-iron chequered plates fixed to strong iron girders at the surface of the ground.

In connection with the shafts he also recommends a cast-iron tank to be erected of the following dimensions, namely, 120 feet long by 60 feet wide, and 5 feet deep. The tank to be constructed of cast-iron plates $\frac{7}{8}$ inch thick, upon strong cast-iron girders, to be fixed on brickwork. The top of the tank to be covered with a light glazed iron roof.

The capacity of this tank would be 225,000 gallons. The cost of the shafts, tank, and pumping machinery, as estimated by Mr. Clarke, is £27,500.

In September, 1854, Mr. Robert Rawlinson and Mr. Pigott Smith reported on the water supply of Birmingham, and condemned in toto the project of sinking wells.

They support their arguments by reference to two deep mines sunk in the new red sandstone at a considerable distance apart—namely, the Pendleton Colliery, near Manchester, and the Monkwearmouth Colliery, near Sunderland. They observe : " After a depth of 900 feet vertical had been attained, the strata were found to be comparatively dry, and water had to be passed down into the workings to water the roads, they were so intolerably dusty. The water obtained at the deepest points to which it was found to penetrate was impregnated with mineral as strongly as brine. Both these shafts yielded water in the upper beds to the depth of several hundred feet. This had to be tubed out by a casing of cast-iron."

They adduced as failures of wells in the new red sandstone the city of New York and other places in America, besides in this country, Manchester, Salford, Liverpool, and Wolverhampton.

The Tettenhall well of the Wolverhampton Waterworks Company is an ellipse of 11 feet by 7 feet 6 inches, with one side flattened. These are the dimensions in the clear, inside the brickwork. The first 16 feet consist of argillaceous marly beds, changing for the next 4 feet into mottled sandy rock, and below this into the soft red sandstone of the district. The first 8 feet of the shaft is steined with 14 inch brickwork, and the next 30 feet with 9 inch brickwork. The whole of the steining is puddled behind in order to keep out land springs and the water from the neighbouring wells. Where the brickwork commences on the top of the solid rock, an elm curb 9 inches wide and 3 inches thick, is introduced below the brickwork. The curb rests on a well-worked bed of white clay, 6 inches thick, which is laid upon the rock, and the puddle behind the steining is worked into and carried up in connection with this white clay, which was procured from a coal pit near Bilston, and cost 3s. 6d. a ton. The depth of the well is about 140 feet, and the lower 106 feet is not steined at all, the rock being sufficiently firm to stand without lining. Several headings were driven at the height of 6 feet above the bottom. These headings are of irregular size, some being 8 feet high, some as much as 14, and some not more than 4 feet. The usual width is 4 feet, and there are in all about 1,100 lineal yards of heading. The usual work of a man in driving the headings was about 1½ cubic yards per day.

The water when not acted on by pumping stands about 20 feet deep in the well.

The Company has at Tettenhall two smaller shafts, each 5 feet diameter in the clear, which were used for air shafts and for pumping out the water during the sinking of the main shaft.

The Company has also two shafts at Goldthorn, on the south side of Wolverhampton. These are each 7 feet in diameter, 300 feet deep, and are sunk within a few yards of each other.

The first 12 feet consist of argillaceous beds, and this part is lined with brickwork. The shafts then pass through about 36 feet of the conglomerate or indurated gravel beds peculiar to this part of the Permian series, and this portion is not steined. The strata below the conglomerate bed are argillaceous and marly, and are lined all the way down to the bottom. The beds dipped across the shaft in a westerly direction, at the rate of about 2 feet per yard. A spring occurred in the gravel bed which drained the neighbouring wells. The strongest spring was met with at a depth of 50 or 60 yards from the surface. The water fell off much towards the bottom of the sinking.

There is a bore-hole 130 yards deep in the bottom of the well. The bore-hole is tubed with iron 3½ inches diameter in the clear at top, and 2½ inches diameter at bottom. The 3½ inch tube extends 25 yards in depth, the remainder being 2 inch.

The first heading was driven to the west in the same direction as the dip, and, therefore, in consequence of the rapid dip, intersecting a great number of beds. This heading was 130 yards in length, and was driven with an in-

clination upwards of about 1 in 20. The principal portion
of the water was obtained by means of this heading. The
other heading was driven 23 yards to the south, in the line of
the strike, and was entirely in one bed of rock. Neither this
nor the boring yielded any water. The headings were 6 feet
square. It was found necessary in places to arch the long heading with two half-brick rings. About 32 yards in length of the
western heading were arched where it crossed beds of marl.

The produce of the red sandstone wells at Wolverhampton
is very inferior to that of any of the Liverpool wells. The
Wolverhampton Waterworks Company spent more than
£60,000, and for this large outlay have only obtained a
supply of about 379,000 gallons a day. They had two
pumping establishments ; one at Tettenhall and the other at
Goldthorn Hill. The Tettenhall works consist of one main
shaft 140 feet deep, of an oval shape 11 feet × 7¼, besides
two other working shafts 5 feet diameter, and about 1,100
yards of heading. This establishment yields only 168,000
gallons a day. The Goldthorn works consist of two shafts,
each 7 feet diameter in the clear, and 300 feet deep, and 153
yards of heading, with a boring 390 feet deep at the bottom.
This station yields 211,000 gallons a day.

No question can exist about the extreme impolicy of sinking this shaft. It is only a quarter of a mile from the great
fault of the coal field, so that the drainage area for supplying
the well is extremely limited. The beds which consisted
chiefly of brown coloured sandstones and red marls, alternating all the way down, are tilted to an angle of more than
80°. A bed of conglomerate or indurated gravel was passed
through at 50 or 60 yards from the surface. This bed
yielded a small quantity of water at the expense of all the
neighbouring wells, which were dried up. It appears that
the marl beds were of much greater thickness than the sandstone beds, and it seems probable that the beds intersected
by the shaft were not permeable water-bearing beds at all,
even had they cropped out at the surface, which is doubtful.

Again, the true water-bearing beds, which may possibly have been reached by the boring, certainly did not crop out at the surface, but were cut off by the fault which also cut off all communication and drainage from the coal field. The miserable failure of this well is therefore not to be wondered at.

Well and boring at Rugby in Lias and New Red Sandstone.

	Feet.
The well is 7 feet in diameter, and in depth	82
In the bottom of the well is a boring 14 inches diameter, lined with cast-iron tubing	61
Then a boring with wrought-iron tubing 12 inches diameter	91½
Then a boring with wrought-iron tubing 10 inches diameter	236⅓
Then a boring with wrought-iron tubing 9 inches diameter	280¼
Then a boring with wrought-iron tubing 7½ inches diameter . . . . :	33
Then a boring with wrought-iron tubing 6½ inches diameter	246
Then a boring with wrought-iron tubing 6 inches diameter	98
Additional boring	13
	1141

The following statement of the strata passed through is condensed from a valuable section presented by Mr. McDougall Smith to the Metropolitan Board of Works :—

	Depth from surface. Feet.	Thickness. Feet.	
Red sand and gravel	10	10	
Blue lias clay and limestone	400	390	Lias.
Dark and brown clays	470	70	
Light hard stone	478	8	
Red clay	780	302	New red sandstone.
„ sandstone	786	6	
„ clay	1045	259	
„ clay sands	1141	96	
Total		1141 feet.	

This well was entirely unsuccessful in yielding a supply of water.

The boring appears to have reached the brine springs of the new red sandstone, and the water was saturated with common salt, and quite unfit for domestic use.

It appears, therefore, that about 470 feet of the sinking were in the lias formation, in which, of course, it was hopeless to expect water, and that the remaining 650 feet were in the argillaceous or gypseous member of the new red sandstone, in which the prospect of water was equally hopeless.

It is true that a very small proportion of the sinking in the new red sandstone is described as sandstone and sandy clay ; but on the whole the character of the beds is clearly argillaceous, and ought to have indicated to a well-informed geologist the extreme improbability of meeting with water. No doubt the engineers were buoyed up with the hope of reaching the lower arenaceous beds of the new red sandstone ; but the remarkable failure at Rugby ought to prove a warning against sinking in the new red sandstone, unless there are circumstances favouring the probability of finding water, and especially unless the geological horizon be well ascertained.

Wells in Red Sandstone rock at Birkenhead.

It appears that Birkenhead is supplied from two wells, one of which is described by Mr. Baldwin Latham, from whose book the following particulars are taken.

The red sandstone rock has been penetrated to a depth of 895 feet, of which the first 95 feet consist of an open well, 9 feet in diameter, executed without lining or steining of any kind. In the bottom of the well is a boring 26 inches diameter and 44 feet deep.

Feet.

Then follows a boring of 18 inches diameter, and 16 feet
   deep, making a total of . . . . , . . . 155
Then a 12 inch boring for 130
Then 7 inch ditto 110
             Total depth of 395

When the pumps are not working the water stands permanently at a level of 93 feet from the surface.

The well yields about two million gallons in twenty-four
hours, the level of the water then being about 134 feet
below surface.

ON THE YIELD OF WATER BY THE CORNISH MINES.

Amongst the valuable information for which we are indebted to the reports of the Cornish engines, must be
mentioned the records which give the average quantity of
water pumped in every month from each of the mines. This
is evidently information of great importance and interest,
as bearing on the subject of supplies from wells and shafts,
and should receive the greatest attention from engineers
who are considering the question of new supplies from these
sources.

From a careful examination of the maximum and minimum
quantities, and the months in which these occur, it appears
that on the average of five years the months stand in the
following order (March being the one in which the greatest
quantity is pumped, and August the least) :—March, February, April, January, December, May, November, June,
September, July, October, August.

It further appears that some of the mines yield double as
much water in certain months as they do in others, but that
the maximum quantity is more commonly about fifty per
cent. in excess of the minimum. The variation from year
to year, however, is much more considerable, and there are
many instances where the highest yield is four times as

much as the minimum yield in some other year. There are two remarkable cases, namely, the Marazion Mine and Cardrew Down, where the maximum is seven or eight times as great as the minimum.

With reference to quantity it must be admitted the yield is generally small considering the great depth of the mines. There are few whose maximum quantity exceeds 2,000 gallons a minute, or 2,880,000 gallons in twenty-four hours, and even in these the lowest yield is occasionally much less.

The remarkable falling off in some mines is also deserving of notice. Thus the Marazion Mine yielded, in the month of March, 1833, no less than 2,180 gallons per minute, whereas in February, 1835, it yielded only 263 gallons a minute, or less than $\frac{1}{8}$th of the former quantity.

It is much to be regretted that the system of reporting the Cornish engines, as to duty performed, water pumped, &c., has not been persevered in during late years. When the Cornish engines were brought some years ago to a high state of perfection, so as to work with an extremely small allowance of coal for each horse power, the wholesome rivalry existing between the principal makers caused the reports issued by Browne and by Lean to be of great value, and the makers and owners of the best engines took great pride in their performance. All this is altered at the present day; and as it is a fact that the best engines are not reported at all, any conclusions drawn from the present reports would only tend to mislead.

PUMPING MACHINERY FOR RAISING WATER.

ON THE PUMPS USED IN WATER WORKS.

The pumps commonly used for raising water from wells may be divided into two classes, lifting pumps and forcing pumps.

The lifting pumps may be again subdivided into two varieties; namely, those with a hollow piston, and those with a solid or plunger piston.

1. Lifting pumps with a hollow piston, called also atmospheric pumps.

This variety, in its simplest form, consists of the following parts :—a cylinder or tube, in which is fixed a valve opening upwards, and above which works a piston provided with a valve, also opening upwards. The part of the cylinder in which the piston works is called the body of the pump, and is the only part which need be bored with any great accuracy. The top of the cylinder may be opened or closed, it matters not which, but somewhere above the level to which the piston ascends there must be an orifice for discharging the water.

The action of the common atmospheric pump is so simple, and is so well known to every school boy, that it will be unnecessary here to dwell upon it. The cylinder is made of various materials, as wood, iron, or copper; and frequently the lower part below the fixed valve is a mere leather hose, furnished with a strainer at its lower extremity. The fixed valve in this kind of pump must be placed at such a level that the depth from it to the surface of the water in the well must never exceed the height of a column of water, which will balance the atmospheric pressure or weight of the atmosphere. This weight is measured in the barometer by a column of mercury, which varies in different parts of the world, and at different altitudes, from 28 to 31 inches. Thus, an atmospheric pump at the level of the sea may have its fixed valve several feet higher than a similar pump working on the top of a high mountain. The height at which the mercury stands in a barometer at any given place affords, in fact, a tolerably practical measure of the height to which water will rise in a vacuum when pressed by the external atmosphere. Thus, in theory, where the mercury stands in the tube of a barometer at a height of 30 inches, the sucker or fixed valve of an atmospheric pump may be placed 30 feet above the surface of water in a well. In practice, however, owing to imperfection of materials, fluctuations of level in the water, and other causes, this difference of level is too great, and should not really exceed 27 or 28 feet. In shallow wells, therefore, which are not more than about 27 feet in depth, the part of the cylinder or pump above the fixed valve need never exceed the length of the stroke or space through which the piston works. In deep wells the ascending part of the cylinder above the body of the pump in which the piston works may be, theoretically, of any height. There are difficulties, however, connected with the valves in the moveable piston, which render it inconvenient to have the lift in this kind of pump much more than 100 feet. Whatever be the height of the column of water above the moveable piston, it is evident that the absolute weight of this whole column has to be lifted at each stroke of the piston; and for this reason atmospheric pumps, which are worked by hand, have scarcely any of the pump above the piston, as otherwise the weight of water to be lifted at each stroke would be too great for the power to be applied. This, practically, limits the height to which water can be raised from wells, by common atmospheric pumps worked by hand, to about 28 feet.

In deep wells, however, when pumps are worked by horse or steam power, this objection does not apply, and if the power be sufficient to raise at each stroke the whole column of water above the piston, the length of the cylinder above this piston is only limited by the practical considerations before alluded to in connection with the valves.

It should be observed that the common atmospheric pump is seldom or never used in water works for the purpose of raising water.

2nd. Lifting pumps with a solid or plunger piston.

In this variety of pump there is the barrel or body of the pump, in which the piston works, and two fixed valves. Beneath the lower of these is the descending pipe, which goes down into the water of the well, and which is frequently, but very improperly, called the suction pipe. Above the upper fixed valve is the ascending pipe going up to the top of the well. The ascending and descending pipes are sometimes in the same vertical line, and sometimes not. The pump barrel is always a little on one side, and has a communication with the descending pipe near the top of the pump barrel, and immediately above the lower fixed valve. It has also a communication with the ascending pipe below the upper fixed valve. The disposition and general arrangement of these various parts will be much better understood from fig. 18 than from any written description. This woodcut represents a lifting pump with a solid piston, as used at the mine of Huelgoat. It was erected by M. Juncker, an able French engineer, and the engraving is reduced from a drawing in M. Combes' work on mines.*

This pump raises the water to the height of 508 feet, and is intended to raise it to a further height of 246 feet, or in all 754, when the mine shall have reached this depth. In fig. 18, *A* is the body of the pump in which works the solid piston *B*. The body of the pump is bored and turned true in the most perfect manner, and is quite open at the bottom. *C* is the piston rod, which works through a stuffing box in

* "Traité de l'exploitation des Mines,' par M. Ch. Combes. Liége, Dominique Avango et Cie., 1846.

the cover of the pump body. *D* is the passage communi-

Fig. 18.

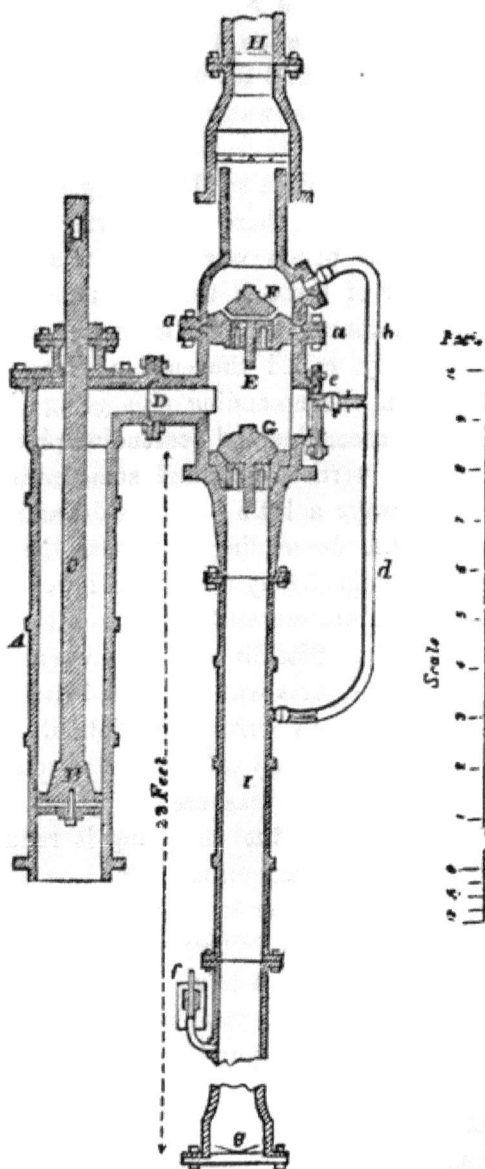

cating between the body of the pump and the valve chest.

E is the valve chest, in which are fixed the two valves F and G. H is the ascending tube, going up to the top of the shaft, and I is the descending pipe, going down to the water at the bottom. The action of this pump will be readily understood. Suppose the piston at the bottom of its stroke in the position represented in fig. 18, the body A filled with water, and all the other pipes also filled with water, as they are in the ordinary working of the pump. This being the state of things, the piston rises and lifts the volume of water resting on it, forcing it through the passage D into the valve chest E, and thence through the valve F into the ascending pipe H. During the rise of the piston, the water can only pass in this way, and can by no means get through the valve G because this only opens upwards, so that the pressure of the water instead of opening it only closes it tighter. When the piston is raised to the top of its stroke it begins to descend, leaving a vacuum behind it, into which the water left in the valve chest immediately passes in addition to some from the pipe I, which is pressed on by the external atmosphere, and in obedience to a natural law rises through the valve G and fills up the body of the pump. The piston is now at the bottom of its stroke, all the pipes being full of water as before, when the same process is again and again repeated.

The upper part of the valve chest fits into an enlargement of the ascending pipe H, and is fitted with a leather collar, screwed down, and kept in its place by a copper ring. The leather collar presses against the interior of the enlarged part of H, and admits of the upper part of the valve chest being raised in order to examine the valves. When this is to be done the flange a a is to be unscrewed.

The small side pipe b c d, which is provided with stop cocks, establishes a communication between the ascending and descending pipes, and between these and the valve chest. This arrangement is necessary for the purpose of filling the descending pipe with water, when the working of the pump has been discontinued for any length of time, and to avoid the shocks and damage which are often experienced when the pump is first set to work in consequence of the confined air compressed in the valve chest between the two valves.

When the pump is to be put in action, the stop cocks in *b c d* are opened, and a communication made between the pipe above the valve *F* and the descending pipe *I*. The latter, as well as the body of the pump *A* is then filled with the water which descends from the pipe *H*. The air contained in the descending pipe *I* passes up through the valve *G*, and escapes by a small orifice *e*, which is closed by a screw as soon as the water issuing from the orifice gives notice that the pump body and the descending pipe are entirely filled.

f is a small side valve fitted to the descending pipe, and loaded with a weight equal to about the pressure of the atmosphere, or about 14 lbs. per square inch. This valve is placed just above the surface of water in the well, and indicates whether the valve *G* is in proper order ; for if the valve *G* does not shut properly the pressure of the water which is raised during the up-stroke of the piston is transmitted to the column of water contained in the descending pipe, and this pressure immediately causes the valve *f* to open.

By means of the same valve *f* also, it can be ascertained if the upper valve *F* closes properly. For this purpose the pump must be stopped and the valve chest put in communication with the descending pipe by opening the stop cocks in the small pipe *c d*. Then, if the valve *F* closes imperfectly, the water from *H* will come through it and fill the descending pipe *I*, raising the valve *f*. This effect will of course not take place unless the foot valve *g* is in order, a fact which can be readily ascertained. These checks upon the perfect working of the pump are excellent, and have been productive of great economy.

The valves in this pump are entirely metallic, and without any leather or hemp packing. They are conical in form being carefully turned and fitted to a corresponding conical opening in the valve seat.

The valves and seats are composed of the following alloy:

85 to 88 parts of copper.
4 to 6 parts of tin.
4 to 6 parts of lead and zinc.

These valves hold water very perfectly, and close with much more readiness and accuracy than any form of leather valve. M. Juncker states that the loss of water in this pump is only one-thirtieth of that due to the diameter and stroke of the piston. This form of valve, however, is now super-seded, in most of the large pumping establishments in England, by Harvey and West's double beat valve, which will be described in a future page.

Fig. 19 represents a front elevation of the pumps in the Tettenhall well of the Wolverhampton Water Works, and fig. 20 represents a side elevation of the same. This may be taken as a good example of the most approved practice, being a recent work executed by Mr. Wicksteed and Mr. Homersham within the last few years.

A A are the air pipes which descend into the water, and which are perforated at the bottom with small openings to exclude gravel, sand, and other impurities. B B are the pump barrels, each containing a solid plunger piston, which here acts as a forcing power, and does not lift the water as in the Huelgoat pump (see fig. 18, page 234).

As soon as the plunger pole begins to ascend, the water, acted on by atmospheric pressure, enters the empty pump barrel through the opening at D. On the descent of the plunger pole a quantity of water, equal to the volume or mass of the plunger pole, is displaced at each stroke and forced up the rising pipes C C. The body of the pump B is considerably greater in diameter than the plunger pole, so that the water readily enters as soon as the down-stroke of the steam piston commences. The plungers are 13 inches

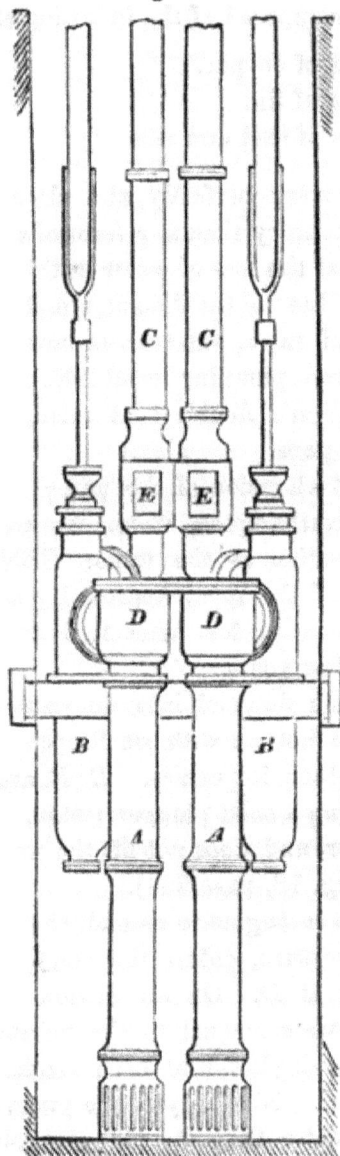

diameter, with a 10 feet stroke.   *C C* are the rising pipes of

Fig. 19.

Fig. 20.

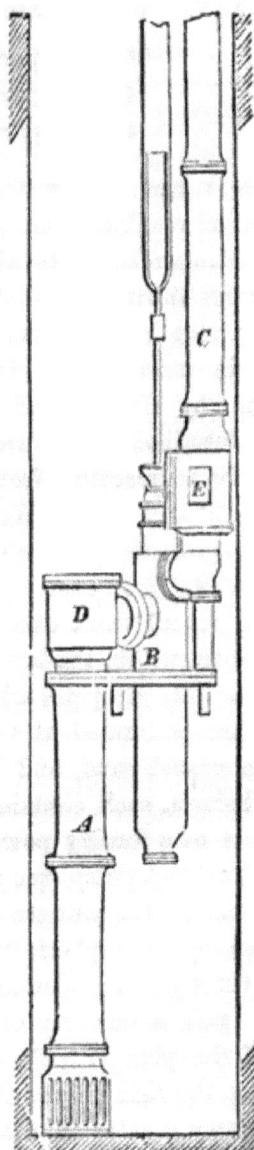

Scale 1 inch = 8 feet.

13 inches inside diameter. *D D* are the valve chests con-

taining the valves at the top of the air pipe, which open and
close the communication between the latter and the body of
the pump. *E E* are the valve chests containing the valves
at the base of the rising pipes, which open and close the
communication between these and the body of the pump.
The two valves here are not inclosed in the same valve
chest, but in separate ones. There is also some difference
between this and the Huelgoat pump in the arrangement of
the several parts, as will be seen more particularly from an
inspection of the two elevations, figs. 19 and 20, and from
the plan of the well, fig. 21. In the Huelgoat pump the
ascending pipe is directly over, and has the same axis as the

Fig. 21.

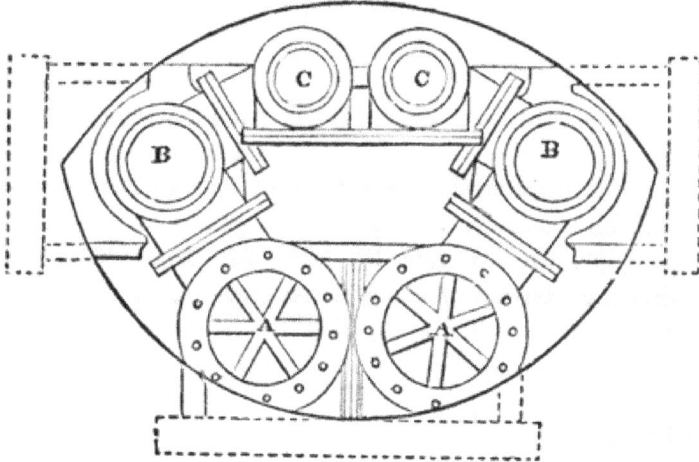

Scale 1 inch = 4 feet.

air pipe, the body of the pump being on one side. In the
Tettenhall pumps, on the other hand, the three parts are
arranged in plan almost as an equilateral triangle, one apex
being occupied by the air pipe, another by the body of the
pump, and the third by the rising pipe. This will be more
clearly seen from fig. 21, in which the three principal parts
are represented by the same letters as in the elevations.
Each of the pumps is worked by what is called a forty horse-
power non-condensing engine, by Kaye of Bury.

The shape of the well is an irregular oval, 11 feet in its longest diameter, and $7\frac{1}{2}$ feet across. *A A* are the two lower pipes, 18 inches diameter, which dip into the water. *B B* are the pump barrels, with a 13 inch plunger and a 10 feet stroke. *C C* are the rising pipes, 13 inches diameter inside. Each of the two pumps delivers 56 gallons a stroke, and makes four or five strokes per minute. The engine is capable of working the pumps at the rate of twelve strokes per minute, and as each pump raises fifty-six gallons per stroke, each would raise in twenty-four hours, $56 \times 12 \times 24 \times 60 = 967,680$, or nearly a million gallons. This is no doubt what the pumps were intended to do, but unfortunately the well does not yield the required quantity of water; the engine can only make about 3000 strokes a day, and works only one pump, so that the quantity raised is only $3000 \times 56 = 168,000$ gallons a day. The engines employed, as we have said, are a pair of forty horse-power each, by Kaye of Bury. The cylinders are 36 inches diameter, and the present consumption of coal is $1\frac{1}{2}$ tons a day of inferior Staffordshire coal for raising 168,000 gallons to the top of a standpipe, the extreme lift being 320 feet. This only requires the exercise of about eleven horse-power during the whole twenty-four hours, so that the consumption of coal is equal to $\dfrac{3360 \text{ lbs.}}{24 \times 11} =$ more than 16 lbs. per horse-power per hour.

The duty of the engine, computed in the Cornish mode, for 1 cwt. of coal will be

$$\frac{168,000 \times 320 \times 10}{30 \text{ cwt.}} = 17,920,000 \text{ lbs. raised one}$$

foot high by 1 cwt. of coal.

The pump rods consist of whole balks of timber, each about 18 feet long and 12 inches square, united to each other with flush scarf joints plated with iron. The single pieces of balk forming each pump rod do not meet in the centre, but are connected by outside balks of similar scantling and about

23 feet long. These outside connecting balks overlap the single balk forming the pump rod rather more than 4 feet at each end, so that a space of 14 feet is left between the abutting ends of the pump rod in the centre. In this central space is fixed the contrivance for adjusting the length of the pump rod, rendered necessary by variations of temperature, &c.

The Wolverhampton Company's engine, at Goldthorn, pumps from a well 300 feet deep into a reservoir close by. This is a Cornish engine of seventy horse-power, with an 8 feet stroke and 45 inch cylinder, working a 15 inch pump and lifting forty-eight gallons at each stroke. Number of strokes 4,400 in twenty-four hours; consumption of coal twenty-three cwt.; quantity of water raised in twenty-four hours $4,400 \times 48 = 211,200$ gallons; lift 300 feet. The duty of this engine is somewhat better than that of the Tettenhall, namely,

$$\frac{211,200 \times 300 \times 10}{23} = 27,547,826$$

℔s. raised one foot high by the consumption of one cwt. of coal.

FORCE PUMPS.

Another kind of pump frequently used in Waterworks for raising water, is of the description called force pumps, with a solid plunger piston, which works through an air-tight stuffing box in the cover of the pump barrel. Pumps of this kind, when used for raising water from wells, consist of the same principal parts as the lifting pump, namely :—1st. The air-pipe below the barrel or body of the pump. 2ndly. The barrel in which works the solid plunger or piston ; and third, the rising or ascending main pipe above the pump body. The air-pipe dips several feet into the water to be pumped, and is usually perforated at the bottom with small holes, which, while they freely admit the water, serve to exclude sand, mud, gravel, and other impurities which might otherwise find their way in.

M

This air-pipe is provided at the top with a valve opening upwards and fixed somewhere between the level of water in the well and the body of the pump. The body of the pump is generally a few inches longer than the stroke as it is called, and longer than the cylinder of iron which constitutes the solid piston. The body of the pump is either placed immediately over the air-pipe, in which case the upper ascending pipe is a little on one side, or the pump body is placed on one side, in which case the upper ascending pipe may be in the same vertical line as the air-pipe. The pump barrel does not require to be turned or bored, as the plunger does not fill up the whole space ; and, in fact, it is usual for the pump barrel to be an inch or so more in diameter than the plunger. The office of the plunger, which is raised by the depression of the piston in the steam cylinder, is merely to force up the water which has risen through the valve in the air-pipe. The upper ascending main or pipe is merely composed of certain lengths of cast-iron, united by flange joints, and at its lower extremity is provided with a valve also opening upwards, to prevent the water which has been forced into the pipe from returning to the bottom of the well.

Fig. 22 is a section of a force pump with solid plunger piston, such as is commonly used for raising water from wells. Here A is the air-pipe, B is the solid plunger piston shown near the bottom of its stroke, C is the ascending pipe going to the top of the well, D is the lower valve chest, and E the upper valve chest. The valves are Harvey and West's patent double beat; that in E should have been shown open in consequence of the descent of the piston having forced up the water and raised it into that position. For the same reason the valve in D is shown closed, and now sustains the whole pressure of the column in C.

Let us now examine the action of the pump. The plunger is raised by the pressure of the steam on the piston in the steam cylinder, and as the cover of the pump barrel through which it works is perfectly air tight, no air can pass in from

Fig. 22.

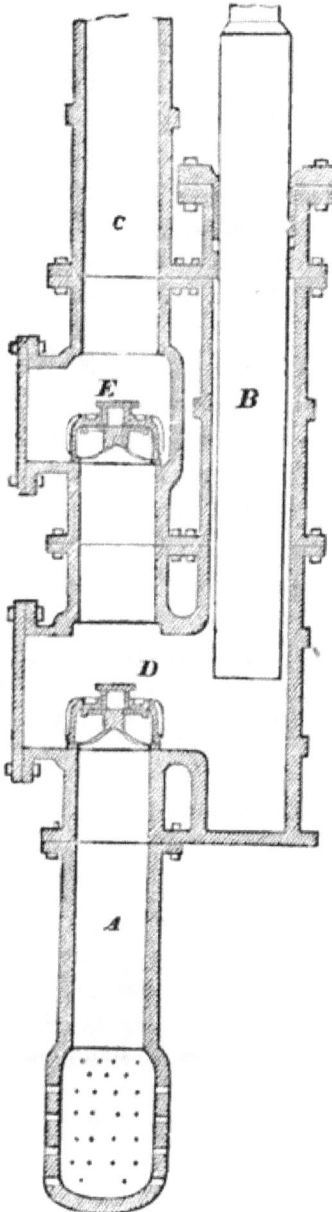

the outside to supply its place. A partial vacuum is therefore formed, which is supplied by the air beneath the lower valve and between this and the water. After each stroke this air becomes more and more rare, and the vacuum more and more perfect; till at length the water in the well, pressed by the atmosphere outside, follows the ascent of the piston, and rises through the foot valve into the pump barrel. When this is full of water the plunger descends and presses on the water, which can neither go back through the lower valve, nor escape through the cover of the pump barrel. In fact, it has only one mode of escape, namely, through the valve of the ascending pipe; and through this it accordingly goes, each successive stroke of the plunger sending into the ascending pipe a quantity of water equal to the volume of the plunger itself. This goes on till the ascending pipe is full, and then, of course, at each stroke it delivers a quantity equal again to the volume of the plunger.

From what has been explained about the rise of water through the lower valve in D, it will at once be seen that the pump would not act if the length of the air pipe exceeds the height of a column of water, equal to the weight of the atmosphere pressing on the same base. In other words, the valve at the top of the air-pipe must not be higher than the corresponding valve in the atmospheric pump. The foot valve, or lowest fixed valve, in water works and mining pumps, is usually placed in practice much nearer to the water level than in ordinary hand pumps, often only a few feet above the water level.

The engines employed to work the pumps for raising water out of wells are essentially the same as those employed in water works for raising water from rivers or storage reservoirs into high service reservoirs, and over stand pipes for forcing it through a train of pipes.

A description of the steam engines will be given under the general head of pumping apparatus, where also the valves used in the pumps of modern water works will be described.

PUMPING APPARATUS USED IN RAISING WATER FOR SUPPLY OF TOWNS.

Some account of the pumps used in raising water from wells will also apply to this part of the subject, as the pumps used in that kind of work contain the addition of a rising main to the ordinary parts of a pump. In the present article we shall treat only of the pumps used for raising water from the level of the earth to a still greater height. We shall also briefly describe the kind of engines usually employed for pumping in water works, whether for raising water from wells or for pumping it from the surface of the ground into elevated reservoirs, for throwing it to the top of stand pipes, or forcing it through trains of pipes.

The same kind of pumping engine is applicable to each kind of work, and therefore may be most conveniently described under the general head of engines.

The two principal varieties of steam engines used for pumping purposes, are condensing low pressure engines, condensing high pressure engines, and ordinary non-condensing high pressure engines.

The first and third varieties are not extensively used in modern water works, and are being very generally superseded by the condensing high pressure engines, working expansively on the Cornish principle.

1st. *Low pressure condensing engines.*—This form is usually made on the pattern of Boulton and Watts' single acting engine. There is little or no expansion of the steam in the cylinder, although the steam is usually cut off when the piston has made from $\frac{3}{4}$ to $\frac{7}{8}$ of the stroke, in order to prevent the danger of breakage from the piston striking the bottom part of the cylinder. This kind of engine is usually applied to work a lifting pump, as at the East London Water Works, but is also sometimes used for working a plunger pole, as at the Birmingham Water Works. When working a lifting pump, the steam being admitted at the top of the cylinder, depresses the steam piston and raises the solid pump piston, which is attached to the other extremity of the beam. The pump rod is only loaded with a sufficient weight to overcome friction, to raise the steam piston, and cause the pump piston to descend to the bottom of the pump barrel. The engine being single acting, only raises water by means of the down stroke in the steam cylinder, corresponding with the up stroke in the pump. The condensing low pressure engine at the East London Water Works, which is a single acting engine, by Boulton and Watt, has been described in great detail by Mr. Wicksteed, who has published valuable plates of this engine, and of the first Cornish engine which was erected for the work of pumping in the metropolis. The Boulton and Watt engine has a steam cylinder of 60 inches, the piston having a stroke of 7·91 feet, and making $11\frac{1}{2}$ strokes per minute; the diameter of the pump is $27\frac{1}{4}$ inches, and that of the pump rod $4\frac{7}{8}$ inches, the stroke being 7·91 feet, the same as that of the piston in the steam cylinder. The water is raised 107 feet high.

The power of the engine is thus calculated by Mr. Wicksteed. He first finds the load on the piston in this manner.

	Inches.					Area in inches.
Diameter of pump	$27\frac{1}{2}$	.	.	.	.	578
Less area of pump rod	$4\frac{7}{8}$	.	.	.	.	19
						559

$$\frac{559}{144} = 3\cdot88 \text{ square feet.}$$

Then 3·88 × 107 × 62·5 = 25947·5 =

the load on the piston—

and 25947·5 × 7·91 × 11·5 = 2,360,314

ifted one foot high per minute,

$$\text{and } \frac{2,360,314}{33000} = 71\cdot5 \text{ horse power.}$$

The power for each stroke is $\frac{71\cdot5}{11\cdot5} = 6\cdot22$ horses nearly.

The boilers of this engine are of the form called waggon headed, and the pressure of the steam in the boilers is about 17 or 18 lbs., its pressure in the cylinder being about $10\frac{1}{4}$ lbs. The duty or useful effect of this engine, according to a great number of experiments tried by Mr. Wicksteed, was equal to 47,718,084 lbs. lifted one foot high by the consumption of 1 cwt. of the best Welsh coal. This, as will be seen hereafter. is less than half the duty performed by the Cornish engine, at the same works and with the same coals.

Fig.23 is an elevation showing the arrangement of the air-pipe, and the pump in the well as worked by this engine. *A* is the air-pipe, *B* the body of the pump, in which works a solid plunger piston, *C* is a blank pipe supporting the upper valve box. *D* is the pump rod, and *E* a counterbalance, consisting of moveable cast-iron weights. *F* is the air vessel into which the water is discharged by the pump. The water usually stands in the air vessel, within 8 feet of the top, the

Fig. 23.

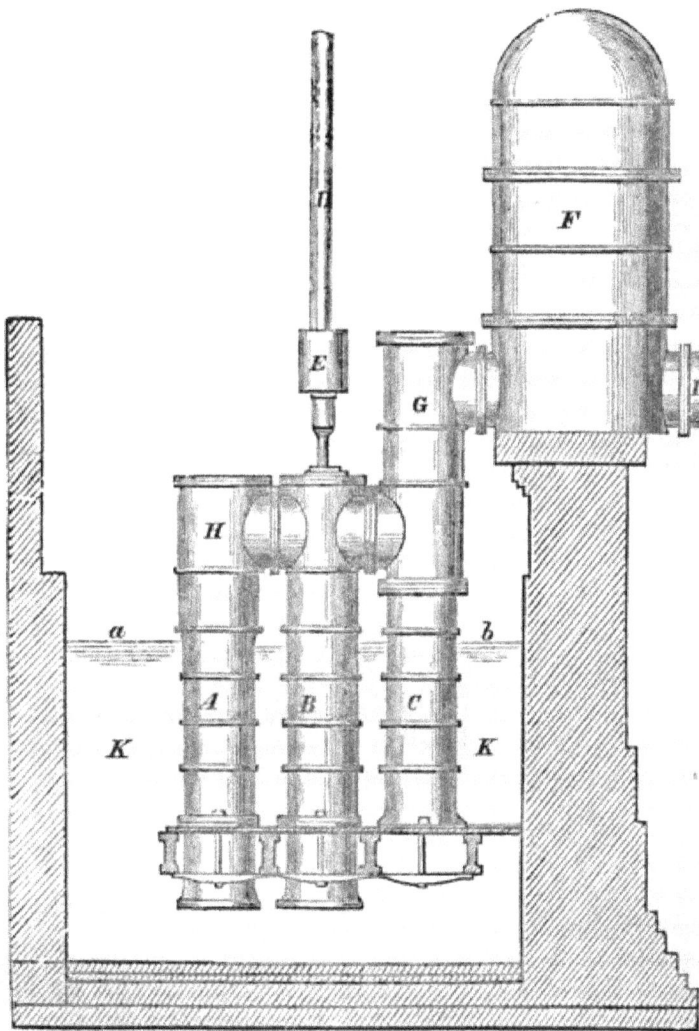

Scale 1 inch = 8 feet.

compressed air in this upper part serving to equalise the
pressure of the water in the main pipe.   G, H are the valve
boxes, and I is the main pipe through which the water
passes for the supply of the district.   K K is the pump well,
and a b the usual surface of the water.

The single acting Boulton and Watt engine has been applied with perfect success in pumping at the Birmingham Water Works. The absolute lift of water here is 252 feet, being considerably greater than at any of the London works, and the pressure is further increased by friction in the pipes to about 285. Two Boulton and Watt engines have lately been erected at these works, each with 10 feet stroke and 72 inch cylinders. Each engine works a plunger pump 23 inches diameter and with 10 feet stroke. The weight of water to be raised at each stroke is equal to 51411 lbs., or rather more than 22 tons; but the total weight upon the plunger required to overcome the load upon the air pump, the friction of the engine, and to maintain a velocity of 10 strokes per minute, is nearly $26\frac{1}{2}$ tons, which is equal to 142 lbs. per square inch upon the area of the 23 inch plunger, and $14\frac{1}{2}$ lbs. upon the steam piston. According to Mr. Wicksteed's mode of calculating the power of this engine, the weight of the column of water would be used to represent the resistance; while, according to others, the whole resistance of $26\frac{1}{2}$ tons would be used. Thus, according to Mr. Wicksteed, the power would be

$$\frac{51411 \times 10 \times 10}{33000} = 156 \text{ horses};$$

but, according to the other mode, the power will be

$$\frac{26\frac{1}{2} \times 2240 \times 10 \times 10}{33000} = 180 \text{ horses nearly.}$$

These engines pump into an air vessel of 7 feet internal diameter, and 18 feet high, or 15 high above the delivery branch into the main; and it is replenished with air by a separate small pump of 6 inches diameter, and 3 feet 6 inches stroke.

An excellent description of these engines was lately read by Mr. Garland of the Soho works, before the Institution of Mechanical Engineers at Birmingham, and as this description contains many particulars which have not been previously noticed with respect to pumping engines, we have thought it convenient to present the principal part of the details as given by Mr. Garland :—

The cylinders have steam-cases, and are enclosed in a covering of felt, having an outside casing of wood, to prevent the radiation of heat ; and the top of the cylinder and upper nozzle are covered in a similar manner.

The steam valve, equilibrium valve, and exhaustion valve, are 13, 15, and 18 inches in diameter respectively, and of the double-beat construction, by which is removed the principal part of the pressure, that the common conical valve is subject to. The steam governor valve is made of the single conical form (there being no necessity for making this valve upon the double-beat principle), and it is regulated by a screw and wheel handle.

The load on these engines is a variable one to the extent of the difference of the dead level of the upper reservoir, and the amount of friction of water in transitu ; and it sometimes happens that the water is being drawn off faster than the engine supplies it, and the velocity of the water beyond where the great draught occurs is consequently decreased, and the resistance proportionably diminished.

To prevent any accident to the engine by going out too suddenly in consequence of this diminished resistance, a throttle valve is placed between the upper and lower nozzle, and in the pipe communicating with the top and bottom of the cylinder, which is regulated in its opening by a screw and wheel handle, and by contracting the passage, or, in other words, wire-drawing the equilibrium, the equalisation of pressure between the top and bottom of the cylinder is more slowly formed during the time the plunger is descending, to the extent which the weight is in excess of the diminished resistance. In these engines this valve has been found of invaluable service, and it will even hold the plunger at the top of the stroke. It acts exactly like putting on a break to a crane when lowering a weight, without absorbing any power or causing any disturbance to the working of the engines.

The opening of the steam injection and exhaustion valve is regulated by a cataract, and the speed of the engine is thus under the control of the engine man. The equilibrium valve is opened by quadrant catches, and is dependent upon closing of the exhaustion valve; the former being opened upon the closing of the latter, and shut in the usual manner by a tappet upon the plug rod.

The injection valve is also made upon the double beat principle, to render the strain upon the exhaustion valve spindle as little as possible, by relieving it of all unnecessary pressure, the underside of it being open to the condenser.

In the event of the bursting of any pipe in the main, and the resistance to the plunger being suddenly removed, a detent is fixed upon the plug rod to prevent the repetition of a blow upon the spring beams by the catch pins. This detent comes into action upon the engine making more than its usual length of working stroke, by holding the steam handle down, and thus preventing the opening of the steam valve. This adjunct to the hand gear, though it may never be brought into operation from such an occurrence, would evidently be of great value in such a case.

The air pump is of 34 inches diameter and 5 feet stroke, and the condenser of similar capacity. The air pump bucket is fitted with a brass annular or ring valve, and the delivery and foot valves are of the usual construction, or what are termed flap valves. A vacuum is obtained varying from 27 to 29 inches, according to the state of the atmosphere. Each engine has its separate condenser cistern formed of cast-iron, which is supplied by a cold water pump of $13\frac{1}{2}$ inches diameter, making a 5 feet stroke. The feed pump is of $6\frac{1}{2}$ inches diameter and 2 feet 6 inches stroke, fitted with an air vessel. The plunger of the main pump is, as before stated, 23 inches diameter, and has the same length of stroke as the steam piston, viz., 10 feet. The suction valves and delivery valves of the pump are of the double-beat kind, and fitted in pairs for the purpose of giving additional security to the action of the pump in the event of one of them sticking or becoming otherwise deranged.

The air vessel is 7 feet internal diameter and 18 feet high, or 15 feet high above the delivery branch into the main, and it is replenished with air by a separate pump of 6 inches diameter and 3 feet 6 inches stroke. An air-cock is fixed upon the suction pipe of this pump, by which is regulated the necessary quantity of air to be supplied. This cock only requires to be partially open, and when closed entirely the pump lifts water only.

The air-vessel is of great importance, as by its equalising action the motion of water in the mains is rendered continuous, and a less weight, in consequence, is required to give the necessary velocity to the descent of the plunger in the out-door stroke. At the top of the pump plunger is fixed the pole case, containing the necessary weights to overcome the load or resistance, and, as before stated, is equal with the plunger and rod to about 26½ tons.

Upon the first delivery pipe joining the air vessel is fixed a safety discharge valve, 6 inches diameter, loaded by a lever and weight a little above the pressure upon the main, to prevent any undue force being thrown upon the pump from the accidental shutting of the sluice cocks between the engines and the town.

The main lever or working beam is 30 feet long, cast in two plates each of 3 inches in thickness, and the depth of it in the middle is 6 feet, and at the ends 2¼ feet. Each of the plummer blocks has saddles of cast-iron between them, and wooden spring beams 30 inches deep and 20 inches wide.

It may be interesting to state that the quantity of water lifted by every stroke of each engine is equal to 180 gallons, or 1,800 gallons per minute, and 108,000 gallons per hour, weighing upwards of 483 tons lifted in each hour.

Mr. Garland further stated, in explanation of his paper, that the pressure of steam was about 12 lbs. per square inch in the cylinder, and that it was cut off at one-third of the stroke, ex-

panding through the remaining two-thirds. The actual duty
had not been ascertained, because the fuel used consisted of
Staffordshire small coal or slack. The evaporative value of
this, as compared with the best Welsh coal (which is commonly used in testing the duty of a pump engine), has not
been ascertained. He stated, in answer to a question, that the
small pump had been found necessary to replenish the air
vessel. It is certain, however, that there are many instances
of air vessels attached to pumping engines in London and elsewhere, without the addition of any such pump to supply the
air vessel.

Mr. Cooper, who was present at the discussion which followed
Mr. Garland's paper, expressed an opinion, that it was preferable to make a pumping engine double acting on the bucket
and plunger plan, with the plunger half the area of the
bucket, so as to pump half the water in the up stroke, and
half in the down stroke, thus enabling an engine and pump of
half the size to do the same work ; also to add a crank and fly
wheel, and work at a higher speed, which further reduced the
size and cost of engine and pump. Mr. Cooper mentioned an instance of some works where there were four 150 horse power
engines working very satisfactorily on this plan from $12\frac{1}{4}$ to
21 strokes per minute, with 7 feet length of stroke. But he
considered the horizontal engine with direct acting pump and
crank, was the most advantageous and economical, when the
water to be pumped was near the engine house floor.

Double acting condensing engines are also occasionally employed for pumping, and so also are direct acting engines
working without a beam, the steam cylinder being placed
directly over the barrel of the pump.

In many of the American water works, as at Pittsburgh,
Alleghany, and Detroit, high pressure non-condensing engines
are used for pumping. The work of these, however, does not
appear to be very satisfactory, as a recent report on the water
works of these cities gives the duty performed by the pumping engines there. The steam is generated by means of wood,

the value of which for evaporating purposes, as compared with bituminous coal, is well known by the American engineers.

The following table shows the duty of several of these high pressure non-condensing engines, from a recent report of the engineer of the Detroit Water Works, reduced to the English standard of lbs., raised one foot high by the consumption of 1 cwt. of coal.

	Duty in lbs.
Pittsburgh Upper Water Works Engines, date 1852 .	19,941,600
Ditto, Lower works	19,112,576
Alleghany city	19,226,700
Detroit	17,397,856

In Cincinnati and other towns, there are both kinds of engines; namely, condensing and non-condensing high pressure engines. This practice of adopting both kinds is worthy of attention. The ordinary and regular work is assigned to a condensing double acting engine, working expansively like our Cornish engines; and the duplicate engine employed for occasional work, and to serve in case of emergency, is a high pressure non-condensing engine, with much smaller cylinder than the other. This gives the advantage of great economy in the regular continuous working of the condensing engine, and as the high pressure engine is much cheaper to erect, it saves a considerable sum in the first outlay. Mr. McAlpine, a celebrated American hydraulic engineer, has adopted this method in most of his recent works, as in Brooklyn, Albany, Chicago, and other places.

Thus, at Albany, where two million gallons have to be raised daily to a height of 156 feet, and another million to a height of 238 feet, exclusive of friction, Mr. McAlpine proposes a double-action condensing Cornish engine for the regular pumping work, with a duplicate non-condensing engine as a reserve.

The Cornish engine, to have a 58-inch cylinder 12 feet stroke, and to be worked at 10 double strokes a minute—Steam in boiler 30 lbs. per inch; ditto in cylinder, effective pressure 20 lbs. The non-condensing engine to be made horizontal, and to work steam at a pressure of 70 lbs. per inch on the piston, cutting off at $\frac{3}{4}$ stroke, making twenty strokes per minute, and calculated to do the whole work by pumping twenty hours a day. The cylinder to be 24 inches diameter and 6 feet stroke, giving an effective velocity of 240 feet per minute. The condensing engine to work two pumps, each 18 inches diameter; the non-condensing engine to do the same work with one 18-inch pump.

At Chicago, where three million gallons a day have to be raised 90 feet high in twelve hours, Mr. McAlpine proposes a condensing engine with a 46-inch cylinder, 9 feet stroke, making $13\frac{1}{3}$ double strokes per minute, effective velocity 240 feet per minute. Steam in boilers and cylinder the same as at Albany.

As the duplicate engine is to be used only for short intervals, cheapness in its construction is more to be regarded than economy in using it. He therefore proposes to make this of the minimum size, necessary to afford the requisite supply of water, by working the whole twenty-four hours. For this engine he specifies a steam cylinder of 18 inches with 6 feet stroke, piston travelling 240 feet per minute, using steam at 80 lbs. pressure at the cylinder, and to be cut off as before at three-fourths of the stroke. This engine to work horizontally and to drive an 18-inch horizontal pump by direct action.

The Brooklyn works are much later and of more recent date than either of the others. Here the work to be done is equal to raising 5 million gallons per day, to the height of 190 feet in twelve hours. To perform this work Mr. McAlpine proposes a double acting 72-inch cylinder Cornish engine, working expansively, and using steam of 20 lbs. pressure per square inch of the piston. The duplicate engine here is to have a 30-inch cylinder, and to use steam as before of 80 lbs. pressure.

The performance of the small high pressure steam engines for farm purposes, exhibited during the last few years at the shows of the Royal Agricultural Society, is calculated to excite some astonishment. It is no uncommon thing to find these small

engines of six or eight-horse power, reported as working with
5 lbs. of coal per horse power per hour, which is equivalent to
a duty of more than 44 millions. Considerable allowances,
however, must be made from this large amount of duty, as it
is almost certain that the whole working power of the engines is
estimated, and no distinction made between the power absorbed
in overcoming friction and that producing useful effect. It is
certain that the recorded performance of those engines does not
represent the actual duty, as we have no instance of high pressure engines anywhere, and, under any circumstances, workirg
with a duty much exceeding twenty million pounds = 11lbs. of
coal per horse power per hour.

CONDENSING HIGH-PRESSURE ENGINES WORKING EXPANSIVELY.

Most of these engines are single acting, like the condensing
Boulton and Watt pumping engines, the principal distinction
being that they work with steam of much higher pressure ; that
this steam is cut off after performing from one-eighth to one-
third of the stroke, and that the cylinder, boiler, and steam pipes
are very carefully clothed with non-conducting material to prevent any loss of heat by radiation. The Cornish engine at the
East London Water Works, which has been described by Mr.
Wicksteed, has a cylinder $80\frac{1}{4}$ inches in diameter, with a stroke
of 10 feet. It works a pump with a plunger pole 41 inches
diameter and stroke of 9 feet. The engine usually makes
eight strokes per minute, and forces the water to the top of a
stand pipe 108 feet high, above the surface of the pumping
well.

Calculation of the power of this engine :—

Area of
plunger in   Lift.   lbs.
sq. ft.

$9 \cdot 168 \ \times \ 108 \ \times \ 62 \cdot 5 \ = \ 61,884$ resistance to plunger.

$61,884 \ \times \ \dfrac{9 \text{ ft. pump stroke.}}{10 \text{ ft. engine stroke.}} \ = \ 55,695 \cdot 6$ lbs. load on steam
piston.

Length　　Strokes
of stroke.　per min.

Then 55695·6 × 10 × 8 = 4,455,648 lbs. lifted one foot high per minute.

and　　$\dfrac{4,455,648}{33,000}$ = 135 horses.

also　　$\dfrac{135}{8}$ = 16·875 horses for each stroke per minute.

This is the power at which the engine actually works, but according to the Cornish method of estimating horse power, an 80 inch cylinder engine would be capable of working at a much higher power than this. The Cornish makers usually construct their engines of this size, capable of working with a pressure equivalent to 15 lbs. per square inch acting uniformly on the piston, and the effective velocity of the piston is taken at 110 feet per minute, which is equal to 11 double strokes of 10 feet each per minute. Hence the power of the engine would be

square ins. area of piston less area of piston-rod.	lbs. pressure per square inch.	Velocity per minute.		
5019	× 15	× 110		

$$\frac{5019 \times 15 \times 110}{33,000} = 251 \text{ horses.}$$

This engine, when making 8 strokes per minute, lifts 660 cubic feet per minute, or 5,940,000 gallons in 24 hours. When making 11 strokes per minute, the quantity would be 8,167,500 gallons in 24 hours.

Fig. 24 is the elevation of the forcing plunger pump worked by this engine, showing also the pumping well and the stand pipe. *A* is the wind bore or air pipe of the pump, *B* is the main pump barrel or pole case, *C* is the plunger pole, *D* is the lower, and *E* the upper valve box, *F* is the delivery pipe leading to the stand pipe *G* ; *H* is the pumping well to which the water is conveyed by a culvert from the settling reservoirs or filter beds, and *a b* is the level of the water in the well.

In the ordinary working of Cornish engines the steam is raised in the boilers to a pressure varying from 35 to 50 lbs. per square inch, and is cut off from the cylinder after the piston has passed through a distance varying from one-tenth to one-third of the stroke. It has been observed that the duty of many large pumping engines in Cornwall has been very great when they were first erected, and that the duty afterwards fell off. Mr. Wicksteed explains and accounts for this by saying, that at first the mine is not deep, and the engine is worked far below her full power, consequently the steam is cut off when the piston has performed only a very small part of the stroke. In proportion as the mine becomes deeper, the work required from the engine increases, so that the expansion of the steam is not carried to the same extent, and the duty falls off.

Mr. Wicksteed made some very interesting experiments on the Cornish engine at the East London Water Works, which amongst other things, show the effect of expansion at different parts of the stroke.

Thus, when the steam was cut off at $\frac{6}{10}$ of the stroke, the duty of the engine with 112 lbs. of Welsh coal of the same quality as used in Cornwall was 76·7 million lbs., and when the steam was cut off at $\frac{3}{10}$ of the stroke the duty amounted to 105·7 million. These experiments were made in the most careful manner, and extended over a period of several weeks.

In these experiments the steam pressure in the boilers varied from 30 to 52 ℔s., the pressure in the cylinder before cutting off the steam varied from 15 to 20 lbs. per square inch, and the mean pressure of the steam on the piston varied from 13·08 to 15·54 ℔s. per square inch.

The large amount of duty reported of the Cornish pumping engines has given rise to a great deal of discussion within the last few years. When Mr. Wicksteed first visited Cornwall thirty years ago for the purpose of examining these engines, he found one of the engines at the Fowey consolidated mines doing a duty of 83,000,000 lbs. with a bushel of coal. The

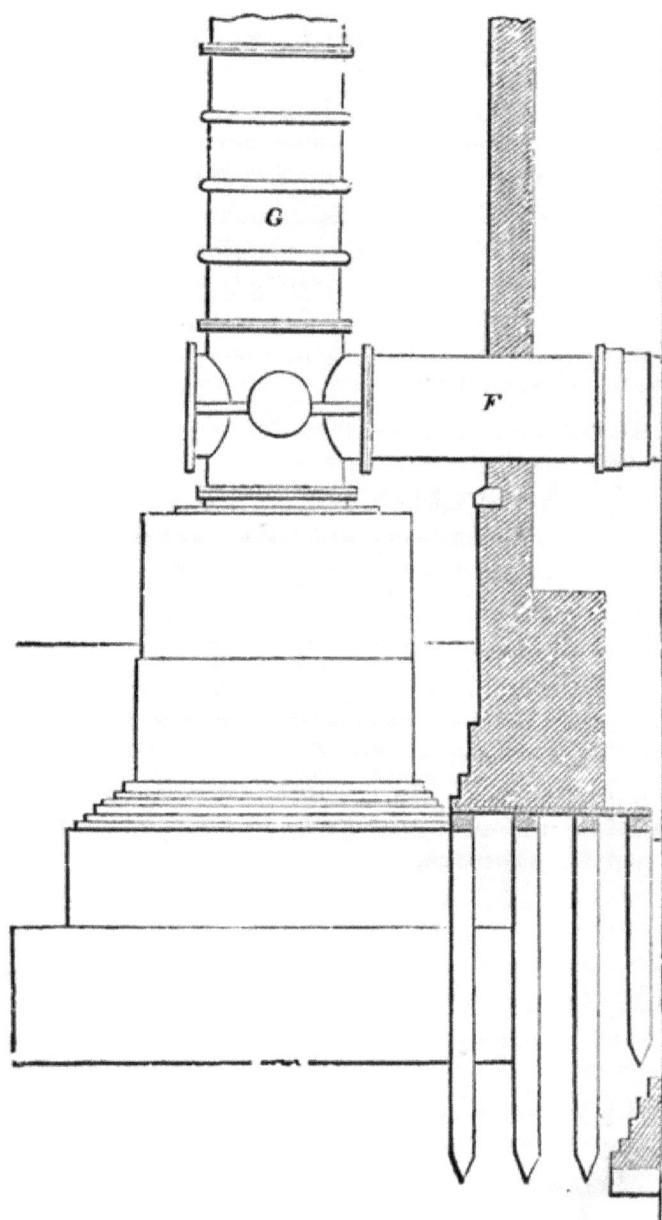

Fig.

G

F

Scale

26.

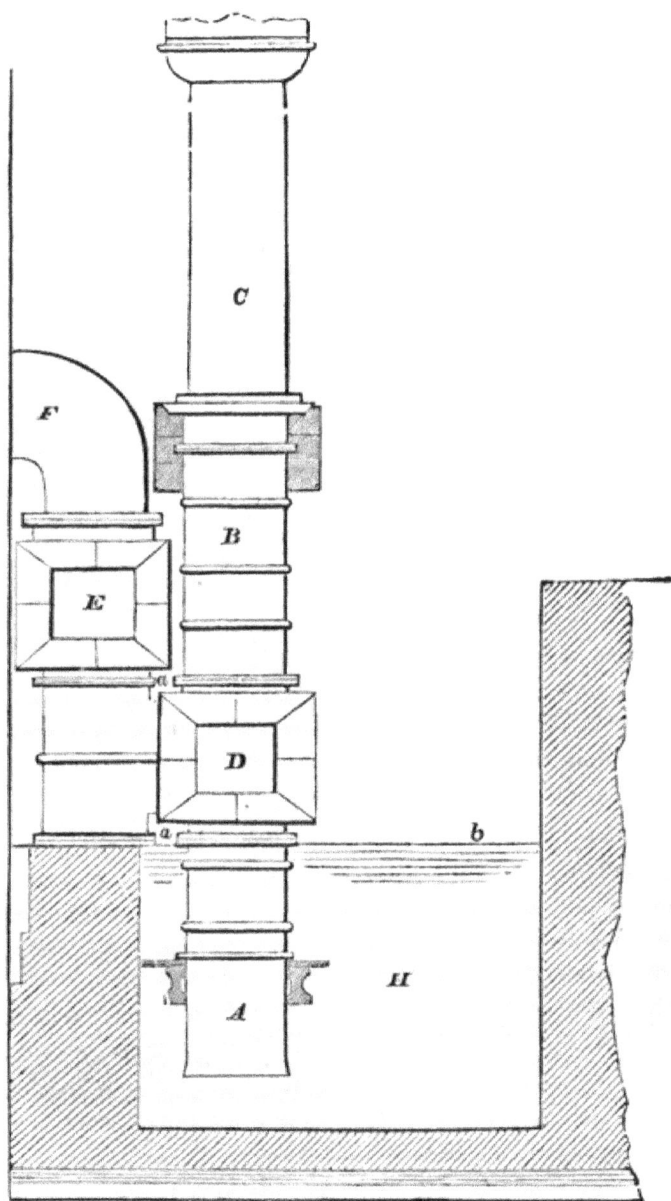

1 inch = 8 feet.

weight of the Cornish bushel of coals was at first supposed to
be 84 lbs., but was since more accurately ascertained to be
94 ibs.   Hence the duty of this engine was at that time very
nearly 100,000,000 lbs. for 1 cwt. of coal.   This duty has
since been considerably exceeded, but Mr. Wicksteed satisfied
himself from the enquiries and observations he was then able to
make, that the Cornish engine would effect a saving of nearly
two-thirds of the coals if used for pumping by the London
Water Companies.   In addition to the use of high pressure
steam and the practice of cutting it off and working it expansively, the Cornish engines differ from the Boulton and Watt
engines in the following particulars.

1st. The boilers are not waggon shaped but cylindrical, having in most cases an internal tube traversing the boiler longitudinally.

2nd. The boiler, the cylinder, and steam pipes are completely cased with non-conducting material, every precaution
being taken to prevent radiation of heat.   The consequence is
that the engine room is at all times perfectly cool, and very
little heat is lost, even when the engine has to stand still for
several hours.

3rd. The steam and exhausting valves, as well as the
pipes leading to the condenser, are of very large capacity, and
the valves are capable of being opened with great facility, in
consequence of which they are worked with much less exer-
tion of strength than in other engines.

4th. The length of the stroke is greater and the number of
strokes per minute fewer than in other engines.

5th. The engine man possesses the most perfect power to
regulate the number of strokes, by means of the cataract.
Mr. Wicksteed observes, that this contrivance is peculiarly
applicable to engines for water works, where the demand for
water increases every year, and where the power must increase
in proportion.

6th. The Cornish engine, being put up at first of greater
power than is actually required, say to work at first with steam

cut off at one-sixth part of the stroke, will continue to be adapted to an increase of work much longer than any other form of engine, because an increase of power will be obtained, both by working steam of higher pressure, and by allowing the steam to act on the piston through a greater proportion of the length of stroke. In this way the expense of additional new engine power may be longer deferred when the Cornish engine is used.

The other advantages which Mr. Wicksteed enumerates have chiefly reference to the pumps being worked by solid plunger pistons, and with an improved form of valves. These advantages, however, are not confined to the adoption of the Cornish engine, as the plunger pump with solid piston is frequently worked by the Boulton and Watt single acting engine.

The celebrated double beat valves of Messrs. Harvey and West, which are used so generally in the pumps worked by Cornish engines, are equally applicable, both to lifting pumps and forcing pumps worked by other kinds of engines.

THE VALVES USED IN PUMPS.

In both lifting pumps and force pumps there must be at least two fixed valves. Although, of course, these valves open and close, they are called fixed, in contradistinction to the valve which is placed in the moveable piston of the common atmospheric pump. The seating of the valves, however, or the part on which they close when shut down, is the only part that is really fixed and immoveable. The two kinds of valves formerly used for pumps in water works were called "butterfly valves" and conical valves. The first kind was so named, from a fancied resemblance to the wings of a butterfly, the valve being composed of two semicircular disks, hinged to the seating along the diameter of the semicircle. The conical valve was of metal, both the valve and the seating being accurately turned and fitted to a true conical form. This kind of valve is shown in the section of pumps employed in the mine of Huelgoat (fig. 18, p. 234).

Both the conical and the butterfly form of valve, however, have been entirely superseded by Harvey and West's patent double beat valve, which is now almost invariably used in all large pumps.

The seating of this valve consists of a circular ring, on which the lower part of the valve closes or beats as it is called, and of a circular plate of somewhat less diameter than the ring. and on the outer edge of which the upper part of the valve beats—hence the name *double beat valve.* Figs. 25 and 26,

Fig. 25.

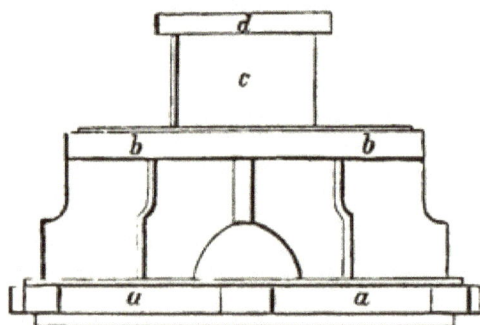

Fig. 26.

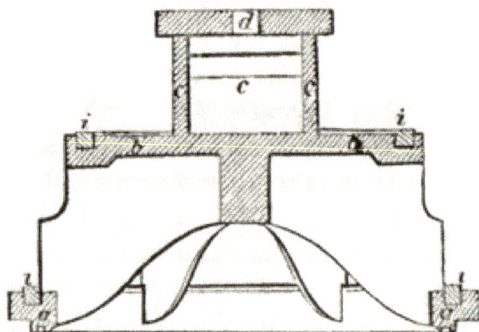

are respectively an elevation and section of the seating. In these figures *a* is the circular ring on which the base of the valve beats, and *b* is the circular plate on which the upper part of the valve beats. *c* is a cylinder cast upon the seat and turned

perfectly true, so as to form a guide for the valve to work
upon and keep it in its place, and *d* is a cap bolted on the
cylinder to stop the upward motion of the valve and prevent
it from rising too high.

The valve itself may be described as a sort of double
cylinder one within the other, forming one piece entirely open
at the bottom and partly so at the top. Figs. 27, 28, and 29,

Fig. 27.

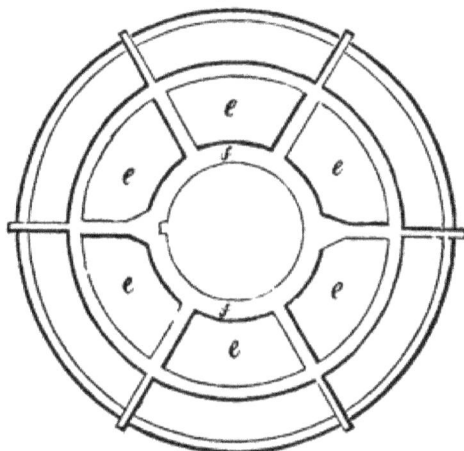

Fig. 28.

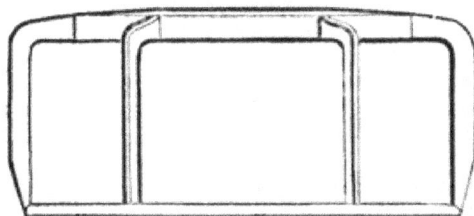

Fig. 29.

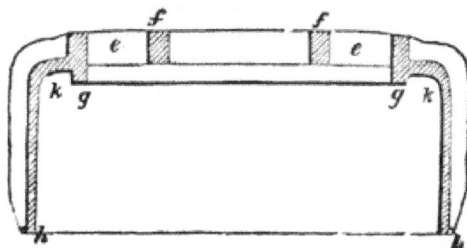

are a plan, elevation, and section of the valve; *e e e* are the
openings in the top of the valve; *f* is the part which embraces
and works on the cylinder *c*. *g* is the upper ring which beats
on the plate *b*, and *h* is the lower ring of the valve, which beats
on the lower part of the seat *a*. The actual beats *i i*, (fig. 26)
on which the parts of the valve rest when closed, are either
formed by a raised ridge cast or wrought upon the seat, and
faced or turned true, or are formed by introducing into circular
grooves, cast in the seat, a ring of wooden wedges or of soft
metal, the top surface in either case being faced or turned true
to receive the valve. The patentees prefer wood or soft metal
for the beating surfaces. The two rings *g* and *h* must also be
faced or turned true, so as to fit accurately to the beating
surfaces when closed. *k* shows the part of the valve which
is exposed to the pressure of the atmosphere, or to the force
created by the motion of the piston, according as the valve is
the lowest or the highest in the pump. This area must be of
such a size, that the pressure acting on it will overcome the
weight of the valve, and cause it to rise. Figs. 30 and 31

Fig. 30.

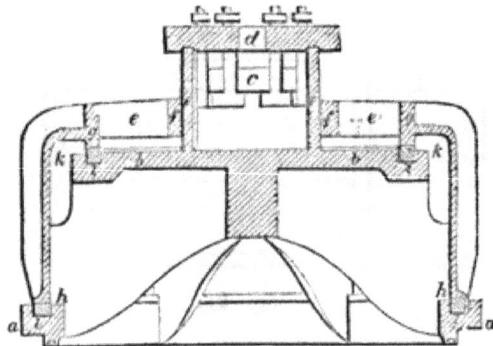

are two sections of the valve and seat, the former showing the
valve closed and the latter open, the arrows marking the
course taken by the water passing through the valve. Figs.
25 to 31, inclusive, are drawn on a scale of 1 inch to 2 feet.

Fig. 31.

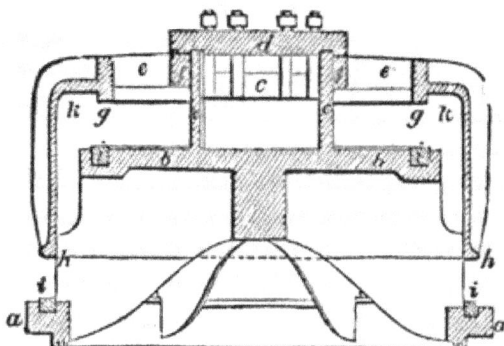

The great advantages of this valve over the old forms of butterfly and conical valves, are the following:—1st, as the area of the valve exposed to the pressure of the column of water, or to the action of the piston upon its return stroke, is considerably less than in any other form of valve, the blow and consequent vibration caused by the shutting of the valves is diminished in proportion, and less costly foundations are therefore required; 2nd, the loss of water upon the shutting down of this valve is considerably diminished.

The testimony of engineers, and all who have ever been concerned in the working of pumps, is universal in favour of these valves, which work without any of the jar, noise, and vibration of the old forms. They open and close quietly, and work for years without appearing to be perpetually trying to destroy and knock themselves to pieces, an idea which the old butterfly valve especially prompted in all who witnessed its performance.

Wood faces were originally used for the valves by the patentees, but these are now frequently superseded by a mixture of tin and lead, forming a composition, which is run into a dovetailed groove in the seat. The valves and the seat are usually of cast iron. At the Birmingham Water Works the pumps are 23 inches in diameter, and each pump is provided with a double system of valves one over the other, in order to

N

give additional security to the action of the pumps, which work under an unusually high pressure. These valves are on Harvey and West's principle. They act so perfectly that the blow when shutting is scarcely perceptible. They were taken out after six months' work, and the beating faces of them were as perfect as when first put in.

Mr. Marten * says, before the introduction of Harvey and West's double beat valves, so great was the objection to the old form of butterfly valve, that the Cornish engine was on the point of being abandoned in despair when first introduced into water works. The concussion caused by the sudden closing of large butterfly valves acting under great pressure, was so severe as to occasion serious alarm for the safety of the machinery and foundations.

The beautiful principle adopted in Harvey and West's valve for regulating the area of the part subjected to the pressure, entirely obviates this serious objection.

Mr. Marten says, for ordinary purposes, as for colliery pumping engines, and where the lifts are small, the butterfly valve is very serviceable and economical; as there are no expensive faces to be ground, the valves are not liable to derangement by grit and other impurities in the water, and they can be readily repaired on the spot.

For a class of work one grade higher than ordinary, Mr. Marten recommends the double beat ring valve, and observes that as large valves from 16 to 20 inches diameter, these work well when made of cast iron with wooden beats. When the valves are smaller, they are better with gun metal beats working face to face.

The following description of a new kind of valve used at the Hull Water Works is given by Mr. Marten in his recent paper :—

"The valve consists of a pyramid of circular seats one above another, on each of which there are a number of small circular beats about 2 inches diameter, into which drop a

* Paper read before the Institution of Mechanical Engineers.

corresponding number of gutta-percha balls. The action of this valve is very simple. It was invented by Mr. William Hosking, and inserted in the place of a double beat valve. It is 22 inches diameter, and works under a head of 160 feet, in connection with a plunger pump, with a direct action steam cylinder. Immediately upon starting, it was found that this valve lightened the burden of the engine about $1\frac{1}{2}$ cwt., and it has since been working with great satisfaction."

This valve is said to act entirely without concussion, and to be almost entirely free from any danger of stoppage or injury by extraneous bodies getting jammed in.

The valve at Hull contains 56 of the small gutta-percha balls, which, being very little heavier than water, are lifted with the greatest ease, and therefore reduce considerably the weight on the engine as compared with any form of metal valve, the latter frequently weighing as much as 5 or 6 cwt. Mr. Marten enumerates many advantages possessed by this form of valve, one of these being the ease with which it can be repaired. Scarcely anything can get out of order, except the gutta-percha balls; and it is only necessary to keep a few spare ones in readiness, while those which are damaged can be warmed and recast in a mould kept for the purpose, when they are again as good as new. This form of valve is to be used for the pumps of the South Staffordshire Water Works (see specification in the Appendix).

ON CALCULATING THE SIZES OR DIMENSIONS OF PUMPS.

The capacity of a pump, unlike that of the cylinder in the engine, is independent of the height to which the water is to be lifted. The dimensions of the pump are regulated simply by the quantity of water to be pumped. Its capacity must be such as to contain the quantity of water to be raised at each stroke of the piston; and hence if we have $Q =$ the gallons to be raised in one minute, and $n =$ the number of strokes per minute, then the capacity of the pump must be equal to $\frac{Q}{n}$ gallons.

Put d = diameter of a pump barrel in inches, l = its length in feet, $\frac{Q}{n}$ = its capacity in gallons. Then as a cubic foot contains 6·25 gallons, we have

$$\frac{d^2 \times \cdot 7854 \times l \times 6 \cdot 25}{144} = \frac{Q}{n}.$$ This being reduced becomes

$$\cdot 03409\, d^2 l = \frac{Q}{n} \quad \cdots \cdots \quad (1)$$

$$\frac{Q}{\cdot 03409\, d^2 n} = l \quad \cdots \cdots \quad (2)$$

and $$\sqrt{\frac{Q}{\cdot 03409\, l\, n}} = d \quad \cdots \cdots \quad (3).$$

On these simple expressions is founded the rule which appears in many elementary works for finding the number of gallons in a yard, a foot, or any other length of pipe of a given diameter.

To find the gallons in a foot of length, take the square of the diameter in inches, strike off one figure, and divide by 3.

To find the gallons in a yard, square the diameter as before, and strike off one figure.

Example: Required the content in gallons of a pipe 15 inches diameter, with lengths of 50 feet and 50 yards.

Here $15 \times 15 = 225$, and $\frac{22 \cdot 5}{3} = 7 \cdot 5$ gallons per foot; then $50 \times 7 \cdot 5 = 375$ gallons; also 22·5 gallons per yard; then $22 \cdot 5 \times 50 = 1125$ in 50 yards.

This rule, it will be observed, is slightly at variance with formula (1). It in fact supposes the factor to be ·033 instead of ·03409. This slight difference is of very little consequence in practice, and in fact the popular rule will agree almost exactly with the formula, if we take the contents of a cubic foot to be 6·23 gallons instead of 6·25, the former quantity being the one commonly assumed by some engineers. It is evident, if we wish to know the *weight* of water in a pump barrel or pipe, we have only to use the factor with the decimal point one place removed, namely, ·3409 instead of ·03409. In the same way, the weight of water in lbs. in a yard length of pipe will, of course, be simply the square of the diameter in inches. The Cornish engine reporters, in calculating the weight of the column of water lifted by their engines, use the factor 2·0454, which is exactly six times ·3409. They then multiply by the lift in fathoms, each of which is, of course, equal to 6 feet.

Required the diameter of a pump with a 10 feet stroke, making 12 strokes per minute, to raise three million gallons in 24 hours.

Here $\dfrac{3,000,000}{1440 \times 12} = 173·6$ gallons per stroke.

Then $\sqrt{\dfrac{173·6}{·03409 \times 10}} = \sqrt{509·2} = 22·6$ inches for the diameter.

The American engineers are in the habit of adding one-third for leakage, so that, according to them, a pump to do the above work would have to be

$$\sqrt{\dfrac{231·5}{·03409 \times 10}} = \sqrt{679} = 26 \text{ inches diameter.}$$

ON CALCULATING THE POWER OF ENGINES.

The work performed by steam engines is commonly expressed in what is termed horse power; that is, an engine is said to be equal to the work performed by a certain number of horses. The standard which has been fixed on to represent the work of one horse, is equal to 33,000 lbs. raised through a space of one foot high in a minute. This is equivalent to saying, that a horse walking at his most effective speed of 2½ miles an hour, or 220 feet per minute, and attached to a weight of 150 lbs. freely suspended over a pulley, will raise this weight at the same rate of 220 feet per minute. Using, then, this standard for computing the work of engines—a standard which has been agreed to by the mechanicians of all countries—we obtain a very ready method of determining the horse power required to raise any given quantity of water to any required height. The data required for this purpose are the quantity to be raised in any given unit of time, and the height to which it is to be raised. The quantity is simply to be reduced to the weight in pounds raised per minute; this weight is to be multiplied by the height in feet, and the product divided by 33,000 in order to find the horse power required to perform the work in question.

A gallon of distilled water, at a temperature of 60° Fahrenheit, weighs exactly 10 ℔s. avoirdupois; so that by adding a cipher to any quantity expressed in gallons, we obtain its weight in pounds. Suppose, now, it be required to find the horse power capable of raising 350 gallons of water per minute to a height of 170 feet. Here we have $350 \times 10 = 3,500$ ℔s. to be lifted per minute, and $3,500 \times 170 = 595,000$ ℔s. lifted one foot high per minute, and $\dfrac{595,000}{33,000} = 18$ horse power.

When the quantity is expressed in gallons to be raised to a given height in 24 hours, it is necessary to divide this quantity by 1,440, in order to bring it into the quantity per minute; and as $33,000 \times 1,440 = 47,520,000$, if we divide the gallons per day of 24 hours by one-tenth of this, or 4,752,000, we obtain the horse power required to lift it.

The table of horse power in the Appendix (Table I.) has been computed in this way, as showing the horse power required to raise any quantity up to 10,000,000 gallons 1 foot high in 24 hours. The first column contains the gallons to be lifted, and the second column gives the horse power required, being simply derived from the first by dividing it by 4,752,000.

The use of this table is so simple as scarcely to demand explanation. Let it be required, for example, to find the horse power necessary to raise 3,550,000 gallons 250 feet high in 24 hours.

Here we have opposite the given quantity $\cdot 7471$: hence, $\cdot 7471 \times 250 = 186 \cdot 8$ horse power.

Suppose the quantity to be raised should not occur immediately in the Table. Let it be required, for instance, to find the power necessary for raising 2,316,500 gallons 234 feet high in 24 hours. Here we have

		Horses.
2,300,000		$\cdot 484$
$16,000 = \frac{1}{100}$ part of $\cdot 3367$		$\cdot 0037$
$500 = \frac{1}{10}$ part of $\cdot 0105$		$\cdot 0010$
		$\cdot 4887$

Then $\cdot 4887 \times 234 = 114$ Horses.

Or the horse power may be derived thus : $2,316,500 \times 234 = 542,061,000$ gallons raised 1 foot high in 24 hours.

			Horses.
Opposite 5 millions is	$1 \cdot 0522$ and this $\times 100 =$	$105 \cdot 22$ for	500,000,000
4 millions is	$\cdot 8418$ „ $\times 10 =$	$8 \cdot 42$ for	40,000,000
2 millions is	$\cdot 4209$	$\cdot 42$ for	2,000,000
50,000 is	$\cdot 0105$	$\cdot 01$ for	61,000
		$114 \cdot 07$ Horse Power	

as before.

Without the use of a table, putting q for the quantity of gallons to be raised in 24 hours, and h the height in feet, we have

$$\frac{qh}{4,752,000} = \text{horses power} \quad . \quad . \quad . \quad (4).$$

When the capacity of engines for waterworks is to be determined according to the horse power, it is not sufficient to take the exact amount of this from calculation, but considerable allowances must be made for the friction of the engine and pumps. Besides the unavoidable friction of machinery, it is also necessary in all engines used at waterworks to provide a considerable amount of surplus power, so that in case of accident and repairs there may be no absolute stoppage of the pumping.

Some engineers are in the habit of doubling the net horse power, and estimating this doubled amount at a fixed price per horse power; others assume the actual friction at one-fourth of the net horse power exerted, and call the united amount the gross horse power. They then divide the gross horse power into two equal parts, and order three engines, each of a power equal to half the gross horse power. According to this mode of viewing the subject, it is assumed that two of the engines—that is, two-thirds of the whole power—will be constantly at work, while the remaining third engine will be in reserve to be used in case of accident or repairs.

It will be seen on examination that these two modes of estimating are not very different in their results. Let P represent the net horse power required by calculation; then, according to the first mode, the gross horse power to be estimated for will be 2P, and according to the second mode it will be $\frac{3}{2}\left(P+\frac{P}{4}\right)=\frac{15}{8}P$, the difference being only the 8th part of the net horse power. For example, suppose the net horse power required were 120 horses; according to one mode of estimating, 240 HP would be specified, and according to the other, $\frac{120+30\times3}{2}=225$ HP would be taken.

It is often convenient, in conveying general and rapid information where minute accuracy is not required, to express the work to be done in millions of gallons raised one foot high. Now, on referring to Table I. in the Appendix, it will be seen that the power required to raise 1,000,000 gallons 1 foot high in 24 hours is ·21 horses. Then, according to the first mode of calculating above alluded to, the gross horse power to be assumed would be ·42 for each million gallons, and according to the other mode it would be $·21\times\frac{15}{8}=·39$.

In general terms, let M represent millions of gallons to be raised in 24 hours, h being the lift in feet, including the friction through the pipes, and let P represent horse power:

Then ·21 h M = P (net)
Also ·42 h M = P (gross) according to first mode,
And ·39 h M = P (gross) according to second mode.

Put c for the coefficient of horse power in the above equations, or the gross horse power required to raise 1,000,000 gallons 1 foot high in 24 hours, also P = gross horse power,

then we have $cMh = P$ and $c = \dfrac{P}{Mh}$.

We shall now examine this coefficient of horse power for several important works, in which the engine power and the quantity pumped are accurately known. And first of the London Companies.

The Southwark and Vauxhall Company, according to their returns in 1865, pumped on the average 6,000,000 gallons a day to a height of 185 feet, and employed for this purpose 355 horse power. Hence, the gross horse power employed to raise each million of gallons 1 foot high is

$$\frac{355}{6 \times 185} = \cdot 32 \text{ horses.}$$

The Grand Junction Company employed a power of 690 horses to raise an average daily quantity of 3·5 million gallons 218 feet high.

$$\text{Here } \frac{690}{3\cdot5 \times 218} = \cdot 90 \text{ horses.}$$

The Chelsea Company employed 221 horse power to raise an average daily quantity of nearly 4,000,000 gallons to a height of 157 feet.

$$\text{Hence } \frac{221}{4 \times 157} = \cdot 35 \text{ horses.}$$

The East London Company employed 568 horse power to raise an average daily quantity of nearly 9,000,000 gallons to a height of 107 feet.

$$\text{Hence } \frac{568}{9 \times 107} = \cdot 59 \text{ horses.}$$

The Lambeth Company employed 222 horse power to raise an average daily quantity of 3,000,000 gallons to a height of 146 feet.

$$\text{Hence } \frac{222}{3 \times 146} = \cdot 51 \text{ horses.}$$

The following Table expresses a summary of the results which have been stated :—

Name of Company.	Horse power employed to raise 1 million galls. 100 feet high.
Southwark and Vauxhall	32
Chelsea	35
Lambeth	51
East London	59
Grand Junction	90
Net horse power required	21
Gross horse power according to the formula 2 P .	42
Gross horse power according to the formula $\frac{15}{8}$ P .	39

It appears that the first two companies on the list have little more than 50 per cent. in excess of the actual net power required exclusive of friction. They each have less than required by either of the formulæ which have been considered above, and are certainly not obnoxious to the charge of having too much surplus power. The Lambeth and the East London have each more surplus power than would probably be adopted for the works of provincial towns, but probably not more than is judicious to provide for the rapid increase of the population which they supply. The Grand Junction is the only company which appears to have a remarkable surplus power.

Looking at the great economy with which engines work when loaded below their full power, and looking also to the constant and rapid increase of the London water companies, it is questionable whether any one of them, except the Grand Junction, could be said to have an extravagant amount of surplus power. Yet we find an Inspector of the Board of Health charging them all in the most wholesale manner with reckless and extravagant expenditure of this kind; and actually attempting to make out that the London water companies in the aggregate employ $4\frac{1}{2}$ times as much engine power as is necessary.

I shall now examine a few instances of recent works of considerable magnitude, in order to show the scale on which the principal engineers of the day have proceeded in fixing the amount of engine power. Here, again, the case of the London companies first presents itself, in their recent application to Parliament for powers to change the site of their works, and take the water from the Thames above the reach of the tide.

At a time when the average daily supply of the Chelsea Company was 5,000,000 gallons a day, Mr. Simpson proposed 600 horse power to raise the water 165 feet high from Seething Wells to a reservoir on Putney Heath.

Hence $\dfrac{600}{165 \times 5} = \cdot 73$ horse power for each million gallons raised 1 foot high.

Mr. Quick, who is engineer for two of the companies at Hampton, proposed 94 horse power for the Grand Junction Works to raise 5,000,000 gallons a day, over a stand-pipe 46 feet high.

He proposed the same power also for the Southwark and Vauxhall Works, in which 8,000,000 gallons a day have to be raised 40 feet high; and for the West Middlesex Company, where the work to be done is nearly the same as for the Southwark and Vauxhall Company, he seems, according to the printed evidence, to have proposed 100 horse power.

Hence the following co-efficients :—

Grand Junction $\dfrac{94}{5 \times 46} = \cdot 41$ $\begin{cases} \text{horse power for each million} \\ \text{gallons raised 1 foot high.} \end{cases}$

Southwark and Vauxhall $\dfrac{94}{8 \times 40} = \cdot 29$ ditto

West Middlesex $\dfrac{100}{8 \times 54} = \cdot 23$ ditto.

ON THE MODE OF CALCULATING THE DIMENSIONS OF
ENGINES REQUIRED TO PERFORM A GIVEN AMOUNT
OF WORK.

This method of calculation is far preferable to that which
simply determines the horse power of an engine, and leaves it
to the maker or contractor to furnish an engine capable of
exerting this power either nominally or really. The calculation of engines according to horse power has led to so many
errors and is capable of so much misinterpretation that it will
be well to abolish it in all commercial transactions of importance.

The principal points required to determine the dimensions
and capacity of a pumping engine are the mean effective pressure of steam in the cylinder, the length of stroke, the number
of strokes per minute, and the diameter of the cylinder.
Amongst the Cornish engineers, and amongst the makers of
their celebrated engines, the former particulars are so well
understood, so generally settled and agreed upon, that the
diameter of the cylinder alone represents with tolerable accuracy the power of the engine.

It seems to be generally agreed amongst the Cornish engineers, that their engines may be made to work with a maximum effective pressure of from 15 to 17 lbs. per square inch
of the piston, and with a velocity of 200 to 240 feet per
minute; but as the engines are commonly single acting, only
half of this velocity is effective.

For example, in "Brown's Engine Reporter," in which not
less than 24 single pumping engines are reported every month,
the following data are assumed with respect to the engines :—

Those with cylinders under 30 inches are assumed capable
of working with a load of 18 lbs. on each square inch of the
piston. Those with cylinders from 30 to 45 inches with a
load of 17 lbs.; between 45 and 60 inch cylinders, with a load
of 16 lbs.; and above 60 inches with a load of 15 lbs. In
making up the horse power the pistons of all single acting
engines are calculated to move with an actual velocity of 220
feet per minute, or an effective velocity of 110 feet per

minute. Tables 2 to 5 in the Appendix show the power of Cornish engines calculated on these data, from 15 inches up to 100 inches diameter of cylinder.

Most of the Cornish pumping engines are single acting; but the double acting engines used in Cornwall for raising the kibbles are usually calculated to work under a load of 10 lbs. per square inch, and the piston is assumed to have an actual and effective velocity of 250 feet per minute.

Mr. John Darlington, the author of a valuable paper which appeared in the first number of the "Engineering Journal," assumes the initial pressure of steam on entering the cylinder of the Cornish engines at 30 lbs. per square inch. He assumes it to be cut off at one-fourth of the stroke, and to have a mean pressure of 17·8 lbs. per square inch. From this he deducts one-fifth for friction, and takes the remaining 14·24 lbs. to represent the effective pressure of the steam. He assumes the same effective pressure for all engines from 15 inch cylinders up to 100 inches. He assumes the length of stroke to be 8 feet in the small sized engines from 15 inch up to 19 inch cylinders, and to be 12 feet in the largest size from 85 to 100 inch. In the same way also he takes the effective velocity (the length of stroke multiplied by the number of double strokes per minute), to vary from 112 to 96 feet per minute, the smallest size having the highest velocity, and the largest having the lowest. These velocities are what Mr. Darlington terms *safe* rates of working, but in his table he gives another column showing the most economical rate of working, and this economical rate is commonly less than *half* of that which is assigned as the safe rate. This is only in accordance with well-established facts—with the opinions also of all practical men, and is borne out by the daily working of the pumping engines in Cornwall.

In proceeding to determine the dimensions of an engine according to the Cornish method, the first and principal thing necessary is to produce an equation between the work to be done in a unit of time and the power of the engine in that same unit. Thus, if we take the work to be done in lbs. raised one foot high per minute, the whole pressure of steam on the piston multiplied by the effective velocity of the piston per minute must be equal to this work.

Putting w for the work to be done in lbs. raised one foot high per minute, $P =$ the whole pressure of steam on the piston in lbs., $v =$ its velocity in feet per minute, then must $w = P\,v$.

It will be most convenient, however, to subdivide P into the two factors which evidently compose it, namely, the area of the piston or cylinder and the pressure p per square inch. We have then the expression $w = a\,p\,v$.

Engineers adopt different modes of calculating the elements or parts of this equation. w, of course, is a determined or given quantity, while p and v are usually assumed at what are known by experience to be reasonable and proper.

Thus if w be the whole work to be done exclusive of friction it will be quite safe to assume, as many of the Cornish engineers do, that p may be 14 or 15 lbs. and v may be 80 or 85. Either of these assumptions will enable us readily to derive a, the area of a cylinder which will do the required amount of work.

Suppose a Cornish engineer prescribes an engine with a cylinder area $= a$ to perform a given quantity of work w, we know immediately by the expression $\dfrac{w}{a} = p\,v$ what he has assumed for the effective pressure of steam multiplied by the velocity of the piston. We know, in fact, what value he has assumed for $p\,v$, and though this may vary slightly among different engineers, there is still a very fair and general uniformity of opinion on the subject. Some may put p a little higher than others, and v a little less, but the product $p\,v$ is usually about the same among the different Cornish engineers.

The following comparison will explain this more clearly:— Three celebrated Cornish engineers were requested to specify separately the kind of engine they would recommend to perform a given quantity of work. They were Mr. William West of St. Blazey, Mr. Samuel Hocking of John Street, Adelphi, and Messrs. Harvey and Co. of Hayle foundry.

Name of Engineer.	Question proposed. Q.	Power required to be exerted per minute in millions raised 1 foot high = W.	Diameter of Cylinder Proposed.	Area of Cylinder = a	Product of steam pressure into effective velocity or value of $\frac{W}{a}$.
Mr. W. West	1½ million gallons raised 314 feet high in 24 hours.	3,270,833	65	3318	986
Mr. Samuel Hocking	Ditto	3,270,833	60	2827	1157
Messrs. Harvey & Co.	1½ million gallons raised 306 feet high in 12 hours.	6,375,000	84	5542	1150
Mr. W. West	1½ million gallons raised 163 feet high in 24 hours.	1,697,917	50	1963	865
Mr. Samuel Hocking	Ditto	1,697,917	45	1590	1068
Messrs. Harvey & Co.	1½ million gallons raised 180 feet high in 12 hours.	3,750,000	64	3217	1166
Ditto	1½ million gallons raised 140 feet high in 12 hours.	2,916,667	58	2642	1104

It appears from the preceding Table, that we shall follow the practice of the most eminent Cornish engineers, in adopting a value varying from 850 to 1200 for the denominator $v\,p$ in the expression $\dfrac{W}{v\,p}=a.$

The following Table shows the value of $v\,p$ in the actual work of the 15 pumping engines, for which particulars are given with sufficient detail in "Brown's Engine Reporter," for 1855

TABLE SHOWING THE VALUE OF vp, FROM THE ACTUAL WORKING OF FIFTEEN CORNISH ENGINES IN 1855.

No.	Name of Engine or Mine.	Average number of strokes per minute.	Length of stroke in feet. ft.	in.	Velocity of piston in feet per minute or value of v.*	Load per square inch on piston $= p$.	Value of vp or continued product of number of strokes per minute, by length of stroke, by load in lbs. per square inch.	Average duty for the year, expressed in millions of lbs. raised 1 foot high by consuming 1 cwt. of coal.
1	Boscundle	6·9	9	0	62·2	13·4	833	52·9
2	Gt. Dev. Consols	3·6	9	0	32·0	29·4	941	49·2
3	Fowey Consols	6·6	10	4	67·7	13·8	934	98·6
4	Great Polgooth	6·3	10	0	63·0	17·7	1115	93·9
5	Ditto	6·4	10	0	63·6	11·4	725	70·1
6	Mary Ann	6·2	7	0	43·7	18·2	795	42·4
7	Par Consols	4·8	12	0	57·3	13·6	779	98·5
8	Ditto	5·2	9	3	48·3	11·9	575	76·2
9	Pembroke	2·9	10	6	30·4	17·6	535	76·7
10	Ditto	4·9	10	0	49·5	19·1	945	76·2
11	South Caradon	6·4	9	0	57·7	12·0	692	49·1
12	Ditto	5·5	9	0	49·3	13·4	661	48·2
13	Trelawny	4·7	10	0	47·1	17·3	815	67·4
14	West Fowey	3·8	10	0	37·8	12·1	457	70·4
15	Wheal Uny	4·1	11	0	44·7	17·0	760	71·3
							Average 771	

* The velocities in this column are the average monthly velocities, and are not always the product of the two first columns, because the average number of strokes is only carried to one place of decimals. The velocities, however, are strictly correct.

It is worthy of observation, that this Table of actual performance does not bear out the idea, that the most duty is invariably performed by engines, which work at the lowest rate of speed, and with the smallest pressure. Of those engines in which $v\,p$ is below the average, only one reaches a duty of 77 millions; while in Nos. 6, 11, and 12, in which $v\,p$ is very small, the duty is also lower than in any of the other engines. In those engines where the duty is highest, as in No. 3, No. 4, and No. 7, which are all 80 inch cylinder engines, the respective values of $v\,p$ are 934, 1115, and 779, or considerably above the average.*

Taking into consideration the practice of the most eminent Cornish engineers, in connection with the results exhibited in this Table, I venture to propose the value $v\,p=1000$ as a constant, in estimating the size of the cylinder for large Cornish engines to be used in waterworks. We shall then have the very simple expression

$$\frac{W}{1000}=a,\text{ the area of the cylinder.}$$

For example, let it be required to find the diameter of cylinder which a Cornish engine should have to raise 3,500,000 gallons 120 feet high in 24 hours.

Here $\dfrac{3,500,000 \times 120}{144}=2,916,667=W.$ ℔s. raised 1 foot high per minute.

Then $\dfrac{2,916,667}{1000}=2917$ area of cylinder in inches $=61$ inches nearly for the diameter of the cylinder.

* In the preceding table it should be observed that the value of p is derived, not from the actual pressure of steam on the piston, but from the actual water load divided by the area of the piston. It follows that the actual pressure of steam must be somewhat more than this, because it has to overcome all the friction of the pumps and parts of the engine, besides raising the actual water load. If, in order to compare the value of $v\,p$ as determined from actual working with that assumed in calculations of power, we add to $v\,p$ in the table $\frac{1}{4}$ of its amount for friction, we shall find the agreement remarkably close.

We have seen that various Cornish engineers assume dif
ferent values for $v\,p$. Thus, Mr. West appears to assume
about 926; Mr. Hocking 1113; and Messrs. Harvey 1140.
According to the practice of calculation followed by each of
these, the engine to do the above work would be determined
thus—

			Area of cylinder.		Diameter of cylinder in inches.
By Mr. West	$\dfrac{2,916,667}{926}$	=	3150	=	$63\frac{1}{4}$
Mr. W. Hocking	$\dfrac{2,916,667}{1113}$	=	2621	=	$57\frac{1}{4}$
Messrs. Harvey	$\dfrac{2,916,667}{1140}$	=	2558	=	57

In the Appendix (Table 6) will be found Mr. Darlington's
Table of horse power, which has been before alluded to. The
pressure of steam which he assumes, corresponding with that
which we have used in speaking of the actual working of Cornish engines, is 14·24 ℔s. per square inch. The Table is
valuable, because it shows in a very simple manner the proportion between the horse power at safe working speed, and
at the economical rate of working.

Mr. Darlington's value for p being constant, namely 14·24,
that of v is variable, being 80 feet per minute for the smaller
sized engines, and increasing up to 96 for those of the largest
size.

Taking a velocity of 80 feet for engines up to 60 inch cylinder, Mr. Darlington's value for $v\,p$ would be $14·24 \times 80 = 1139$.
Taking a velocity of 84 feet for engines from 60 to 70 inches,
the value of $v\,p$ would be $14·24 \times 84 = 1196$. Taking a velocity of 88 feet for engines between 70 and 80 inches, the value
of $v\,p$ would be $14·24 \times 88 = 1253$. Taking a velocity of
92 feet for engines between 80 and 85 inches, the value of $v\,p$
would be $14·24 \times 92 = 1310$. Finally, taking a velocity of 96
feet per minute for engines with cylinders between 85 and 100
inches, the value of $v\,p$ would $14·24 \times 96 = 1367$.

These latter values appear to be greater than those which

obtain in practice, and are not in accordance with those of other engineers. It will readily be seen, that in adopting Mr. Darlington's values, we should fix an engine of smaller size than by using any of the constants before given. On the whole, there appears no reason to vary the opinion already expressed in favour of the expression

$$\frac{W}{1000} = \text{area of cylinder.}$$

STEAM WORKED EXPANSIVELY.

When steam is admitted throughout the whole of the stroke, its efficiency is of course equal to that of the uniform pressure or unity. When cut off at any part of the stroke as $\frac{1}{n}$ its efficiency is equal to $1 \times$ Hyp. Log. n. Hence the following table, showing the efficiency of steam at different degrees of expansion :—

Steam * admitted throughout the stroke					1·000
„ cut off at $\frac{3}{4}$	.	.	.	.	1·287
„ „ $\frac{2}{3}$	.	.	.	.	1·405
„ „ $\frac{1}{2}$	.	.	.	.	1·693
„ „ $\frac{1}{3}$	.	.	.	.	2·099
„ „ $\frac{1}{4}$	.	.	.	.	2·386
„ „ $\frac{1}{5}$	.	.	.	.	2·609
„ „ $\frac{1}{6}$	.	.	.	.	2·792
„ „ $\frac{1}{7}$	.	.	.	.	2·946
„ „ $\frac{1}{8}$	.	.	.	.	3,079
„ „ $\frac{1}{9}$	.	.	.	.	3.197
„ „ $\frac{1}{10}$	.	.	.	.	3·303

Now, if steam be admitted with any given pressure p, and be cut off at the nth part of the stroke, it will have an equivalent pressure throughout the whole stroke $= p \times \dfrac{1}{n} \times \overline{1 + \text{Hyp. Log. } n}$. Suppose steam of 35 lbs. pressure cut off

* The greater part of this scale is taken from Lean's Historical State-ment of the Steam Engines in Cornwall, a work which has been before referred to.

at $\frac{1}{7}$ part of the stroke. Then we have $\frac{35}{7} \times 2\cdot946 = 14\cdot7$ for the mean effective pressure throughout the stroke.

The pumping engines used in the American waterworks are usually double acting engines, and they are commonly calculated to work with an effective mean pressure of about 14 lbs. per square inch. Thus, in several engines designed by Mr. McAlpine for Brooklyn, Albany, Chicago, and other works, the pressure of the steam is thus estimated—

	lbs. per square inch.
Pressure in boiler	30
Do. when admitted to cylinder . . .	20

This being cut off at $\frac{1}{4}$ of the stroke, we have (by Table in page 284) $\frac{20}{4} \times 2\cdot386 = 11\cdot93$ lbs. per square inch for the mean pressure on the piston throughout the stroke.

To this must be added, say, 9·5 lbs. per square inch for the additional pressure due to the vacuum produced by condensation. From the total pressure so derived the American engineers deduct one-fifth the initial pressure of the steam, or 4 lbs. per square inch for the friction of the engine, and one-half this quantity, or 2 lbs. per square inch, for the friction of the air-pump piston. Hence the effective pressure will be derived as follows :—

	lbs. per square inch.
Mean pressure of steam at 20 lbs. per inch, cut off at $\frac{1}{4}$ of the stroke	11·93
Addition for vacuum	9·5
	21·43
Less for friction of engine $\frac{20}{5} - 4$ lbs.	
,, ,, air-pump 2 . .	6·
Mean effective pressure	15·43

or nearly $15\frac{1}{2}$ lbs. per square inch.

This mean effective pressure is then to be multiplied by the effective velocity of the piston, which, in the double acting engines, is frequently as much as 300 feet per minute, and from this is deducted the actual resistance of the air-pump, which is obtained by multiplying the area of the air-pump by 9·5, the vacuum pressure as before, and by the velocity of the air-pump piston. This last deduction reduces the actual effective pressure to about 14 lbs. per square inch, which is a pressure commonly assumed by the American engineers in their calculations.

The mode of calculation will, perhaps, be better understood oy putting the equation into the algebraical form given by Haswell :—

Let P be the mean effective pressure on the whole area of the piston, due to the expansive action of the steam usually assumed at 12 lbs. per inch.

v = pressure on the whole area of piston due to the vacuum produced by condensation usually assumed at 9·5 lbs. per square inch.

f = pressure on the whole area of piston necessary to overcome the friction of the engine usually = 4 lbs. per inch.

m = pressure on the whole area of piston necessary to overcome the friction of the air-pump usually = 2 lbs. per square inch.

S = velocity of steam cylinder piston, usually from 240 to 300 feet per minute for double acting engines.

n = velocity of air-pump piston, usually assumed at 80 to 100 feet per minute.

b = resistance of the vacuum against the air-pump piston = area of air-pump piston × 9·5.

W = weight in lbs. lifted one foot high per minute, then

$$\overline{S\ (P + v)\ (f + m)} - n\,b = W.$$

EXAMPLES.—The condensing engine proposed for Chicago waterworks has the following dimensions.

Diameter of steam cylinder, 46 inches.

Length of stroke, 9 feet.

Effective velocity of piston, S = 240 feet per minute.

Pressure of steam on entering cylinder, 20 lbs. per square inch.

Steam cut off at ¼ of stroke.

Pressure due to vacuum v = 9·5 lbs. per square inch.

Diameter of air-pump, 34 inches.

Also value of f = 4 lbs. per square inch.

    ,, m = 2 lbs. per square inch.

    ,, n = 80 feet per minute.

Then S= 240

      P=1662 × 12 = 19944

      v=1662 × 9·5= 15789

                35733

Also f = 1662 × 4 = 6648

     m = 1662 × 2 = 3324

              9972

25761 Total pressure acting
on piston.

Then 25761 × 240 = 6,182,640 lbs.

Less area of air-pump

34 in. = 907·92 × 80 × 9·5   690,019

                                5,492,621

The work to be done by this engine is equivalent to raising three million gallons 90 feet high in twelve hours =
$\frac{3,000,000 \times 90}{72}$ = 3,750,000 lbs. raised 1 foot high per minute. To this Mr. McAlpine adds one-fifth for the friction of the pumps and machinery, making $4\frac{1}{2}$ millions of pounds. The friction of the water in passing through the pumping main will increase the duty on the engine to $5\frac{1}{4}$ million pounds, raised 1 foot high per minute. Now, a single acting Cornish engine, according to the formula $\frac{W}{1000} = a$, must have a cylinder equal in area to $\frac{5,250,000}{1000}$ = 5250 = a diameter of 82 inches. We have seen that the American engineers adopt a double acting engine, with a 46-inch cylinder, having an area of only 1662 inches, or less than $\frac{1}{3}$ that of the single acting engine.

At the Albany works the engine is required to raise in 12 hours 2 million New York gallons (of 8 lbs. each) to a height of 156 feet, and 1 million gallons to a height of 238 feet.

Hence W =   6,111,111

To this add for friction in main   .   .   .   733,270

Also $\frac{1}{5}$ for friction of pumps and machinery . 1,222,222

                                          8,066,603 lbs.

to be raised one foot high per minute.

For this Mr. McAlpine proposes a steam cylinder of 58 inches and 12 feet stroke, the piston making 10 strokes per minute, and having an effective velocity of 240 feet per minute.

The assumed pressure of steam, the vacuum, &c., being the same as in the Chicago engine, the power calculated by the same formula $S\overline{(P + v)} - (f + m) - nb$ is equal to 8,667,600 lbs.

Now to do this work, a single acting Cornish engine, computed as before, would require a cylinder with an area = 8667 inches = a diameter of 105 inches, or considerably more than 3 times the area adopted for the American engine.

In the Brooklyn works the duty required is much greater than in either of the preceding, being equal to the elevation of 5 million gallons in 12 hours, through a main 6,000 feet long and 30 inches diameter, to a height of 190 feet. This duty, including the friction of the machinery, and that of the water passing through the rising main, is equal to 17 millions of pounds, raised one foot high per minute, or about 515 horse power.

To effect this work Mr. McAlpine proposes an engine with a steam cylinder of 72 inches diameter, and 12 feet stroke, working steam at 20 lbs. pressure, to be cut off as in the other cases at $\frac{1}{4}$ of the stroke. The effective velocity of the piston or value of S in Haswell's equation, is here 288 feet per minute. The diameter of the air-pump is 48 inches, and the velocity of the air-pump piston is 96 feet per minute, all the other values being

the same as those which are described for the Chicago
engine. The power of this 72 inch cylinder engine, when
computed by the formula $S \overline{(P + v)} - (f + m) - nb$, is equal
to 16,524,912 lbs. raised one foot high per minute.

Now, if this work were to be done by Cornish single acting
engines, it would require 2 engines, each with cylinders exceeding 100 inches in diameter, whereas the American engineer proposes only one engine of 72 inches.

We have seen, that in the practice of the most eminent
Cornish engineers, the value of $v\,p$, or the product of effective
velocity of piston by mean pressure of steam, is equal to 1,000.
If the American engines were simply double acting, all other
things remaining the same, the value of $v\,p$ would of course
be double, or 2,000. This, however, is not so, because the
actual velocity of the piston is considerably greater.

If we take in each case the value of W or work to be done in
one minute, and divide it by the area assumed for the cylinders
of the American engines, we shall have the value of $v\,p$ for
the purpose of comparison with the single acting Cornish
engines.

$$\text{Thus at Chicago} \quad \frac{5,250,000}{1,662} = 3,159$$

$$\text{,, at Albany} \quad \frac{8,066,603}{2,642} = 3,053$$

$$\text{,, at Brooklyn} \quad \frac{17,000,000}{4,071.5} = 4,174$$

Thus the value of $v\,p$ being in the American engines from 3
to 4 times as great as in the Cornish single acting engines,
and the area of the cylinder requiring to be inversely as $v\,p$,
it follows that the American double acting engines require
cylinders only one-third or one-fourth the area of the single
acting engines.

The mode in which the American engineers provide for the
auxiliary or surplus power has been already alluded to. The
double acting condensing engine is designed with sufficient
power to do the whole work in 12 hours, and in addition, a

high pressure non-condensing engine is erected, capable of doing the whole work in 24 hours, and sometimes in 20 hours.

In order to make the comparison complete between the American system and our own, I shall briefly notice the non-condensing engines proposed in the three works which have been taken as examples, namely those of Chicago, Albany, and Brooklyn.

The work to be done by the non-condensing engine at Chicago is that of raising 3 million gallons 90 feet high in 24 hours. When the friction of the pumps and machinery is added to this, and also the friction of the water passing through the pumping main, the work to be done is equivalent to raising 2,600,000 lbs. one foot high per minute, or W = 2,600,000.

To effect this, an engine is proposed with an 18-inch cylinder and 6 feet stroke, with a piston travelling 240 feet per minute, using steam at 80 lbs. pressure per square inch at the cylinder, and cut off at one-fourth of the stroke.

According to the table at page 284, the mean effective pressure or value of p will be $\dfrac{80}{4} \times 2\cdot386 = 47\cdot7$ lbs. per square inch.

The value of P will be $254 \times 47\cdot7 = 12116$.

Then $12116 \times 240 = v$ P the whole power $= 2,907,840$ lbs., or about 300,000 lbs. in excess of the power actually required.

At Albany the work to be done is equivalent to raising 1,600,000 imperial gallons 156 feet high, and 800,000 gallons 238 feet high in 20 hours;

$$\text{Then } \frac{1,600,000 \times 156}{120} = 2,080,000$$

$$\text{And } \frac{800,000 \times 238}{120} = 1,586,667$$

$$W = 3,666,667$$

To this is added the same amount for friction of water in mains, in

pumps, machinery, &c., as for the
condensing engine, namely . . . 1,955,492

Total 5,622,159 lbs. raised one
foot high per minute.

The engine proposed for this has a steam cylinder of 24
inches' diameter and 6 feet stroke, making 20 strokes per
minute, and working steam of 70 lbs. pressure to be cut off at
$\frac{3}{4}$ stroke.

$$\text{Here } v = 240$$
$$a = 452$$

$p = 70 \times \frac{3}{4} \times 1\cdot287 = 67\cdot57$ lbs. for the effective mean
pressure.

Then $67\cdot57 \times 452 \times 240 = 7,329,994$ lbs. raised one foot
high per minute for the power of the engine, which is considerably in excess of the power actually required.

At Brooklyn, where the work to be performed by the non-
condensing auxiliary engine is equal to raising 5 million
gallons 90 feet high in 24 hours, the power required, including the friction of machinery and the friction of the water
passing through the mains, is equal to 8,067,476 lbs. raised
one foot high per minute. For this work an engine is proposed with a 30 inch cylinder and 6 feet stroke, using steam of
80 lbs. pressure at the piston, with an average effective pressure
of 48 lbs. and an effective velocity of piston = 240 feet per
minute. Hence the power of the engine will be

Area of Cylinder.	Pressure per inch.	Velocity.	
706·86	× 48	× 240	= 8,143,027 lbs.

The following table gives at one view the particulars relating to the American non-condensing engines which have
been noticed in the preceding pages :—

Name of Work.	Diam of Cylinder ins.	Velocity of piston in feet per minute.	Pressure of Steam at Piston. lbs.	Part of Stroke at which Steam is cut off.	Average pressure of Steam.	Value of $v\,p$.
Chicago	18	240	80	$\frac{1}{4}$	48	11520
Albany	24	240	70	$\frac{3}{4}$	67·6	16224
Brooklyn	30	240	80	$\frac{1}{4}$	48	11520

The Chicago and Brooklyn are more recent works than the Albany, and may be taken to represent the most modern practice amongst the American engineers.

ON THE COST OF ENGINES FOR PUMPING PURPOSES.

This is often estimated at a price per horse power of actual work to be performed, but the practice has been productive of serious errors and misunderstandings.

Engineers have been heard gravely calculating the net horse power, at £50 per horse for condensing engines, a sum which is far too low, and which in reality represents about the value of each horse power of the gross instead of the net amount.

The following table presents some examples of the cost of engines according to horse power of actual work to be performed. The unit of work assumed here is the usual one, commonly called Watt's standard, namely 33,000 lbs. raised one foot high per minute = 1 horse power. The last column contains the cost per net horse power reduced to this standard. The price of the engine in each case includes the boilers and pumps, and where no note occurs to the contrary, it also includes the erection of engine-house, boiler-house, and all necessary masons' and bricklayers' work.

Name of Work, Engineer who Estimates, and description of Engine.	Work to be done.	Millions of Gallons raised 1 ft. high in 24 hours.	Cost.	Horse-power, or actual work to be done in units of 33,000 lbs. raised 1 foot high per minute.	Cost per actual net horse-power, or cost for each unit of 33,000 lbs. raised 1 foot high per minute.
			£		£
Mr. Quick's estimate for Grand Junction Works, Evidence, 1851—either one 64-inch or two 45-inch cylinder single-acting Cornish engines.	5 million galls. raised 46 ft. high	230	7000	48·40	144·6
Ditto in Southwark and Vauxhall	8 million galls. raised 40 ft. high	320	7000	67·34	103·9
Ditto for West Middlesex	5 million galls. raised 46 ft. high	230	10,000	48·40	206·6
Homersham, in Watford supply	8 million galls. raised 174 ft. high	1392	35,357	292·92	120·7
Daglish's estimate for Liverpool works —one engine and pumps, exclusive of buildings.	2 million galls. raised 300 ft. high	600	8470	126·26	67·1
Daglish's estimate for two steam engines and pumps, exclusive of buildings.	2 million galls. raised 300 ft. high	600	9850	126·26	78·0
Ditto for one steam engine and pump, exclusive of buildings.	1 million galls. raised 75 ft. high	75	3400	15·78	215·4
Sandys, Vivian, & Co.'s estimate for Liverpool engines—for two engines, exclusive of buildings.	2 million galls. raised 300 ft. high	600	8000	126·26	63·4
Sandys, Vivian, & Co.'s estimate for one engine, exclusive of buildings.	2 million galls. raised 300 ft. high	600	7000	126·26	55·4
Ditto for one engine. exclusive of buildings.	1 million galls. raised 75 ft. high	75	1800	15·78	114·1

Name of Work, Engineer who Estimates, and description of Engine.	Work to be done.	Millions of Gallons raised 1 ft. high in 24 hours.	Cost.	Horse-power, or actual work to be done in units of 33,000 lbs. raised 1 foot high per minute.	Cost per actual net horse-power or cost for each unit of 33,000 lbs. raised 1 foot high per minute.
			£		£
Forester & Co.'s estimate for Liverpool engines—for either one or two, engines, exclusive of buildings.	2 million galls. raised 300 ft. high	600	9700	126·26	76·8
Ditto for either one or two engines, exclusive of buildings.	1 million galls. raised 75 ft. high	75	3200	15·78	202·8
Hurry & Bateman's estimate, Wolverhampton Waterworks.	1½ million galls. raised 314 feet high, and 4½ million gallons raised 264 feet	471 } 1188	35,400	349·11	101·4
Seward & Capel's estimate for Wolverhampton Corporation Waterworks, exclusive of buildings.	1½ million galls. raised 566 ft. . .	849	22,000 to 25,000	178·65	{ 123·2 139·9
Mr. William West's estimate for Wolverhampton Corporation Waterworks, including buildings.	1½ million galls. raised 314 ft. by a 65-inch cylinder engine.	471	8400	99·11	84·8
Mr. William West's estimate for the same, exclusive of buildings.	1½ million galls. raised 314 ft. by a 65-inch cylinder engine.	471	6000	99·11	60·5
Mr. William West's estimate for Wolverhampton Corporation Waterworks.	1½ million galls. raised 163 ft. by a 50-inch cylinder engine.	244½	7100	51·45	138·0

Ditto exclusive of buildings . . .	1½ million galls. raised 163 ft. by a 50-inch cylinder engine.	244½	5000	51·45	97·2
Mr. William West's estimate for Wolverhampton Corporation Water-works.	1½ million galls. raised 22 ft. high in 6 hours, by a 35-inch cylin-der engine.	132	4600	27.78	165·5
Ditto, exclusive of buildings . . .	1½ million galls. raised 22 ft. high in 6 hours, by a 35-inch cylin-der engine.	132	3100	27.78	111·6
Mr. William West's estimate for Wolverhampton Corporation Water-works, including duplicate engines.	1½ million galls. raised 566 ft. in 24 hours.	849	38,200	158·66	240·8
Harvey & West's estimates for Wol-verhampton Corporation Water-works, exclusive of buildings.	1½ million galls. raised 306 ft. by an 84-inch cylinder engine.	459	11,000	96·59	113·9
Ditto, ditto	1½ million galls. raised 180 ft. by a 64-inch cylinder engine.	270	7000	56·82	123·2
Ditto, ditto	1½ million galls. raised 140 ft. by a 58-inch cylinder engine.	210	6500	44·19	147·1
Messrs. Hawthorn's (of Newcastle) estimate for double-power expansive condensing beam engines for Wol-verhampton Corporation Water-works, exclusive of buildings.	1½ million galls. raised 314 ft. in 24 hours.	471	6055	99·11	61·1
Ditto	1½ million galls. raised 163 ft. high in 24 hours.	244½	2520	51·45	48·9
Messrs. Hawthorn's estimate for high pressure double acting horizontal engines.	1½ million galls. raised 22 ft. high in 6 hours.	132	1170	27.78	42·1

Name of Work, Engineer who Estimates, and description of Engine.	Work to be done.	Millions of Gallons raised 1 ft. high in 24 hours.	Cost.	Horse-power, or actual work to be done in units of 33,000 lbs. raised 1 foot high per minute.	Cost per actual net horse-power, or cost for each unit of 33,000 lbs. raised 1 foot high per minute.
			£		£
Messrs. Hawthorn's estimate for high pressure double acting horizontal engines.	1½ million galls. raised 22 ft. high in 9 hours.	108	780	22·73	34·3
Ditto	1½ million galls. raised 22 ft. high in 12 hours.	66	570	13·89	41·0
Mr. Hocking's estimate for Wolverhampton Corporation Works (single acting Cornish engines).	1½ million galls. raised 500 ft. by one 60-inch cylinder, one 45-inch, one 36-inch.	750	20,800	157·83	131·8
Ditto, exclusive of buildings.	1½ million galls. raised 500 ft. .	750	13,200	157·83	83·6
Mr. Hocking's estimate for the same works.	1½ million galls. raised 500 ft. high in 12 hours by two 60-inch cylinders, two 45-inch ditto, and one 36-inch ditto.	1500	32,600	315·65	103·3
Ditto, exclusive of buildings . . .	Ditto	1500	21,800	315·65	69·0
American Works.					
Mr. McAlpine's estimate for double acting expansive high pressure condensing engines for Brooklyn Works, exclusive of buildings.	5 million galls. raised 190 ft. high in 12 hours by 72-inch cylinder engine.	1950	18,000	399·83	45·0
Ditto for non-condensing engine and pumps to perform the same work by working 24 hours, exclusive of buildings.	5 million galls. raised 190 ft. high in 24 hours by 30-inch cylinder non-condensing engine.	1035	3000	199·92	15·0

			£		£
Ditto for the whole engine power required to raise 10 million gallons a day, and including a surplus of non-condensing engine power, exclusive of buildings.	10 million galls. raised 190 ft. high in 12 hours.	3800	39,000	799·66	48·7
The same for a supply of 20 million gallons a day, exclusive of buildings	20 million galls. raised 190 ft. high.	7600	53,000	1599·23	33·1
The same for a supply of 30 million gallons a day, exclusive of buildings	30 million galls. raised 190 ft. high.	11,400	85,000	2398·99	35·4
Mr. McAlpine's estimate for the Albany Water Works (engines to work only in the day time), including every expense of land and buildings	2 million galls. raised 156 ft. · 1 million galls. raised 238 ft. · One 58-inch condensing beam engine, and one duplicate not condensing.	624 } 476 }	13,320	231·48	57·5
Mr. McAlpine's estimate for the Chicago Works (engines to work only 12 hours), including, as before, every expense of land and buildings.	3 million galls. raised 107 ft. high by 46-inch cylinder condensing engine, and duplicate non-condensing engine, with 18-inch cylinder.	642	11,258	135·10	83·3
Same works	3 million galls. raised 116 ft. high (same kind of engine, but with 48-inch cylinder).	696	11,642	146·46	79·5

The statement of work to be done by the American engines is in each case somewhat understated, as it is derived from the actual height to which the water is to be lifted without any addition for friction through the pumping main.

It follows that the price per horse power is in each case somewhat higher than it would be if based upon the full amount of work including this friction, as in the estimates quoted from the Cornish engineers. Notwithstanding this, however, the American estimate is in every instance considerably below the price of single acting Cornish engines.

The highest American estimate is that for the Chicago works, where the horse power is 135 horses, and where the cost of entire engine power, including duplicate engine, land, and buildings, is only £83 per horse power.

For other works, the cost per horse power appears to diminish nearly as the magnitude of the work increases, till we have for the large pumping power at Brooklyn, only from £33 to £35 per horse power.

This great difference of price, as against the single acting Cornish engines (see first part of the Table), is in some measure due to the engines being made double acting, and thus requiring much smaller cylinders, and a proportionate reduction of other parts ; and also, in some measure, to the use of the non-condensing engines as auxiliary power. We feel bound also to call attention to the fact, that Messrs. Hawthorn's double acting expansive engines will bear comparison even with the cheapest of the American.

Experiments and recorded observations are still required as to the working and actual duty of these double acting expansive engines. Their first cost is certainly much less than that of the single acting Cornish engines; and unless they are more expensive to work—in other words, unless they perform less duty—they ought to be preferred to the single acting engine. The Table here given, and the remarks here made, are not ventured as conclusive or satisfactory even to the mind of the author; but are merely given to draw attention to a subject of great importance, and one which is daily becoming of more moment with reference to the supply of towns having moderate command of capital. In new waterworks it should of course be the aim of the engineer to effect as much as possible with the means at his disposal. It is dangerous to be led away by theories in favour of any particular kind of engine; but the whole subject, on the other hand, requires the exercise of calm and deliberate judgment, based on the most accurate information which can be procured.

As an example of large double acting pumping engines in this country, a specification is given in the Appendix, of the engines now being erected by Messrs. Boulton and Watt, for the South Staffordshire Works, under the direction of Messrs. M'Clean and Stileman, of London, and Messrs. Marten, of Wolverhampton, Civil Engineers.

Opinions are much divided, as to the comparative merits of beam and direct acting engines for pumping purposes. The single acting Bull engine, with the top of the cylinder closed, the piston-rod working through a stuffing-box in the bottom, and the steam admitted also at the bottom, was introduced many years ago in Cornwall, and several engines on this plan have been erected for American waterworks, and more recently by three of the London companies for their new works at Hampton. The advocates of the direct acting engine contend that, the cylinder being placed immediately over the pumping well, and the piston-rod being in fact also the pump rod, there is much less friction than in the beam engine. Mr. Marten, who has under his management both kinds of engines, gives a preference to that with a beam. He observes that, as a rule, direct acting engines when working under a high initial pressure are apt to start off at a speed which jars and strains the whole of the machinery throughout.

"The speed attained by the piston as driven indoors at the beginning of the stroke, is many times greater than the average velocity per minute; and consequently unless all the parts are made extra strong in proportion, the bearings wear out with great rapidity, and the machinery is soon loose at every joint.   In a beam engine, on the other hand, a very large proportion of the initial force is absorbed in overcoming the inertia of the heavy beam, which thus becomes a reservoir of surplus force in the earlier portion of the stroke, to be given out during the latter; and the result is, that a comparatively steady velocity is maintained throughout the stroke, much to the advantage of the whole of the machinery; indeed it is only with this adjunct that expansion can be carried safely to a very high degree.   The beam in fact is a reciprocating fly-wheel, and is attended with precisely the same action and the same beneficial results.   The writer is acquainted with a case of two large expansive engines of nearly the same size working near together, one of which has an open network beam of about 30 tons, and the other a strong heavy beam of 45 tons' weight.   The difference in the working of the two engines is very perceptible, and nearly 5 million pounds duty in favour of the heavy beam.   In many cases where a jar is perceived in pumping engines working with a high expansion, it may be cured by increasing the weight or inertia of the beam."*

For pumping a large quantity of water through a great length of main pipe, under a heavy pressure, Mr. Marten's experience has led him to prefer the double acting beam engine erected in duplicate, the two engines being coupled together at right angles to one large fly-wheel.   Such is the style and combination of pumping engine adopted by Mr. Marten, in conjunction with Mr. M'Clean, for the South Staffordshire Works.   See specification of these engines in the Appendix.

DOUBLE CYLINDER ENGINES.

Ever since the double cylinder engine was first introduced by Woolf, this form has been in favour with some engineers. These are not much used in waterworks, but there are many

* From Marten's Paper on Pumping Engines already referred to.

combined cylinder engines on Simms' principle in the Cornish mines. The French use extensively double cylinder engines, and contend that they obtain by their means a more useful and economical effect from the expansion of the steam. The monster pumping engines, with 144 inch cylinders, erected for draining the Haarlem Mere, from the designs of Mr. Gibbs and Mr. Dean, have double cylinders, one within the other, the outer being fitted with an annular piston. Mr. Marten, while admitting the advantages of double cylinder engines in some cases, where uniformity of power throughout the stroke is a desideratum, yet for large pumping engines prefers single cylinder double action engines. He remarks that the arrangements with a double cylinder are much more complicated, and he finds that all useful degrees of expansion can be carried on sufficiently with a single cylinder.

PUMPING INTO A MAIN.

Where more than one pump is used there is often only one air-vessel. Mr. Marten, however, recommends a separate air vessel and back flap valve to each pump, also a blow-off valve loaded with a certain weight, so that in case of any recoil in a great length of main the pumps would not be burst.

In the main pipe, when the pumping lift is considerable, he recommends the insertion of a back flap valve at every 50 feet of elevation above the pumps, so that in case of any pipe bursting, the whole main shall not be run dry.

The leading point to be kept in view in the design and construction of engines under these circumstances is the maintenance of a constantly uniform flow of water through the main pipe from the pumps. This is provided for by the compound double acting pumps, by large air-vessel accommodation, and by the coupling of the engines at right angles.

PUMPING INTO A RESERVOIR.

At Wolverhampton, the reservoirs are prevented from being overfilled by a self-acting check valve, which shuts against any supply beyond a certain limit, so that the man working his engine at a distance knows when his work is done. The valve is so arranged, that immediately the engine ceases to work the supply to the town is maintained from the reservoir through flap valves, underneath the self-acting stop-valve, opening immediately as soon as a supply is required for the town.*

STAND PIPES.

These appear to be an unnecessary addition to the expense of a pumping establishment. They were first introduced, not so much to give the required pressure in the main pipes supplying a town, as to equalize the weight on the engine, and cause it to pump always against a uniform load. An air-vessel however costs only about one-tenth as much as a stand pipe, and is thought by some engineers to answer the purpose equally well.

The Tettenhall engine, at the Wolverhampton works, pumps from a well 140 feet deep, over a stand pipe 180 feet high, making a total lift of 320 feet. At the time the stand pipe was erected the Company had no summit reservoir, and the stand pipe was thought necessary to give the pressure in the mains.

Mr. Marten, the engineer of the works, in his recent paper read before the Institution of Mechanical Engineers, appears to be of opinion that the stand pipe was unnecessary. He observes, that all the requisite safety can be secured by pumping into an air-vessel with a check valve on the delivery side, so that in case of a pipe bursting, or any sudden diminution of pressure taking place, it would be impossible for the engine to "go out-of-doors" at more than a certain regulated speed. Mr. Marten says, "Unless the stand pipes are carefully cased in winter they are in great danger of being frozen, and very serious consequences have arisen from this cause. There is also a drawback with them on account of the great weight of the column of water, which has to be set in motion from a dead stand at each stroke of the engine."

* From Marten's Paper on Pumping Engines.

ON THE DUTY OF PUMPING ENGINES.

This term was first explained in a definite and precise manner by the learned and accomplished Davies Gilbert, President of the Royal Society, in a paper read before that body in 1827. "The criterion of the efficiency of ordinary machines is force, multiplied by the space through which it acts; the effect which they produce, measured in the same way, has been denominated *duty*, a term first introduced by Mr. Watt in ascertaining the comparative merit of steam engines, when he assumed one pound raised one foot high, for what has been called in other countries the dynamic unit; and by this criterion, one bushel of coal has been found to perform a duty of thirty, forty, and even fifty millions."

Mr. Wicksteed* says, "As regards the term 'duty,' I understand it to mean the useful effect, or actual weight of water raised by a given weight of coals, the same weight of coals also generating a sufficient quantity of steam to work the engine and overcome the friction of the pit or pump work."

It is clear, from these definitions, that the duty is not an expression of the work done, as this would include the power to overcome friction and other resistances, but is the actual useful effect expressed in lbs. weight of water actually raised.

To the enterprise and enlightened spirit which have long distinguished the mining interests of Cornwall, we are chiefly indebted for those vast improvements, of various kinds, which have absolutely increased the power of pumping engines to the extent of five times that which they possessed forty years ago, when the Cornish engines were first reported. To them, also, we are indebted for that valuable series of annual reports which have recorded, year by year, the gradual and successive improvements of the engines. Mr. Lean, in his historical statement of the steam engines in Cornwall—a work compiled

* Wicksteed's Experimental Inquiry concerning Cornish and Boulton and Watt Pumping Engines, p. 32. London, Weale, 1841.

at the request of the British Association for the Advancement of Science, by the well-known registrars and reporters of these engines—makes a statement which shows, in the clearest manner, that the improvements made in the engines of Cornwall, up to 1835, were then saving the country £80,000 a year in coals alone, as compared with the cost of working the same engines in 1814, or twenty-one years before. All the statements in Mr. Lean's book are characterized by moderation and truthfulness, and appear to be thoroughly worthy of confidence.

Something more than mere praise and simple admiration are due to the labours of the men who quietly and unostentatiously, without parade of any kind, have been thus steadily promoting the substantial and vital interests of their country. Without these improvements, and without the exertions of the men to whom they are due, it is probable many of the mines of Cornwall would have become unprofitable, and must have been abandoned, on account of the expense required to keep them free from water.

Mr. Taylor says, in his records of mining, that in early times the duty of atmospheric engines was equal to 5 million pounds raised one foot high by a bushel of coals $= \dfrac{5 \times 112}{94} =$ nearly 6 millions for 1 cwt. of coal.*

During the ten years from 1770 to 1780, it appears that Smeaton's atmospheric engines were doing an average duty of 7 to 11 millions, and that Boulton and Watt's engines were doing about double this amount. About 1785 Boulton and

---

* The duty given here and in the following pages is always expressed in lbs. raised by 1 cwt. or 112 lbs. of coal. In all the earlier reports and writings on the subject of duty the unit was a *measured* bushel of coal, which has been variously estimated at 84 to 100 lbs. in weight. It is now generally considered, however, that the bushel is equivalent to 94 lbs., but both Lean's and Brown's reports now also give the duty reduced to a cwt. of 112 lbs. As this standard is more convenient, and will be better understood than the other, I have adopted it throughout, and whenever necessary in extracting from the old reports, have reduced the duty to this standard.

Watt introduced the improvement of working steam expansively in Cornwall, and at this time the duty somewhat increased, although the steam was not raised to any higher pressure than before.

In 1800, when Boulton and Watt's patent expired, the best of their engines in Cornwall were doing an average duty of about 24 millions. After this time Mr. Murdoch and other skilful and experienced agents having left the country, a great deterioration took place in the Cornish engines, and it is said that in the following year several of the largest engines with 63-inch cylinders on Bull's mode of construction were working with an average duty under 12 millions. Soon afterwards, however, owing to the able exertions of Captain Lean, the duty began to improve at several of the mines, and the example set by these produced a beneficial result also in others.

The following figures show the regular successive improvements of the engines as recorded in the earlier years of Lean's reports :—

Year.	No. of Engines.	Average Duty.	Average Duty of the best Engine.	Year.	No. of Engines.	Average Duty.	Average Duty of the best Engines.
1812	21	23·0		1828	57	44·1	91·4
1813	29	23·2	31·4	1829	53	49·6	91·6
1814	32	24·5	38·1	1830	56	51·5	92·8
1815	35	24·4	34·1	1831	58	51·6	84·6
1816	35	27·4	38·6	1832	59	52·6	101·1
1817	35	31·5	49·5	1833	56	55·4	100·3
1818	36	30·2	46·8	1834	52	56·9	108·1
1819	40	31·3	47·6	1835	51	56·9	109 1
1820	46	34·1	49·1	1836	61	55·4	101·6
1821	45	33·6	50·9	1837	58	55·9	102·8
1822	52	34·4	50 6	1838	61	58·0	100 2
1823	52	33·6	50·0	1839	52	65·4	92 6
1824	49	33·7	51·8	1840	54	64·3	97·2
1825	56	38·1	54·0	1841	56	65·1	121·3
1826	51	36·3	53·8	1842	49	64·0	127·9
1827	51	38·2	71·0	1843	36	71·4	114·4

It will be observed, that up to the year 1827, the duty of the best engine is seldom more than 50 per cent. above the average duty of all the engines. In 1827, however, the average duty of the best engine is nearly double the general average, and in the following year is rather more than double. This increase was due to the improvements made by Samuel Grose, and to the introduction of a 90-inch cylinder on Woolf's principle, which performed a duty considerably exceeding that of any former engine. In the following years the duty of the best engine never appears to double the average duty, although it more nearly approaches 100 per cent. than 50 per cent. in excess of this. The duty of the best engine in 1842 appears to be the largest ever recorded for any continuous period, being nearly 128 millions. This duty was performed by Taylor's 85-inch cylinder engine at the United Mines in Gwennap.

The engine* was erected in 1840 by Messrs. Hocking and Loam, and was especially intended to work more expansively than had hitherto been practised. The boilers were made smaller in diameter than usual, and of stronger plate, so as to stand a higher pressure of steam, the working elasticity being fixed at 40 lbs. per square inch above the atmosphere. Also an extra number of boilers was provided, in order to give an increased proportion of heating surface, and the strength of the working parts of the engine and machinery was augmented to withstand the strain caused by the great force of the steam on the piston at the commencement of the stroke. In this engine (on a visit being made in 1841) the steam was cut off at about one-tenth or one-twelfth of the stroke, thereby carrying out the principle of expansion to a greater extent than had ever before been attempted, except by Woolf in his combined cylinder engines, where he expanded it above twenty times.

The following is the monthly performance of the engines from Lean's Report for the year 1854 :—

* Engineer's Pocket Book for 1849.

	No. of Engines reported.	Average Duty. Millions.	Duty of best Engine.	Name of best Engine.
January	17	57·2	72·0	Leed's 60-inch.
February	22	54·8	74·0	Mitchell's 85-inch.
March .	23	53·6	69·0	Penrose's 85-inch.
April . .	23	54·8	70·0	Ditto.
May . .	22	53·6	70·0	Leed's 60-inch.
June . .	22	53·6	68·0	Mitchell's 60-inch.
July . .	20	52·4	65·0	Ditto.
August .	20	52·4	72·0	Leed's 60-inch.
September	19	53·6	77·0	Ditto.
October .	19	52·4	75·0	Ditto.
November	20	53·6	74·0	Ditto.
December	20	52·4	73·0	Ditto.
Average .	..	53·7	72·0	

This Table shows that the duty of the engines reported by Lean in 1854 is much less than in 1843 and the preceding years, the average in 1854 being 53·7 millions against 71·4 millions in 1843, and the best engines having only a duty of 72 millions instead of 114 to 127 millions. The table for 1854 also shows that all the engines are more nearly on an equality than formerly, as the performance of the best is only 34 per cent. in excess of the average, instead of being 50, and even 100 per cent., as in some former years.

I have been informed, however, that the best engines are not now reported by Lean, the proprietors of some of the best engines not caring to pay the expense of having them reported, although the duty is regularly recorded for their own satisfaction, and for the purpose of comparison with other engines.

The following is the performance of the engines from Browne's Reports for the year 1855 :—

	No. of Engines reported	Average Duty in Millions.	Duty of best Engine.	Name of best Engine.
January .	15	69·9	100·7	Austin's 80-inch.
February .	15	70·1	97·9	Treffry's 80-inch.
March .	15	68·7	97·4	Ditto.
April . .	15	68·4	101·3	Ditto.
May . .	15	68·9	100·3	Ditto.
June . .	15	69·4	98·0	Austin's 80-inch.
July . .	15	69·1	99·1	Ditto.
August .	15	69·4	100·1	Ditto.
September	14	69·7	101·4	Ditto.
October .	13	71·6	100·4	Ditto.
November	13	71·3	98·8	Ditto.
December	12	70·2	100·0	Treffry's 80-inch.
Average .	..	69·7	99·6	

It will be observed that the engines reported by Browne work with a considerably greater amount of duty than those now reported by Lean.

The average working of Browne's engines is very nearly equal to the highest average of former years, but the duty of the best engine is somewhat less than in those years in which Taylor's 85-inch cylinder engine was reported. This engine does not appear in either of the reports published at the present time. Many of the engines reported by Browne were constructed from the drawings of Mr. West, and most of the others are under his superintendence.

The following Table contains the monthly duty of each engine reported by Browne during the year 1855. It shows the small fluctuations in the amount of duty for each engine, and the average duty of each during the whole year.

DUTY OF CORNISH ENGINES IN 1855, FROM BROWNE'S CORNISH ENGINE REPORTER.

	BOSCUNDLE. Matthews' 33 inch cylinder, 46 horse engine.	GREAT DEVON CONSOLS. Morris' 40 inch cylinder, 71 horse.	FOWEY CONSOLS. Austin's 80 inch cylinder, 251 horse.	GREAT POLGOOTH. Darlington, 80 inch, 251 horse.	GREAT POLGOOTH. Tom Bell, 67 inch, 176 horse.	MARY ANN. 46 inch cylinder, 89 horse.	PAR CONSOLS. Treffry, 80 inch, 251 horse.	PAR CONSOLS. Puckey's 72 and 36 inch, Simms' combined, 204 horse.	PEMBROKE AND EAST CRINNIS. 80 inch cylinder, 251 horse.	PEMBROKE AND EAST CRINNIS. Carlyon's 70 inch, 192 horse.	SOUTH CARADON. Pearce's 45 inch, 85 horse.	SOUTH CARADON. Jope's 40 inch, 71 horse.	TRELAWNEY. 50 inch cylinder, 105 horse.	WEST FOWEY CONSOLS. 60 inch, 141 horse.	WHEAL UNY. 50 inch, 105 horse.	Average duty for each month.
Jan.	53·4	47·9	100·7	91·5	72·0	44·1	99·8	78·2	74·4	74·6	54·0	47·9	70·2	72·8	66·4	69·9
Feb.	54·0	48·0	96·7	95·5	71·5	43·2	97·9	79·1	76·2	75·7	53·5	48·6	70·5	72·8	67·7	70·1
March	53·2	46·8	96·5	94·0	72·3	43·1	97·4	79·5	70·1	76·2	46·4	46·3	66·9	73·5	69·4	68·7
April	52·3	46·9	96·9	92·2	67·8	43·5	101·3	79·5	67·5	73·6	45·2	47·3	69·1	72·4	69·9	68·4
May	54·5	50·4	95·9	92·0	67·5	43·5	100·3	76·6	70·0	75·1	46·4	46·7	70·1	72·1	73·6	68·9
June	52·2	50·9	98·0	93·5	68·1	45·7	97·4	72·8	76·1	76·0	48·5	48·9	69·8	70·1	73·4	69·4
July	51·8	50·4	99·1	93·2	68·4	41·4	97·3	72·7	78·7	77·2	49·0	49·0	63·5	69·0	75·5	69·1
August	51·6	50·7	100·1	94·5	68·7	41·2	97·2	74·4	80·2	77·1	48·5	48·4	64·8	68·8	74·7	69·4
Sept.	51·9	51·1	101·4	96·8	69·2	42·0	98·0	75·2	80·5	77·9	48·8	49·6	66·1	68·1		69·7
Oct.	53·0		100·4	94·5	72·1	40·4	97·8	76·3	83·4	79·2	49·6	48·7	66·6	68·6		71·6
Nov.	53·3		98·8	94·0	72·0	40·8	98·1	74·1	84·9	78·0	49·6	48·5	67·1	68·6		71·3
Dec.	54·1		98·3	94·7	72·1	40·8	100·0		78·8	73·5	49·4	47·9	64·5	68·5		70·2
Averages	52·9	49·2	98·6	93·9	70·1	42·4	98·5	76·2	76·7	76·2	49·1	48·2	67·4	70·4	71·3	69·7

RELATION BETWEEN DUTY AND CONSUMPTION OF FUEL.

If we know the consumption of fuel in an engine, it is easy to convert this into duty. For example, suppose an engine requires a lbs. of coal per horse power per hour. The effect produced by one horse power is that of raising 33,000 lbs. 1 foot high in a minute, or $33,000 \times 60 = 1,980,000$ lbs. raised 1 foot high in an hour. This result is produced by a lbs. of coal, hence the duty for 1 cwt. of coal is

$$a : 1,980,000 : : 112 : \frac{1,980,000 \times 112}{a}$$

Hence we have this rule: Divide 221,760,000 by the consumption of coal per horse power per hour, the quotient is the duty of the engine expressed in lbs. raised 1 foot high by 1 cwt. of coal. The following table is calculated to show the duty of engines consuming from 1 to 12 lbs. of coal per horse power, per hour.

Consumption of Coal per horse power per hour. lbs.	Duty in lbs. raised 1 foot high by 1 cwt. of coal.	Consumption of Coal per horse power per hour. lbs.	Duty in lbs. raised 1 foot high by 1 cwt. of coal.
1	221,760,000	7	31,680,000
2	110,880,000	8	27,720,000
3	73,920,000	9	24,640,000
4	55,440,000	10	22,176,000
5	44,352,000	11	20,114,545
6	36,960,000	12	18,480,000

Put D = duty in millions of lbs. raised 1 foot high per minute,

Then $\dfrac{221 \cdot 76}{D}$ = lbs. of coal per horse power per hour.

Also $\dfrac{221 \cdot 76 \times 365 \times 24}{2240} = \dfrac{194261 \cdot 76}{2240} = 867$

Then $\dfrac{867}{D}$ = tons per annum for each horse power.

DUTY OF CONDENSING ENGINES, FROM HOUGHTON'S REPORT ON
DETROIT WATERWORKS, p. 12, AND OTHER SOURCES.

	Millions of lbs. raised 1 ft. high by 1 cwt. of coal
East London Waterworks, single acting Cornish Engine, 1836	105·7
Ditto ditto, Boulton and Watt . .	46·6
Haarlem Mere, Holland	89·4
Average of 36 Cornish Engines, 1843 . . .	71·5
Cincinnati direct action	53·6
Buffalo Cornish Bull Engines, 1852 . . .	37·0
Boulton and Watt's non-expansive Rotative Engine Albion Mills, London, 1786	25·8
Spring Garden, Philadelphia, 1832 . . .	24·6
United States Dry Dock Engine, Brooklyn . .	22·4
Non-condensing 30-inch Cylinder Engine at Kingston, United States	9·8
40-inch Cylinder Condensing Engine at ditto .	28·7

DUTY PERFORMED BY VARIOUS OTHER ENGINES.

Direct Acting Non-condensing Engine at Tettenhall Station of Wolverhampton Works, using small Staffordshire coal	17·9
Cornish Engine at Goldthorn Station of Wolverhampton Works, using small Staffordshire coal . .	27·5
Grand Junction Waterworks Engines in 1849 .	46·8
Southwark and Vauxhall Engines in 1849 . .	69·6
Beardmore's Exemplar Cornish Engine . .	83·4
Beardmore's Duty of Cornish Engines at London Waterworks	56·25
Fowey Consols 50-inch Cylinder Engine of 103 horse power, experimented on by Mr. Wicksteed, and worked at a power of 26¼ horses . . .	130·2
Holmbush 80-inch Cylinder = 251 horses, worked at a power of 62 horses	122·4
Estimated duty of 72-inch Cylinder Condensing Engine for Brooklyn Works	35·5
Engines at the East London Works in 1850 from Mr. Wicksteed's evidence	63·8
Engines at the same works before the use of the Cornish Engines	26·1

Average duty of Cornish Engines from Lean's Re-
    porter, 1854 . . . . 53·7
Duty of the best Engine from ditto . . . 77·0
Average duty of Cornish Engines, Browne's Reporter
    1855 . . . . 69·7
Duty of the best Engine from ditto . . . 101·4

I shall now invite attention to two remarkable statements which have been made with reference to the duty of Cornish engines by two writers who have both contributed much valuable information on this interesting subject. The first is a statement by Mr. Lean in his work already referred to— "That the duty of the best pumping engines in Cornwall far exceeds what could be effected, were it even possible to apply the force of the steam immediately to the water, unencumbered by the friction and imperfection of machinery, or the loss arising from accidental condensation." The other statement is made by Mr. Pole in his valuable treatise on the Cornish engine, and is to the effect that the steam generated in the boilers of the Cornish engine is capable of performing a much higher duty than that which has ever been reported of any engine. Thus he shows that when the steam is cut off at one-sixth of the stroke, and the steam expanded through the remaining five-sixths, a motive power is produced which is capable of raising 154,000,000 lbs. 1 foot high, by the consumption of 94 lbs. of coal; and that, with eight times expansion, a motive power is developed equal to 170,000,000. Now although these two statements may appear at first sight at variance with each other, we shall find on examination that this is really not so, but that in fact they are perfectly consistent one with the other.

Mr. Lean's statement is founded on the quantity of work which the steam will do when worked altogether without expansion; and doubtless if we calculate the motive force developed by the steam when used at a simple initial pressure without expansion, we shall find it much less than that which is really produced in the Cornish engine.

Mr. Lean shows, from actual recorded measurements and observations made during six months at the United Mines, that 100 lbs. of coal will convert 15 cubic feet of water into steam, and that 500 cubic feet of steam, at a pressure of 50 lbs. on the square inch, are generated from each cubic foot of water.* It follows that 100 lbs. of coals would produce $500 \times 15 = 7500$ cubic feet of steam at a pressure of 50 lbs. Hence, "if the steam could be applied immediately to the water at the bottom of the pump without loss by condensation, while a perfect vacuum was constantly maintained at the top, the weight lifted 1 foot high by the consumption of 100 lbs. of coal, would be $7500 \times 144 \times 50 = 54,000,000$ lbs.: while the duty of the best steam engines in Cornwall (although encumbered by much machinery) has been known to exceed double that number." He then proceeds to show that the steam when worked expansively developes a great addition of power beyond that which is due to its first action on the piston at its full pressure. The mode of calculating the increased effect due to the expansion of the steam when cut off at various parts of the stroke, is explained in a very simple and familiar manner, both by Mr. Lean and Mr. Pole. It is also to be found in many books of a more popular character, where a small table of hyperbolic logarithms is inserted for the purpose of making the necessary calculation. It would be going beyond the limits of this work to enter on this subject, and we must be content, therefore, with referring to the scale at page 284, which shows the efficiency of an engine at different degrees of expansive working.

If then the steam admitted at full pressure during the whole of the stroke produce an effect of 54,000,000, it will produce, when cut off at any part of the stroke, an effect equal to

---

* This experiment on the evaporative power of coal agrees remarkably with those made by Mr. Wicksteed at the East London Waterworks. Thus $\frac{15 \times 62.5}{100} = 9.37$ lbs. of water evaporated by each lb. of coal. Mr. Wicksteed found that the best Welsh coal would evaporate 9.493 times its own weight of water.

P

54,000,000 multiplied by the corresponding number in the above scale. Thus, if cut off at one-fifth, the effect will be 54 × 2·609 = 141,000,000.

Mr. Pole takes up the subject just where Mr. Lean leaves off, and shows that the effect due to the steam when worked expansively, is much greater than that ever reported as the duty of any Cornish engine whatever.

He takes for example a cylinder 70 inches in diameter, to which steam of 45 lbs. pressure is admitted during one-sixth of the stroke. Using the hyperbolic or Naperian logarithm of 6, in the same way as already explained, he finds the effect which would be developed by steam worked at this rate of expansion is 154,000,000; and in a similar manner he finds that when cut off at $\frac{1}{8}$ of the stroke the effect produced is 170,000,000, and with ten times expansion the effect would be 180,000,000. According to Mr. Lean the effect produced by expanding six times would be 54 × 2·792 = 151,000,000; at eight times, it would be 54 × 3·079 = 166,000,000; and at ten times, 54 × 3·303 = 178,000,000. The slight difference between the two is accounted for by Mr. Pole assuming the relative density of water and steam at 45 lbs. pressure as 1 to 608, whereas Mr. Lean takes the relative densities at 50 lbs. pressure as 1 to 500. The densities being inversely as the pressures, if one cubic foot of water expands into 500 feet of steam at 50 lbs. pressure, it ought to expand into $\dfrac{500 \times 50}{45} = 556$ cubic feet at 45 lbs. Mr. Pole's rate of expansion is taken from De Pambour's tables in his Theory of the Steam Engine. On the whole, it agrees remarkably well with the experiments described by Mr. Lean at the United Mines; and the statements made by the two authors we have quoted, must be held essentially to confirm each other.

COST OF RAISING WATER BY STEAM POWER.

Mr. Wicksteed has recorded the fact, that the expense of pumping by Cornish engines at the East London Waterworks is ·150 pence per 1000 gallons raised 100 feet high. This is equal to 150 pence, or 12s. 6d. for a million gallons raised 100 feet high.

According to some recent returns by Mr. Duncan, engineer of the Liverpool Waterworks, the cost at the most expensive of their pumping stations—namely, at Hotham Street—was nearly £4 for a million gallons raised 100 feet; but at Windsor station the cost was only 18s. 1d. per million, and at Green Lane only 15s. 9d. per million. As these two are the only establishments that have really good engines and machinery, it is, perhaps, only fair to reject the others, and take the mean of these two. This, accordingly, is 16s. 11d., or say 17s. per million gallons raised 100 feet high.

The work done by the engines of the Wolverhampton Waterworks Company is equal to raising annually 426 million gallons 100 feet high; and the cost of this, in a district where coal is exceedingly cheap (only about 7s. 6d. per ton), appears from the Company's published accounts to be about £750 a year, or about 35s. per million gallons. This is double the cost of the same work at Liverpool.

The East London Company, according to a return made to the General Board of Health, used in 1849 a quantity equal to 2,121¼ tons of coal at 10s. 6d. per ton delivered, and employed an average steam-engine power of 372·6 horses, working 12 hours per day.

Hence $\dfrac{2121\cdot5 \times 2240}{372\cdot6 \times 12 \times 365} = 2\cdot9$ lbs. per horse power per hour.

The Southwark and Vauxhall Company in 1849 employed four engines, whose united power was equal to 355 horses, and the coals consumed amounted to 8 tons per day on the average of the year. This is equal to nearly 3 cwt. per day

P 2

for each horse power. The coal had cost for some years
13s. 3d. per ton, but in 1849 the coal was only 10s. per ton.

Supposing the whole supply of this company pumped over
their stand-pipe, which is 185 feet high, the quantity in
millions of gallons raised 100 feet high will be

$$\frac{2195 \times 185}{100} = 4{,}061 \text{ millions raised 100 feet high.}$$

The coals used for this are 2,920 tons, costing £1,460, or
only about 7s. for the coals used to raise one million gallons
100 feet high. If we add to the cost of the coals the
large sum of £1,000 for labour, repairs, oil, tallow, wear
and tear, &c., we shall have the whole cost of a million
gallons only 12s.

The establishment of the Southwark and Vauxhall Company, under the able and skilful management of Mr. Quick,
compares very favourably in point of economy with any
other that can be cited. The duty of the engines appears,
from the preceding figures, to be very nearly 70 million lbs.
for 1 cwt. of coal, the coal consumed being barely 3 lbs. per
horse power per hour. This work and the East London,
according to the information now before us, appear to be
about on a par as to economy of working.

The Grand Junction Company in 1849 lifted 1,289 million
gallons, the greater part of it passing over a stand-pipe 218
feet high, and consumed 3,170 tons of coal, which cost
£2,285.

$$\text{Hence } \frac{1289 \times 218}{100} = 2{,}810 \text{ million gallons}$$

raised 100 feet high at a cost of rather more than 16s. per
million gallons for coals alone. Adding, as before, £1,000
for labour, &c., the cost per million gallons raised 100 feet
high would be rather more than 23s. per million. This is
considerably higher than the cost either in Liverpool, East
London, or Southwark and Vauxhall, but is in some measure
accounted for by the higher price paid for the coal.

The following estimates of working expenses are taken

from Mr. Hack's evidence before Mr. William Beckett's Committee in 1852 :—

Estimate for working 100 horse power engine at Hampton, for the West Middlesex Works :—

	£	s.	d.
3¼ tons of Welsh coal per day=1,277½ tons per annum, at 26s., in the stoke-hole . . .	1,660	15	0
Two engine workers, two stokers, and two labourers, at £9 per week	468	0	0
Tallow, oil, hemp, and sundry stores. . .	150	0	0
Repairs to engine and boiler, and wear and tear	100	0	0
	£2,378	15	0

The work to be done here is to raise 4,000,000 gallons a day 46 feet high, which is equal to a power of 39 horses working during twenty-four hours ; so that the annual cost is about £61 per horse power of actual work.

The cost per million gallons raised 100 feet high will be very great, according to the above estimate : thus,

$$\frac{4 \times 46 \times 365}{100} = 671 \cdot 6 \text{ millions raised 100 feet high,}$$

and $\dfrac{£2,378 \ 15s.}{671 \cdot 6} = £3 \ 11s.$ nearly,

for each million gallons raised 100 feet high.

Both this and the following estimate by Mr. Hack are evidently extravagant. It should be observed that these are not in the same category as the estimates usually brought by engineers before parliamentary committees. The object here was not to show the committee how cheaply the work could be done, but rather to impress them with an idea of the great sacrifice which the West Middlesex Company was going to make for the public benefit by removing their works to Hampton, and erecting an additional engine at Barrow Hill. No one can fail to be struck with the amount put down in these estimates, both for coals and labour.

Estimate for working a 75 horse power engine for the West Middlesex Works at Barrow Hill :—

		£	s.	d.
Coals, 75 H.P. × 24 × 4 lbs. per H.P. per hour, = 3 tons 4 cwt. 1 qr., say 3¼ tons per day, at 24s. per ton=£3 18s.; and this for 365 days		1,423	10	0
Two engine workers, at 40s. per week . .		208	0	0
Two stokers, each at 26s. per week . . .		135	4	0
One labourer and engine cleaner, at 21s. . .		54	12	0
Tallow, 10 lbs. per day, at 6d.		91	5	0
Oil, 1 pint per day=46 gallons, at 5s. . .		11	10	0
Yarn, hemp, flax, and sundry stores . . .		25	0	0
Wear and tear of machinery		50	0	0
		£1,999	1	0

or about £26 10s. per horse, reckoned on the full power of the engine.

Using the formula $\frac{867}{D}$ (see page 310) to derive from the duty the coals consumed per annum for each horse power, I find that the smallest consumption of coal in any engine reported by Browne in 1855 was at the rate of 8 tons 16 cwt. for each horse power of actual work. This is the consumption by Austin's 80-inch cylinder engine at Fowey Consols. The largest consumption by any engine in the same report is 20 tons 9 cwt. per annum for each horse power; and the average of the whole fifteen pumping engines is 12 tons 9 cwt., or a little more than a ton per month for each horse power. The coal used by the Cornish engines is understood to be the best Welsh coal, capable of evaporating from 9 lbs. to 10 lbs. of water by each pound of coal.

Let us examine for a moment the expense of the best Cornish engine as compared with the worst. Taking an engine working at 250-horse power, and assume that the cost of labour, oil, tallow, and small stores is the same for each, but that the difference is in the coal alone, and assume the latter to cost 16s. per ton. According to the above figures, we have the worst engine costing for coal alone

Tons.	Tons. cwt.			£	s.	d.
$250 \times 20 \cdot 9 = 5112$	10, at 16s.	.	.	4,140	0	0
And the best, $250 \times 8 \cdot 16 = 2200$, at 16s.	.	.	.	1,760	0	0

Annual difference against the worst engine   £2,380   0   0

The true expense of an engine is the original price added to its annual cost capitalized, say at twenty years' purchase.

Suppose each engine to have cost originally £25,000, then we have the following comparison :—

WORST ENGINE.

	£	Total.
Original cost	25,000	
Annual expenses, 4140×20 . .	82,800	
		107,800

BEST ENGINE.

	£	
Original cost	25,000	
Annual expenses, 1760×20 . .	35,200	
		60,200

Difference in favour of the best engine   .   . £47,600

So that the one engine will cost in the end more than 50 per cent. above the other.

Many comparisons of a similar nature might be made ; but the above is sufficient to draw attention to the fact of the immense consequence involved in what is called the duty of an engine and its corresponding consumption of coal.

The following table shows the quantities of oil and tallow used for eleven Cornish engines during the year 1855, compiled from " Brown's Engine Reporter " :—

Average Horse Power employed.	Oil consumed in quarts.	Tallow consumed in lbs.	Oil used per horse power, per annum, in quarts.	Tallow used per horse power, per annum, in lbs.
33·6	64	767	1·9	22·8
140·5	329	3994	2·3	28·4
169·1	351	2886	2·1	17·1
77·6	104	1560	1·3	20·1
33·3	51	920	1·5	27·6
124·6	240	2450	1·9	19·7
76·5	98	980	1·3	12·8
81·5	86	1963	1·1	24·1
111·3	137	2000	1·2	18·0
49·3	109	1620	2·2	32·9
37·3	105	540	2·8	14·5
		Average	1·8	21·6

Hence it appears that the consumption of oil by the Cornish engines is equal to about 1·8 quarts per annum for each horse of working power, and that the consumption of tallow is equal to 21·6 lbs. per annum for each horse.

In order to show how this mode of calculation will work out, and to compare it with the known expense of raising water at certain establishments, let us take the case of a 150 horse power engine, working night and day throughout the year, and compute everything at the most moderate price possible. The annual working cost of such an engine will, on the preceding data, be about as follows :—

Estimated expense of 150 horse power Cornish engine, taking coals at 10s. per ton :—

	£	s.	d.
150 × 12 = 1800 tons of coal at 10s.	900	0	0
Wages, say £9 per week, for 52 weeks	468	0	0
Oil 150 × 2 = 300 quarts at 1s. 6d.	22	5	0
Tallow 150 × 22 = 3300 lbs. at 6d.	82	10	0
Yarn, hemp, flax, &c.	50	0	0
	£1522	15	0

Work 150 × 33,000 = 4,950,000 lbs. raised one foot high per minute, or $\frac{495 \times 1440 \times 365}{100}$ = 2,602 million gallons, raised 100 feet high in a year, then $\frac{1523}{2602}$ = 11s. 8d. per million gallons, the price at Liverpool being 16s. according to Mr. Duncan, and 12s. 6d. at the East London Works according to Mr. Wicksteed. Of course, if we take the price of coal at a higher rate than 10s. per ton, the cost of pumping will be proportionately increased. There are few places in the kingdom where coal of such quality as that used in the Cornish engines can be procured for 10s. per ton ; and if we have an inferior coal, of course the consumption will be more than has been assumed in the above calculation.

It is remarkable that the average duty of the Cornish engines, as reported by Browne for the year 1855, and on which the preceding calculation is founded as to consumption of coal, corresponds almost exactly with what was understood to be the duty of the best pumping engines then employed at the London Waterworks—namely, about 70 millions. It will not be safe, therefore, in calculations for waterworks engines, to reckon on a higher duty than this: and as a consequence, it must be assumed that the consumption of coal, equal in quality to that used in Cornwall (namely, capable of evaporating about 9 times its own weight of water), will be about 12 tons per annum for each horse power.

The following estimates of working expenses were made by Mr. Hocking for the Wolverhampton Waterworks, in which the quantity to be pumped was 1¼ million gallons a day.

The estimate was required in two forms, namely for raising 1¼ millions in 24 hours, and in 12 hours. In either case however a part of the work was to be done in 6 hours, namely the pumping of the whole 1¼ million gallons a height of 22 feet on to the filter-beds. For this Mr. Hocking recommended a 36-inch cylinder engine and a 38-inch pump; stroke of engine and pump each 9 feet, and 9½ strokes per minute. The other two lifts 299 feet and 148 feet, making together 447, exclusive of the friction through the pumping main, which was estimated at 15 feet additional for each lift supposing the work done in twenty-four hours, and at four times this quantity or 60 feet, supposing it done in twelve hours.

For one of these stations Mr. Hocking recommended a 60-inch cylinder engine and 19-inch pump, with 10 feet stroke for each, and making 8½ strokes per minute. For the other station he recommended a 45-inch cylinder engine and 19-inch pump; length and number of strokes as before. This was for doing the work in 24 hours; if to be done in 12 hours, the engines and pump were to be in duplicate.

MR. HOCKING'S ESTIMATE.

	1½ million gallons in 24 hours.	1½ million gallons in 12 hours.
	One 36-inch cylinder and 38-inch pump. One 60-inch ditto, and 19-inch pump. One 45-inch cylinder and 19-inch pump.	One 36-inch cylinder and 38-inch pump. Two 60-inch cylinders and two 19-inch pumps. Two 45-inch cylinders and two 19-inch pumps.
	£	£
Cost of Engines and Pumps .	13,200	21,800
„ Buildings . . .	7,600	10,800
	20,800	32,600
WEEKLY WORKING EXPENSES.	£ s. d.	£ s. d.
Coal at 8s. per ton . .	16 8 0	20 0 0
Enginemen's wages . .	8 10 0	8 0 0
Firemen's wages . .	3 16 0	2 0 0
Occasional assistance . .	1 10 0	2 0 0
Tallow, oil, hemp, and yarn for packing, waste and small stores	3 10 0	4 18 0
	33 14 0	36 18 0
Per annum . . .	£1752 8 0	£1918 16 0

The work to be done for raising the water in 24 hours is equal to 158 horses, and for raising it in 12 hours about 163 horses; so that the annual expense per horse power in the one case is about £11 2s., and in the other £11 15s.

The price is also rather less than 13s. per million gallons raised 100 feet high when the work is done in 24 hours, and about 13s. 2d. when the work is done in 12 hours. These results agree remarkably with the price of pumping given by Mr. Wicksteed for the East London Works, namely, 12s. 6d. per million gallons, and also with the cost of pumping at other well-managed establishments in London.

The cost of working engines of course will vary with the

price of coal, which may range even for the same quality
from 7s. to 30s. per ton. The price of labour, oil, tallow,
yarn, &c., will also influence the working expenses. The
usual price of oil is from 5s. to 6s. per gallon, and that of
tallow 6d. to 7d. per lb.

WATERWORKS OBTAINING A SUPPLY FROM RIVERS AND STREAMS.

Some of the principal towns in this country are supplied from the rivers in their immediate neighbourhood.

The daily supply to the metropolis now exceeds 100 million gallons, and about one-half of this quantity is pumped by five companies from the River Thames. The remaining three companies supply as follows:—

The New River Company, from the River Lea about	15	mil. gals.
,, ,, from wells and springs .	9	,,
The East London Company, from the River Lea .	20	,,
The Kent Company, from chalk wells . . .	6	,,
	50	

So that about 85 million gallons, or 85 per cent. of the whole, are supplied from rivers.

Amongst the principal towns in England which derive the greater part, and in most cases the whole of their supply from rivers, are York, Penrith, Darlington, Newark, Derby, Nottingham, Chester, Worcester, Norwich, Exeter, and Plymouth.

The South Staffordshire Works afford an example of a modern work on a very large scale taking a supply from a river. The works are designed to afford water to a congeries of towns grouped together in South Staffordshire, forming what is called the Pottery District.

The system of pumping water from rivers has also been much adopted in many foreign works, especially in France, Prussia, and the United States.

Much has been said in favour of a supply from large rivers on sanitary grounds. The water is usually softer than that derived from wells, springs, and small streams, and contains a less amount of mineral salts than either of these, at the same time that it is commonly more impregnated with organic matter. A large river flowing over many geological formations and many different varieties of soils, may be naturally expected to take up in solution a variety of mineral matters, and therefore to present a greater number of ingredients than water derived from a more limited area; and this we generally find to be the peculiar character of river water.

It seldom happens that the water of a large river can be made available to supply a town by gravitation. Rivers usually occupying the lowest levels of a country have in most cases to be pumped to a considerable height, and it seldom pays to take the water at a point so far above the town that it will flow by gravitation, without the necessity for pumping at all.

So small a proportion of the New River water now flows by gravitation to the New River Head, that we shall not be far wrong in saying that the whole supply of the metropolis —about 100 million gallons a day—is now pumped to a height of 200 feet, in order to afford sufficient pressure for the supply of London.

VOLUME OF RIVERS.

The volume of water carried off by rivers is exceedingly various, depending on many conditions, such as the basin which they drain, the rainfall, the nature of the soil over which they flow, &c. In the larger rivers of England the volume of water is so great as to render them adequate to supply any quantity likely to be required for many years to come by the whole population seated on their banks. From 300 to 360 million gallons per day may be taken as the lowest summer discharge of each of the three principal rivers in England, the Thames, the Severn, and the Trent, and there are several others which nearly approach this quantity. Of course the discharge in winter and in seasons of flood is much greater. The following very valuable table has been published by Mr. Beardmore, showing the summer discharge of various rivers, streams, and springs, when uninfluenced by any immediate rain. The table also contains many other useful particulars, as the geological formations through which the rivers flow, their height above the sea, and the proportions which their total discharge bears to the basin or drainage area, and the proportion which the volume of discharge bears, in certain cases, to the total rainfall of the district.

TABLE OF ORDINARY SUMMER DISCHARGE OF VARIOUS RIVERS, STREAMS, AND SPRINGS, AS UNINFLUENCED BY ANY IMMEDIATE RAIN, WITH THE DRAINAGE AREA, AND AMOUNT RUN OFF THE SURFACE REPRESENTED IN DEPTH OF RAIN.

RIVERS.	Height above sea.		Drainage Area.	Total Discharge.	Discharge per square mile.	Representing rainfall per annum.	Total average rainfall per annum.
	Valley.	Hill.					
	feet.	feet.	Square miles.	cubic ft. per minute.	cubic ft per minute	inches.	inches.
Thames, at Staines—chalk, green-sand, Oxford clay, oolites, &c.	40 to	700	3,086	40,000	12·98	2·93	24·5
Severn, at Stonebench—silurian	400 to	2,600	3,900	33,111	8·49	1·98	...
Trent, at its mouth—oolites and Oxford clay	100 to	600	3,921	...	...	...	...
Loddon (Feb. 1850), greensand	110 to	700	221·8	3,000	13·53	3·01	25·4
Nene, at Peterborough—oolites, Oxford clay and lias	10 to	600	620·0	5,000	8·45	1·88	23·1
Mimram, at Panshanger—chalk	200 to	500	29·2	1,500	51·4	11·58	26·6
Lea, at Lea Bridge—chalk. (Rennie, April, 1796.)	30 to	600	570·0	8,880	15·58	3·53	...
Wandle, below Carshalton—chalk	70 to	350	41·0	1,800	43·9	9·93	24 0
Medway, driest seasons (Rennie, 1787)—clay	...	...	481·5	2,209	4·59	1·04	...
Ditto, ordinary summer run (Rennie, 1787)—clay	...	...	481·5	2,520	5·23	2·19	...
Verulam, at Bushey Hall—chalk	150 to	500	120·8	1,800	14·9	3·37	...
Gade, at Hunton Bridge—chalk	150 to	500	69·5	2,500	36·2	8·19	...
Plym, at Sheepstor—granite	800 to	1,500	7·6	500	71·4	15·10	45·0

Some of the results in this table are very striking, and are worthy of observation. It appears the Thames, the Severn, the Loddon, the Medway, and the Nene, which flow over a great variety of surface, many of them little absorbent, all carry off in the middle of summer less than one-eighth part of the average annual rainfall. Contrasted with this are two chalk rivers, the Mimram and the Wandle, which each discharge at lowest summer level nearly half the total average rainfall. This shows very clearly the influence of the springs by which such rivers are mainly fed. Rivers flowing in a clay basin are only fed by the rain falling within the actual basin ; and as this rain evaporates very slowly in winter and very rapidly in summer, such rivers are subject to great winter floods and to severe summer droughts. The flow in chalk districts is, however, much more uniform, because the rivers are fed by springs as well as by surface drainage ; hence the water stored up in subterranean reservoirs is discharged by chalk rivers even in the driest seasons. In fact, they draw their supplies from areas beyond their actual basin, and their discharge is much more uniform throughout the year than in most other rivers.

It is probable that the two other chalk rivers in Mr. Beardmore's table, namely the Verulam and the Gade, are not so largely fed from springs as the Mimram and the Wandle, the summer discharge of these not much exceeding that of the clay rivers.

WORKS OBTAINING A SUPPLY FROM DRAINAGE AREAS.

There are certain geographical and physical considerations connected with this subject which it will be proper to allude to before noticing some of the most remarkable works of this kind. A flat, low-lying country is seldom well adapted for the impounding of water by embanking across the valleys. In such a district long and shallow embankments would be required, and these would cause the water to spread out over a great area with a very shallow depth. Under these circumstances the water is apt to vegetate and become highly impure.

Again, in the low-lying districts of flat countries the rainfall is seldom nearly so great as in upland districts, so that much larger drainage areas must be sought. The water power is also much more valuable as the falls of rivers become less, and a much larger compensation is claimed by millers for the water abstracted for economical purposes.

Nearly all these conditions are reversed in the elevated districts and among the older rocks, considered geologically, in contrast with those of the secondary and tertiary formations. It follows from these considerations that most towns situate in low and flat districts are supplied from neighbouring rivers, while those in more elevated districts, such as the great manufacturing towns of Yorkshire, Lancashire, and some in Scotland, are chiefly supplied from water impounded by embankments across the valleys.

In addition to the general configuration of the valleys, which ought to be deep, and with precipitous sides flanked by lofty hills, there are several other points which require attentive examination in projects for collecting water from drainage areas. These are,

1. The area of the watershed.

2. The geological character of the soil as affecting its capacity to absorb rain and to allow the infiltration of water through it.

3. The character of the surface soil or covering of the district, as affording soluble ingredients, which may be taken up by the water and serve to contaminate its quality. In this point of view districts of decomposing peat, districts of arable agricultural land richly manured, and places thickly covered with population, are often highly objectionable.

4. The rainfall of the district, and especially the minimum fall in any one year.

5. The nature of the surface soil as affording facilities for procuring puddle and constructing retentive reservoirs.

The consideration of compensation to millowners, and \ly to landowners, where the water is used for irrigation.

The geological structure is extremely important in estimating the capacity of a drainage area. It is not alone the rain which falls on the sloping surface of the hills and finds its way by gravitation to the lower levels, but the effect of springs is also often very great in augmenting the quantity of water. Mr. Beardmore relates an instance where an oolitic district was found discharging a very large volume of water with scarcely any drainage area lying above or beyond it. In this case the porous strata, with a very small dip cropping out on the sides of the valley, were delivering the water which filtered into them far beyond the limits of the drainage area, as indicated by the levels of the surface. In fact, many districts will be found to have a geological drainage area as well as a surface drainage; and it often happens that the former is far the most important of the two.

Many drainage areas are also valleys of elevation, in which the strata dip in opposite directions on opposite sides of the valley. In this case it is evident that much of the rain falling on a porous surface will insinuate itself between the partings of the strata, and flow off in a direction contrary to that of the surface drainage. In this case we shall have the converse of that quoted by Mr. Beardmore, namely very large drainage areas yielding a very small volume of water, and therefore very inadequate as storage ground.

In the actual examination and investigation of drainage areas every possible case will be found. Some will yield only the quantity due to the surface drainage, and will be uninfluenced by springs either one way or the other. Some will yield in addition spring water which has been absorbed within the true drainage area, and other districts will yield spring water from a more distant drainage. A fourth watershed will yield far less than the quantity due to its own area and slopes, owing to the peculiar dip of the strata, a peculiarity which has been already alluded to. All these points require accurate and minute investigation in every special case. Nor is this sufficient. The actual discharge of water by the streams must be gauged day by day for several years, and compared with the indications of a rain gauge, kept as near the site of the drainage area as possible. If the quantity of rain which has fallen during a series of years be known, it is not essential to have the streams gauged during each of those years. The gauging of one, two, or three years may be sufficient to establish the proportion between rainfall and actual discharge. This proportion must then be applied to the minimum rainfall in any one year in estimating the capability of the reservoir.

It will be seen in the following table, chiefly taken from Mr. Beardmore's valuable work, which has been so often alluded to, how variable are the proportions which obtain between the whole rainfall and the available quantity which can be collected in storage reservoirs.

NAME OF DRAINAGE AREA.	Registered rainfall per annum.	Depth of rain per annum flowing off the surface as ascertained by gauging of the streams.	Proportion or percentage of rainfall which flows off the surface.
	inches.	inches.	per cent.
Bann Reservoirs (moorland) .	72·	48·0	66
Greenock (flat moor) . . .	60·	41·0	68
Bute (low country) . . .	45·4	23·9	53
Glencorse (Pentland Hills) . .	37·0	22·3	60
Belmont (moorland) 1843 . .	63·4	50·7	80
„ „ 1844 . .	50·0	33·3	67
„ „ 1845 . .	55·0	41·2	75
„ „ 1846 . .	49·8	33·2	67
Rivington Pike	55·5	24·25	44
„ from Stephenson's Report (1847 & 1848) . .	63·6	40·0	63
Turton and Entwistle 1836 .	46·2	41·0	89
„ „ 1837 .	48·2	39·0	81
Ashton	40·0	15·5	39
Drainage areas on south side of Longridge Fell, near Preston, May 1852, to April 1853	54·	15·5	29
		18·0	33
	...	22·0	43
Bateman's Evidence on the drainage area of Longdendale .			
First half of 1845. very dry .	21·2	13·5	64
Second half of 1845 . .	38·6	27·25	71
First half of 1846 . . .	22·5	17·5	78
Oct., Nov., & Dec., 1846. .	10·2	8·67	85

The proportion appears to range from one-third to four-fifths of the whole rainfall.

The preceding table represents the depth of rain falling per annum on certain drainage areas, and in the next column the depth of rain which will produce the actual quantity flowing in the streams and rivulets of the district. The difference between the two depths in each case is composed of the following :

1. The loss by evaporation and the moisture entering into vegetable life.

2. The amount absorbed by the soil, sinking into the ground, and not afterwards given out by springs within the drainage area.

The third column of the table shows the percentage of the whole rainfall which can be collected.

Mr. Hawkesley is said to have made experiments on an area of 100 square miles, which showed that 43 per cent. of the whole rainfall could be collected in reservoirs.

Mr. Stirrat found, as the result of three years' experiments at Paisley, that 67 per cent. of the whole rainfall could be collected and delivered in the town.

Some very accurate experiments were made in America, to ascertain the proportion between the rainfall and the depth which could be collected to supply the reservoirs of the Chenango Canal. These experiments were made in Madison County, New York.

One experiment was made on the watershed of Eaton Brook, an area of 6,800 acres, with a steep slope, and a compact soil underlaid by hard greywacke rock, elevated 1350 feet above the sea. The quantity of water flowing off this drainage area was accurately gauged every day for a period of two years, and was found to amount to 66 per cent. of the whole rainfall.

Another experiment was made on the watershed of Madison Brook, an area of 6,000 acres, 1,200 feet above the sea. The slopes here are not so steep as in Eaton Brook valley, and the soil is gravel, resting on greywacke. It was found in this case that 50 per cent. of the whole rainfall was carried off by the streams.

Experiments were made at two stations on the drainage ground of the Albany Waterworks. At the first station, having a watershed of 2,600 acres, it was found that from May till October inclusive only 41½ per cent. of the rainfall was carried off by the streams, but in the other six months, from November till April inclusive, 77·6 per cent. was so carried off. This was in the year 1850; but in the very next year, 1851, the streams carried off, between May and October inclusive, no less than 82·6 per cent.

On another area of 8,000 acres the streams carried off, from July to December inclusive, 33·6 per cent., and from January to June inclusive, 53·6 per cent.

The following table, showing the capacity of reservoirs in proportion to drainage area, is taken partly from Beardmore and partly from other sources.

NAME OF RESERVOIR.	Drainage area. Square miles.	Reservoir Room per square mile in millions of cubic feet.	Total capacity of reservoir in millions of cubic feet.
Greenock	7·88	38	300
Glencorse (Edinburgh) . .	6·00	7·66	46
Belmont	2·81	26·8	75
Rivington Pike. . . .	16·25	29·6	481
Turton and Entwistle . .	3·18	31·43	100
Bolton	·80	25·6	20
Sheffield	1·42	36·5	52
Ashton	·59	21·0	12
Longdendale	23·8	12·3	292
Proposed reservoir for Wolverhampton Works . . .	22½	·7	16
Albany Works, U.S. . . .	29	1·1	32
Dilworth reservoir of Preston Works, Lancashire . .	·092	54·0	5
Homersham's estimate of 24,000 cubic feet of reservoir to each acre of drainage . . .	1	15·36	15·36

CAPACITY OF IMPOUNDING RESERVOIRS IN PROPORTION TO THE SUPPLY TO BE AFFORDED.

The Rivington Pike reservoir was to contain 3,156 million gallons, and was intended to supply Liverpool with an average of 13 million gallons a day, besides 8 million gallons a day to millers. Hence it is calculated to hold 150 days' average supply.

The compensation reservoir of the Gorbals Gravitation Works for the supply of part of Glasgow, contains 12 million cubic feet, and covers an area of 30 acres.

The other reservoir covers an area of 40 acres, and contains 38 million cubic feet of water, or about 80 days' supply.

The reservoir of the Bolton Works contains nearly 21 million cubic feet, and has to supply 900,000 gallons a day, so that it holds 146 days' supply.

The Belmont reservoir contains 75 million cubic feet, and has to supply nearly 3½ million gallons a day, so that it contains about 136 days' supply.

The Longdendale reservoirs for Manchester were to contain 292 million cubic feet, and were intended to furnish a supply for Manchester (including the compensation to millers) equal to 74 days.

The reservoirs of the Preston Works are four in number, at different levels above the town, varying from 448 feet to 50 feet. They contain when full 167 millions of gallons, or about half a year's supply for the population of 80,000 persons.

DIMENSIONS OF EMBANKMENTS FOR IMPOUNDING RESERVOIRS.

Bateman's Compensation Reservoir at Longdendale has an area of 123¾ acres, and contains nearly 155 million cubic feet. The embankment is 27 feet wide at top, and 4 feet high above top water ; inside slope 3 to 1, outside 2 to 1.

His Crowden reservoir has an area of 18 acres, contains 18,493,600 cubic feet, the embankment is 15 feet wide at top,

4 feet high above surface of water, and the same slopes as the Compensation reservoir.

The embankment for the impounding reservoir of the Great Croton Waterworks for supplying New York is composed of earthwork, with a base of 275 feet. Behind this is a mass of masonry, 8 feet wide at top and 65 feet at the base. The masonry is built upright on the upstream side, or that adjoining the earthen embankment, but with occasional offsets. The outer face of the masonry has a curved form, so as to pass the water over without giving it a direct fall on the apron at the foot. The apron is formed of timber, stone, and concrete, and extends some distance from the base of the masonry, so as to afford protection at the point where the water has the greatest force.

A lower dam has been built at the distance of 100 yards below the masonry of the main dam in order to pen up and form a basin of water setting back over the apron at the toe of the main dam, in order to break the force of the water falling on it. This lower or secondary dam is formed of round timber, brushwood, and gravel.

RESERVOIRS OF THE GORBALS GRAVITATION-WORKS.

The Ryatt's Lynn, or upper reservoir, which holds the compensation water for the millers, covers 30 acres of ground, and has a capacity of 75 million gallons, the surface of water being 300 feet above the Broomielaw quay at Glasgow. It is formed by an embankment about 450 feet in length, with a height in the centre of 40 feet.

The lower or Waukmill reservoir covers an area of 40 acres, and has a capacity of 38 million cubic feet. It is formed by an embankment about 550 feet in length, with an extreme height of 55 feet.

AMERICAN WORKS.

The dams constructed by Mr. McAlpine for the reservoirs of the Albany Works were 10 feet wide at top, and were carried up 8 feet above top water line. The slope on the outside is 2 to 1, and the inside has the same slope to the bottom of the conduit, where a berm of 5 feet wide is made, and thence to the bottom of the dam is a slope of 3 to 1. The inner slope of the bank is pitched with stone to the level of the bottom of the conduit, and the top and outer slopes are covered with turf. Through the bank is carried up a puddle wall, 8 feet wide at the top, and increasing in width at the rate of 4 feet for every 10 feet of depth.

His dams for the reservoirs of the Brooklyn Works are 20 feet wide at top, and carried up 5 feet above top water line. The slope on the outside is 2 to 1, and on the inside 3 to 1.

In the centre of the bank is a puddle wall of clay, 8 feet wide at top, which is 3 feet below the top of the dam, and increasing in width at the rate of 4 feet for each 10 feet of depth. The top and outer slope of the dam are covered with turf, and the inner slope is protected by stone pitching to the level of the conduit.

Great precautions are necessary in the construction of large embankments for the purpose of impounding water. An accurate examination of the ground is essential, to determine whether the seat of the embankment requires puddling, in which case the puddle of the seating should be perfectly joined, and worked into the puddle trench carried up through the middle of the bank.

The bank should be formed in layers not exceeding 2 or 3 feet in thickness, and should be kept higher at the sides, especially the outer side, than in the centre, and every means should be taken to consolidate the materials and prevent slips. A judicious examination of the material to be used is also necessary, as any kind of clay approaching in its nature to 'fullers' earth' would be highly objectionable, owing to its property of being acted on by water.

Mr. Thom, of Glasgow, who has had great experience in the construction of these works, recommends that the embankments should have a slope of not less than 3 to 1 on the water side. He does not approve of puddled trenches in the bank ; but after excavating the foundation to such a depth as to be firm, and to prevent the passage of water, he forms the bank by spreading alternate layers of puddled peat or alluvial earth and gravel, beating them well with wooden dumpers till they are completely mixed. He then covers the slopes with a puddle made of small stones or furnace cinders mixed with clay, so as to prevent the possibility of moles or other vermin penetrating into the embankment. Mr. Thom refers to many reservoirs he has constructed in this way without puddle trenches, at Greenock, Paisley, and elsewhere.

COST OF IMPOUNDING RESERVOIRS.

This has been very variable, owing to the great range of prices, which in some cases have not exceeded 6*d*. per cubic yard for the bank, while in others the price has been 1*s*. The following are some examples showing the cost of large reservoirs, including every expense of earthwork, puddling, pitching, waste-weirs, valves, &c., but exclusive of land.

Bateman's Crowden reservoir, to contain 18,493,600 cubic feet of water, to cost £10,100, or £561 per million cubic feet.

His Armfield reservoir, containing 38,755,556 cubic feet, to cost £17,065, or £438 per million cubic feet.

His Hollingworth reservoir, containing 12,348,100 cubic feet, to cost £5,500, or £458 per million.

His Armfield Moor reservoir, containing 13,072,581 cubic feet, to cost £8,100, or £623 per million.

His Tetlow Fold reservoir, containing 8,849,310 cubic feet, to cost £6,500, or £722 per million.

All the above are connected with Mr. Bateman's Longdendale Scheme for supplying Manchester, and appear to be estimated at fair prices, the embankments being taken at 1*s*. per cubic yard, including puddle

His large compensation reservoir for the millers is to contain 154,573,420 cubic feet, and to cost £11,250, or at the rate of only £73 per million feet. This is so much at variance with all the others, that there must be something peculiar to account for it. One must either conceive a remarkably favourable configuration in the valleys to admit of such an enormous volume of water being dammed up by a comparatively small embankment, or assume that a lake or body of water already exists at the place in question, and that the proposed embankment will increase its volume to the extent indicated by the figures.

The Spade Mill reservoir, on the Preston Waterworks, contains 20,934,824 cubic feet, and was estimated at £5625, or £268 per million feet.

The Knowl Green reservoir, on the same Works, contains 7,256,869 cubic feet, and was estimated at £4,288, or £612 per million.

In the preceding examples the prices appear to range from £268 to more than £700 per million feet; this great difference being chiefly caused, not by the difference of price, which is nearly the same in each case, but principally by the variation of shape and form in the valleys, some of which admit so much more readily than others of water being stored up. Most of these examples are taken from districts of millstone grit, where the valleys are deep and the sides precipitous. When embankments are made across valleys of a more open character, and with flatter slopes, the cost of water stored up will evidently be much greater.

IMPOUNDING RESERVOIRS WITH DAMS OF MASONRY.

These have not been much used in this country, but have been extensively adopted in France by eminent engineers of that country. Mr. Conybeare, in his Report on the supply of water to Bombay, quotes three examples of large stone dams executed on the Canal du Midi and other canals in France. These dams vary in height from 40 to 70 feet, and are constructed according to a formula given in the *Aide-mémoire*, and which is commonly followed by the French engineers. In this formula the thickness of the wall is made as follows :—at bottom seven-tenths of the height, at middle five-tenths, and at top three-tenths.

These dams of masonry, it is believed, would be much more expensive than earthworks in this country; and from a comparative estimate made by Mr. Conybeare for his own case of the Bombay Works, he found that while a dam of masonry would cost £235 per yard forward, one of earthwork would only cost £100.

SERVICE RESERVOIRS.

This is the name given to the small reservoirs holding from one to three days' supply, which are constructed in the immediate neighbourhood of a town. The most useful kind of service reservoir is one situate upon an eminence at a sufficient height to give high pressure over the tops of all the houses. Sometimes however the service reservoir is at such a low level that the water has to be pumped up from it. Under all circumstances, however, the contents of the service reservoir will be available in case of any accident happening to the main which supplies it, or to any more distant part of the works.

Service reservoirs as formerly constructed were commonly mere open ponds, either with upright or sloping sides, lined either with concrete, brickwork, or masonry. The sanitary principles of the present day, however, seem to require that, at all events in the neighbourhood of large towns, the service reservoirs should be covered over with a roof either of brick or stone. The advocates of covered reservoirs not only claim in their favour the advantage of preserving the water free from soot and the atmospheric impurities of large towns, but they also insist strongly on the superiority which the water possesses in other respects when stored in covered reservoirs. The principal of these are the uniform temperature and the freedom from vegetation, which is found to be a serious annoyance, especially in water from the New Red Sandstone, when exposed to the action of air and light.

The construction of covered service reservoirs is **extremely** simple. A series of parallel walls or piers is built throughout the reservoir, and between these arches are turned, of half a brick or one brick in thickness, according to the span or distance between the piers. The arches are either semicircles or flat segments of a circle. Other covered reservoirs are constructed by building parallel rows of brick, iron, or stone columns, which support cast-iron girders, and from these girders the arches spring, as before.

The cost of covered reservoirs varies from 30s. to £3 per thousand gallons of capacity. One of the cheapest covered reservoirs which have been constructed is that of the Wolverhampton Waterworks at Goldthorn Hill, from the designs of Mr. Marten. This reservoir is in two parts, containing together 1,500,000 gallons, and the cost exclusive of land is said not to have exceeded £2,200.

The covered reservoirs, with solid brick piers, appear to be somewhat less expensive than those built with brick arches supported on iron columns and girders.

The following trial estimates, made in my own office for various reservoirs at the same schedule of prices in each case, will show the difference :—

Contents of Reservoir.	With brick arches on brick piers and cross walls.	With brick arches on iron columns and girders.
1 million gallons	£ 2696	£ 2920
1½ million do.	3588	3937
2 million do.	4162	4469

The above prices do not include the cost of land, sluices, or culverts.

One of the best and most recent examples of elevated service reservoirs is that erected by the late Mr. Simpson on Putney Heath, for the new Works of the Chelsea Company.

In the works designed by Mr. Simpson for this Company there is an obvious advantage over the schemes of the other London Companies, namely that the water is pumped up to an intermediate summit reservoir on Putney Heath, capable of containing 10 million gallons, and from this reservoir the

water gravitates to all parts of the Chelsea district. The
spot selected for the reservoir is the highest ground on
Putney Heath, being situate immediately on the west side of
the Old Portsmouth Road, and also adjoining the west side
of the road from Wimbledon to Fulham. This part of the
heath is distant six miles from Seething Wells near Thames
Ditton, where the new supply is taken from the river Thames,
and 165 feet above the river at that point.

The Works at Putney Heath consist of a double covered
reservoir, to contain filtered water for the domestic consumption of the district, and of a smaller open reservoir, to contain
unfiltered water for the supply of the Serpentine, and to fill
a main pipe for the purpose of watering the streets. The
covered reservoir is in duplicate, each part having an area at
the water surface of 310 feet by 160 feet, and a depth of
20 feet, the sides all round having a slope of 1 to 1. This
gives a mean area of 290 feet by 140 feet, and a capacity of
5,075,000 gallons for each reservoir, inclusive of the space occupied by the piers. Hence the whole capacity may be taken,
as stated by Mr. Simpson in his Evidence, at 10 million gallons. The sides of the reservoir are cut out in the form of
steps, which are filled up with concrete to a uniform slope of
1 to 1. A bed of concrete one foot in thickness is also laid over
the whole bottom. Each half of the reservoir is covered with
8 brick arches, averaging rather less than 20 feet span, the
side arches being each 20 feet span, and the others 18 feet 8
inches. The piers, supporting these arches, are built length-
ways, and are each 310 feet long at top and 270 feet at base.
The arches are each one brick in thickness, and are covered
over with a layer of puddle, the haunches being filled up with
concrete. The piers are carried up 14 inches thick, but the
division wall between the two parts is rather more than 4 feet
thick, with a concrete slope of $1\frac{1}{2}$ to 1 on each side. The 14-
inch piers supporting the arches are built with large circular
hollows $17\frac{1}{2}$ feet diameter. The centres of these circular hollows are 40 feet apart, so that solid brickwork 23 feet long is
left between the circular hollows, supposing a horizontal sec-

tion taken through the centres of the hollows. Each of the 23-feet spaces has a 14-inch counterfort carried out at right angles. These counterforts occur at intervals of 26 feet and 18 feet alternately, and project 6 feet wide at the base on each side of the pier, and run out to nothing at the top or springing of the arches. In each of the 18-feet spaces between the counterforts there is a smaller circular hollow of 5½ feet diameter. The arches spring from skewbacks formed of carefully moulded bricks, perforated longitudinally with inch hollows. These bricks are made at Kingston-upon-Thames by machinery expressly designed for the purpose. The versed sine, or rise of the arches, is 4 feet 3 inches, or rather more than one-fifth of the span. Each arch is provided with two openings in the centre, communicating with a line of 12-inch earthenware tubular pipe, which passes through the spandrils, and communicates with perforated iron tops in the division wall between the two parts of the reservoir. By this contrivance the space above the water in the covered reservoirs is effectually ventilated.

The supply-pipe from Thames Ditton is 30 inches diameter, and comes into each part of the reservoir at the level of top-water, which is a few inches below the springing of the arches. At this level a waste weir or overflow is fixed, to prevent the reservoir from being filled too full. The exit mains to London consist of two 24-inch pipes, and they pass off from the bottom of the reservoir which has an inclination in one direction of 1 in 20, and a fall across of 6 inches. The surface of top-water in the reservoirs will be 163 feet above Trinity high-water mark, the highest part of the district to be supplied being Queen's Road, Kensington, where the houses stand on ground 135 feet above Trinity high-water mark, and where the ground immediately beyond the houses is 145 feet high. To supply the tops of these houses a stand-pipe, 45 feet in height above the surface of the reservoir, will have to be erected.

Open Reservoir at Putney Heath.

Closely adjoining these covered reservoirs, and on the north side of them, is the open pond to contain the unfiltered water for flushing sewers, watering streets, and supplying the Serpentine in Hyde Park. The area of this reservoir is 194 feet by 104 feet at top. The depth is 12 feet, with concrete slopes of $1\frac{1}{2}$ to 1. Hence the mean area is 172 feet by 86, and the capacity, when full, something more than a million gallons. The supply-pipe into this reservoir is 15 inches diameter, and enters at the bottom, discharging through two openings fitted with valves, which open only to admit the water, but will not allow it to run back. The entering pipe is continued vertical to a height above the surface of water, in order to admit of the escape of air, &c. The surface of water in this reservoir is $6\frac{1}{2}$ feet above that in the covered reservoir: this is to produce a more efficient discharge through the 12-inch pipe which leads to London.

The 12-inch discharge pipe goes off from a circular well sunk 5 feet below the floor of the reservoir, which has a fall or slope of 18 inches. The depth of water is therefore 12 feet at the upper part, and $13\frac{1}{2}$ feet at the lowest.

FILTER-BEDS.

There are two distinct methods of constructing these, in one of which the various kinds of filtering material are placed in compartments side by side, while in the other kind of filter-bed the materials are placed in successive layers one above the other. The first is the method commonly adopted in Scotland, as used for the Gorbals Works, and for the Works at Paisley, Kilmarnock, and other towns. The Scotch system of filtration is also that which has been adopted by Mr. Wrigg in the new Works which he is executing for the town of Preston in Lancashire. The other mode, namely that of filtration by descent through successive horizontal layers, was first adopted in England by Mr. Simpson for the Chelsea Works, and has since been followed in all those numerous English Waterworks in which filtration is practised at the present day.

Scotch System of Triple Filtration.

The filters of the Gorbals Gravitation Works of Glasgow afford, perhaps, the best example of the filtration through compartments. They were designed to filter about 3 million gallons per day, but the quantity is gradually increasing, having been 2,904,000 in 1852, and 3,274,000 in 1854, with a further increase during the last year.

The filters are situate on the brow of a hill about 330 yards distant from the Regulating House, and the water is carried along the sloping surface of the ground in an arched stone culvert, which is laid on a dead level all the way. The culvert is flat-bottomed, 4 feet wide and 4 feet high.

The filters occupy a rectangular space 360 feet long by 80 feet wide. The length of 360 feet is separated into two compartments by a division wall, and the breadth of 75 feet, is divided into three spaces on each side of this division. The filter may be described, therefore, as a double range of three compartments, each range being 180 feet long and 75 feet wide. The first compartment, or that nearest to the delivery culvert, is 15 feet wide and 4½ feet deep, being filled with broken freestone. The second compartment is nearly 24 feet wide, and is filled with gravel to a depth of 3 feet. The third compartment is 34 feet wide, and is filled with coarse sand to a depth of 2 feet. The bottom of all the three compartments in each range of filters is on the same level, and was thus prepared:—the bottom, after being excavated to the proper depth and well levelled, was filled with one foot of puddle, and on this was placed a layer of small stones or very coarse gravel, which was well beaten in, to form a surface proper for the reception of a layer of cement one inch in thickness.

On this layer of cement rows of brick on edge are laid, one inch apart longitudinally, and 10 inches apart from centre to centre, measured transversely. On this open groundwork of bricks rests a close-set flooring of tiles one inch thick perforated all over with holes one-eighth of an inch in diameter. This foundation is the same for each of the filter-beds, but the material with which they are filled is different, as already stated. The broken freestone in the first compartment is in pieces about the size to which road-metal is commonly broken, namely such as will pass through a 2½-inch ring. The second compartment contains screened gravel, and the third contains coarse sand procured from the larger Cumrae Island, in the Frith of Clyde. The culvert which conveys the water to be filtered approaches at one corner of the compartment filled with the freestone, and passes along the whole length of the two ranges at the back of this compartment. This part of the culvert is a rectangle, 4 feet wide, 2½ feet high, and is covered over by 6-inch flags, through which pass the screwed rods of a series of sluices with adjusting nuts at the top. Each range of filters is provided with 10 sluices. The openings of the sluices are long and narrow, so as to admit the water in a thin sheet on to the surface of the first or coarse filtering compartment.

Between each of the filtering compartments is a passage two feet wide, and extending the whole length of the two ranges of filters, in which the water rises up to its original level, after having passed down through the filtering medium. There is also a similar passage between the third or final filtering medium and the pure water basin. Each of these passages is provided with twelve sluices, to admit the water from under the filter-beds. The process of filtration will now be readily understood. The water passing from the regulating valve-house through the culvert already described, proceeds along at the back of the coarse filtering medium, and the sluices being open, it spreads gently over the surface of broken freestone, through which it percolates with considerable freedom. Having passed through this, it finds its way, by means of the perforated tiles and the open brick channels on which they rest, into the first passage, where it rises nearly to the level of the first filtering bed. At this level it pours over in a thin film on to the surface of the second or gravel filter, the top level of which is 18 inches lower than that of the freestone. Here it goes through the same process as before, and then enters the sand filter, which again is 18 inches below the level of the gravel. Finally, having passed through the sand filter, the water rises in the third passage and falls over into the pure water basin, which consists of two compartments, each 180 feet long and 66 feet wide. The side walls of the pure water basin have each a batter of 3 feet on the inside, so that the breadth at bottom is reduced to 60 feet. The depth of water is commonly about 16 feet. The duplicate arrangements of the filters and pure water basins afford convenient facilities for emptying and cleansing at any time, without interfering with the progress of the Works.

An ingenious arrangement is adopted for cleansing the filter-beds by means of an upward current of water, which carries the impurities that have been deposited up to the surface of the bed, where they are floated off to the drains. This cleansing current is brought by a 6-inch pipe from the level of the culvert before it discharges into the first filter-bed. The 6-inch pipe passes under the division wall between the two ranges of filters, and a stop valve branches off into each of the first compartments. When it is desired to cleanse by means of the upward current the first compartment of the filter-bed, the sluices communicating between it and the first or adjoining passage are closed. The same movement of the spindles which closes these bottom sluices opens those at the top, and the upward current of water, carrying with it the deposited matter mixed with the broken freestone, is taken off by a drain which passes under the division wall. In the same way the second and third filters are cleansed by the ascending current, which is admitted to them by opening the bottom sluices, and allowing the current from the 6-inch pipe to enter the filtering bed at the bottom.

Filter-beds of the Chelsea Waterworks.

The filter-beds at the old Works of this Company on Thames

Bank were constructed about the year 1839, and have served as a model for most of those which have since been constructed.

The old filter-beds were two in number, the southern one being rectangular, 240 feet long by 180 feet wide, and the northern one 851 feet in extreme length by 180 feet wide. These filter-beds had an area of 9,000 yards; and taking the average daily quantity filtered in 1853, the area appears to be at the rate of one square yard to 626 gallons. The sides of the filter-beds were embanked to a height of 12 feet above the natural surface of the ground. The slopes were covered with turf. The bottom was puddled with clay 18 inches in depth. In the northern filter-bed were nine brick tunnels, and in the southern eleven, which were laid upon the clay puddle, and extended from one end of the filter-bed to the other. Each of the tunnels was 8 feet in diameter and 18 inches thick, built with every alternate brick left out, so as to form a free passage for the water into the tunnels. The brick tunnels were then surrounded on all sides, and covered over to the depth of 2 feet with coarse gravel. Above this was a layer of 6 inches of shells, upon this a bed of coarse sand, and then a bed of fine sand. The depth of the beds above the gravel was about 5 feet. A deposit of 2 or 3 inches in depth was formed on the sand, and required to be washed off at intervals. This operation was performed in a few hours, and the intervals at which it was required depended on the action of the wind and tide. The water was admitted into the filtering ponds by a number of openings, corresponding with the valleys or hollows in the filtering material, the brick tunnels being also laid in these hollows. The sediment deposited on the surface of the sand required to be scraped off twice a week in the summer-time, and about once in ten days in the winter. From a quarter to half an inch of sand was scraped off each time with the sediment. When the depth of the sand was reduced by this process to about a foot, a fresh supply of sand was added. Nearly 8,000 cubic yards of fresh sand were annually required for this purpose.

The surface of water in the filtering beds was about a foot above the bottom of the settling ponds.

The water after filtration passed through a culvert to the engine well, whence it was pumped into the mains for distribution.

The filter-beds were provided with weirs about 4 feet wide, to let off the water into a culvert in case it should rise in the filtering bed faster than it can be filtered.

The Works at Thames Bank, however, were abandoned in 1856, when the new works at Seething Wells came into operation.

NEW WORKS AT SEETHING WELLS.

The depositing reservoirs and filter-beds are constructed close to the side of the river, in a long narrow strip of ground which borders the river between Raven's Island and Thames Ditton. The Lambeth Works occupy the southern extremity of this strip, and the Chelsea Works are being made close to them, and occupy the northern extremity. The reservoirs and filter-beds are bounded on the west side by the river, and on the east by the turnpike road leading from Thames Ditton to Kingston.

Depositing Reservoirs.

A substantial concrete wall has been built alongside the river for the whole extent of the Works. This wall is about 2 feet 6 inches wide at top, with a batter on the face of about 2 inches per foot; at the back of the wall are substantial counterforts, about 14 feet apart from centre to centre. The two depositing reservoirs are each 272 feet by 226 feet at top, while at bottom the dimensions are 24 feet less in each direction, namely 248 by 202 feet. The reservoirs are 12 feet deep, but the usual depth of water in them will be 8 feet; the mean area of each being 255 by 214 = 55570 square feet, the capacity of each at 8 feet deep will be 2,778,500 gallons.

The level of water in each of these reservoirs is the same

as that of the river, the water being simply admitted through a pipe opening in the concrete wall, so that there is no lift into the depositing reservoir, as at the old works at Thames Bank. The sides of the reservoirs are formed of substantial concrete walls, having a slope of 1 to 1 towards the inside, while the outside is formed in steps, so arranged as to make the average thickness of the mass of concrete about 3 feet. The bottom is also lined with a small thickness of concrete. The work is situate in a diluvial covering of the London clay, and the excavation extends several feet into the solid blue clay containing septaria. No puddle is used for the walls, the concrete being merely placed upon the steps cut out of the clay.

Filter-Beds.

Owing to the peculiar shape of the site to which these Works are confined, the two filter-beds are not each in contact with one of the settling reservoirs, but the whole four are placed in a row thus:—Beginning at the north is, first, reservoir No. 1, then reservoir No. 2, filter-bed No. 1, filter-bed No. 2. The filtering beds are each 300 feet by 150 feet at the surface, having an area therefore of 45,000 square feet each, or in all 10,000 square yards—about the same capacity as the filters at the old Works. The sides of the filters are formed, like the reservoirs, with concrete slopes of 1 to 1. The surface of water in the filtering beds is about 3 feet above the bottom of the settling ponds, in order to allow a certain depth for deposit. The depth of water above the filtering material is generally 4 or 5 feet. The filtering material is composed as follows, beginning at the top :—

	ft.	in.
Fine sand, the surface of which will occasionally be scraped off, so that the thickness will be reduced, but at first it will be formed of a thickness equal to	2	6
Coarse sand		6
Shells, consisting of perfectly clean bleached cockle, tellina, and other shells, from Shellness, or some similar bays or havens	0	6
Fine gravel	0	4
Coarse gravel, varying in depth from a few inches where laid over the perforated pipes to about two feet where laid between them.		

Below the filtering material the water drains off by means of perforated tubular pipes, stretching across the filter-bed, and communicating with a central inclined channel. Each filter contains on either side of the longitudinal centre drain five rows of transverse pipes. The pipes forming these transverse lines are of three sizes, namely 6 inches, 9 inches, and 12 inches diameter. Each row begins at the side of the filter-bed with 6-inch pipe, and terminates at the centre with 12-inch, an arrangement by which the size of the pipe is rudely proportioned to the quantity of water which it has to carry off. The arrangement, in fact, represents an arterial system of drainage, in which the area of the pipes at successive points is proportionate to the quantity of water passing through them.

The tubular drain-pipes are made in lengths of 2 feet, with butt joints, and are bedded on small earthenware chairs, which serve to keep them in position, in the same way as the socket or half socket joint, which are sometimes used for the same purpose. The pipes are perforated all over with $\frac{3}{4}$-inch round holes; and when laid in the chairs the butt ends are not laid close together, but with a space between them of half an inch or so.

The pipes were made at Bristol by Messrs. Pountney and Goldney, from a description of fire-clay found in the neighbouring coal-field.

The pumping station is on the opposite or north side of the road leading from Thames Ditton to Kingston, so that the pipe conveying the filtered water to the engine well is carried under and across the road. The engines for pumping the filtered water are four in number, of a united horse power equal to 600 horses. They have thirteen single-flued Cornish boilers, each 31 feet 9 inches long, and 5 feet 8 inches in diameter. The flues are 3 feet 3 inches in diameter.

In addition to these four engines, which were erected in 1856, two new auxiliary engines were put up in 1866. These are similar to the four older engines, and are provided with seven similar boilers:

Two separate engines of 25 horse power each are also employed at Seething Wells for pumping the unfiltered water to the open reservoir on Putney Heath.

The numerous filter-beds constructed by Mr. Hawkesley at Leicester and other places, are usually designed with an area of 1 square yard for each 700 gallons of water to be filtered in twenty-four hours.

The kind of filter adopted by Mr. Hawkesley is similar to those which have been described in Mr. Simpson's works, the principle being that of descent through horizontal beds.

We have seen that the surface of filter-bed at the old Chelsea Works was at the rate of 1 square yard to 626 gallons. Mr. Bateman is understood to adopt a proportion of 1 square yard to every 675 gallons of water filtered in twenty-four hours. In the filter of the Works for Chester, however, Mr. Bateman proposes only 953 square yards to filter 960,000 gallons a day, or at the rate of 1 yard to 1007 gallons. Mr. Lynde, at Norwich, has adopted the rate of 1 square yard to 513 gallons. In the Gorbals Works at Glasgow, where the filtering material is in three separate compartments, the proportion of the whole surface is 1 square yard to 1135 gallons; while Mr. Wrigg, at Preston, adopting the same system of triple filtration, has the proportion of one square yard to 456 gallons.

The cost of constructing filter-beds ranges probably from 18s. to 30s. per square yard. The following is the result of detailed estimates made in my own office for filter-beds in connection with the Wolverhampton Waterworks, calculated from a uniform scale of prices :—

	£.
Filter-beds containing an area of 5,625 square yards, calculated to pass 3 million gallons in 24 hours, would cost	7302
Filter-beds containing an area of 2813 square yards, calculated to pass 1½ million gallons in 24 hours, would cost	3791

The cost of working filter-beds, comprising the attendance of labourers, scraping, cleaning, and restoring sand where necessary, is usually estimated at about 10s. per million gallons of water passed through.

ON GAUGING THE DISCHARGE OF RIVERS AND STREAMS.

This is an operation which comes frequently within the province of the hydraulic engineer, sometimes to determine the quantity to be relied on for his own works at various seasons of the year, and frequently as involving and connected with questions of compensation to millers, landowners, and others, for water to be abstracted from existing streams.

Large volumes of water are frequently conveyed in artificial open channels for the supply of towns; and the method of determining the relations between the quantity delivered, and the section and inclination of such channels, is of course one which claims particular study and attention.

There are several methods of gauging, the nature of which may be briefly noticed.

1st. The method of determining the discharge from the area and inclination of uniform channels. This method is not applicable to ordinary rivers, but has often to be adopted in finding the discharge of new cuts, or channels made to convey water either to impounding reservoirs, or from these to the town for distribution. The New River is an example, on a large scale, of a work of this kind.

2nd. Gauging by means of the surface velocity in the centre of the stream.

3rd. Gauging by means of sluices or orifices.

4th. Gauging by means of current-meters or other instruments for observing velocities at various depths; and

5th. Gauging by means of weirs.

1. *Motion of Water in Uniform Open Channels.*

The French writers Coulomb and Prony have shown that the velocity of water moving in channels of this kind may be derived from the equation

$$R I = a U + b U^2 \quad . \quad . \quad . \quad (5);$$

in which R represents the *hydraulic mean depth*, or the quotient derived in dividing the sectional area of the water by the wetted perimeter or border. I represents the inclination or fall of the surface-water* per foot run, and U the velocity in feet per second, a and b being coefficients which have to be determined by experiment.

During three-quarters of a century this problem has engaged the attention of some of the most able intellects of France, Germany, and England.

Among those who have experimented on the subject are Du Buat, Coulomb, Prony, Bossut, Couplet, Dr. Young, Girard, Woltmann, Funck, and Brünning. Some German writers, as Eytelwein and Weisbach, have chiefly contented themselves with reducing and comparing the experiments of others, and founding upon them formulæ of the highest value for the purpose of hydraulic calculations.

Eytelwein especially, in his 'Handbuch der Mechanik und der Hydraulik' (Berlin, 1801), has rendered essential service in this department of practical science. After reviewing all the experiments that had been made on the subject of water flowing in uniform channels, he shows that the mean velocity is very nearly $\frac{10}{11}$ of a mean proportional between the hydraulic mean depth and twice the fall in feet per English mile. Let H be the fall in feet per mile, then

$$U = \tfrac{10}{11} \sqrt{2\,R\,H} \quad . \quad . \quad (6);$$

or the velocity in feet per minute will be $54\frac{1}{2}$ times the square root of the mean hydraulic depth multiplied by twice the fall in feet per mile. This is the rule by which Mr. Beardmore

---

* Du Buat, Prony, and the other French writers commonly use what is called the hydraulic inclination, or surface slope of the water, in their formulæ. In short lengths of canal or pipes this is not to be confounded with the inclination of the pipe or channel, as it may vary considerably from this. In long lengths however, say upwards of 100 feet, the hydraulic inclination may be taken to be the same as that of the bottom of the channel. Mr. Neville has pointed out the distinction between the inclination of pipes and the term "hydraulic inclination" in the introduction to his Hydraulic Formulæ. He there shows that serious errors have been made by applying Du Buat's formulæ to short lengths of pipe, in which the slope of the pipe has been confounded with the hydraulic inclination.

has computed the table of velocity and discharge for arterial drains and rivers in his excellent work entitled 'Hydraulic Tables.' The only difference is, that in order to facilitate calculation, he has taken 55 as the multiplier instead of $\frac{600}{11} = 54.54$, which would be the exact multiplier derived from Eytelwein's valuable formula.

Let it be required to compute the velocity and discharge of a channel having the following dimensions :—

Width at bottom, 12 feet.

Side slopes of channel, 2 to 1.

Depth of water, 7 feet.

Inclination or fall, 3 inches per mile.

Here the area of water is $\overline{12 + 14} \times 7 = 182$ feet.

The wetted perimeter is $12 + 2\sqrt{14^2 + 7^2} = 43.3$.

Then $R = \dfrac{182}{43.3} = 4.2$, the hydraulic mean depth.

Also $H = .25$.

Then $55\sqrt{4.2 \times .25 \times 2} = 55 \times 1.414 = 78$ feet per minute for the velocity, and $182 \times 78 = 14196$ cubic feet discharged per minute.

Mr. Neville, in his Hydraulic Formulæ, gives the following modification of Eytelwein's equation :—

$$U = 5604\sqrt{RS} . \quad . \quad . \quad . \quad . \quad (7),$$

where S is the natural sine of the inclination or $\dfrac{H}{L}$. This formula gives the same result as Equation 6, which latter however has the advantage of being more easy to calculate. Hence we may take the velocity in feet per second to be $\frac{10}{1}\sqrt{2\,RH}$; and per minute $U' = \frac{600}{11}\sqrt{2\,RH} . \quad . \quad (8).$

This formula is also applicable to pipes and culverts both circular and oval, whether running full of water or with any less quantity; but is not applicable to pipes under pressure, such as those employed for carrying water from a steam-engine or from a stand-pipe for the supply of towns.

2. *On Gauging Rivers by means of the Surface Velocity in the Centre of the Stream.*

The relation between the mean velocity of water in a channel and that of the surface velocity in the centre, is another subject which has long engaged the attention of hydraulic engineers. Among those who have experimented on the subject Du Buat and Prony are again in the first rank. Du Buat operated on a canal whose section was tolerably uniform, sometimes rectangular and sometimes trapezoidal, whose greatest breadth was about 18 inches, and depth from 3 to 9 inches.

Prony, using the results derived by Du Buat in this experimental channel, established the empirical formula $U = \dfrac{V (V + 2 \cdot 37187)}{V + 3 \cdot 15312}$, where U and V are in metres per second, U being the mean velocity per second and V the surface velocity per second. This equation, reduced into English feet, becomes $U = \dfrac{V(V + 7 \cdot 783)}{V + 10 \cdot 345}$ . . . (9).

Mr. Neville has used this formula for calculating the velocities in small channels. Prony states, that for surface velocities less than 10 feet per second U may be taken simply
$$= \cdot 816 \, V . \quad . \quad . \quad . \quad (10),$$
but that the formula is only to be taken as a simple empirical rule, subject to modification according to the section and slope of the channel to which it is applied.

Recent experiments which have been made with great minuteness by M. Baumgarten on the river Garonne, and on the canal between the Rhone and the Rhine, are scarcely more satisfactory. In these experiments, which were made in channels varying from a foot to 25 feet in depth, it was sought to determine the surface, bottom, and mean velocities at a number of vertical lines across the stream.

Putting v for the mean velocity in any vertical line, and V for the surface velocity, M. Baumgarten found for a navigable canal, with earthen slopes varying from 2 to 6 feet in depth, that $v = \cdot 903 \, V$, and for smaller canals he found $v = \cdot 901 \, V$.

M. Boileau * is of opinion, after reviewing these experiments of Baumgarten, that for canals of all kinds, provided the section be uniform, the mean of these, or more simply $\frac{9}{10}$ of the surface velocity, may be taken as the mean velocity in the same vertical line.

In applying this rule to a wide river or canal, it is of course necessary to determine the maximum or surface velocity at a number of points across the stream; so that it does not dispense with the use of hydrometrical instruments. It may however serve much to diminish the number of observations which would otherwise be required in deep canals.

With reference to the mean velocity for the entire section of a stream, Boileau, in his work already quoted, gives some experiments, in which he finds the values of U for small depths not exceeding a foot vary from ·785 V to ·865 V.

In the 'Annales des Ponts et Chaussées' for 1848 M. Baumgarten gives a table of mean and observed velocities at various depths for twenty-two sections across the Garonne. The author compares his mean velocities with those calculated by the formula of Prony, and finds the latter too great,—a result directly opposite to that established by Boileau for his largest canal of one foot deep. Baumgarten concludes, that for large sections 'such as those of rivers) with irregular banks, the velocity is only four-fifths of that determined by Prony, so that instead of U = ·816 V, it should be U = ·653 V.

Mr. Beardmore, in his work already quoted, has a table showing the mean velocity corresponding with surface velocities, varying from 5 feet per minute up to 950 feet. This table is calculated from the formula $U = V + 2\cdot5 - \sqrt{5V}$. For a surface velocity of 5 feet per minute he gives a mean velocity of 2·50, so that here U = ·5 V; and for his highest surface velocity of 950 feet, the mean velocity is 883·6, so that here U = ·930 V; and so in proportion for intermediate velocities.

Mr. Neville has calculated two sets of mean velocities, both

---

* Traité de la Mesure des Eaux Courantes. Par P. Boileau. Paris, 1854.

of which differ considerably from Beardmore's. One of these, for large channels, is based upon experiments by Ximenes, Funck, and Brünning, and is calculated from the formula $U = \cdot 835\, V$. The other, as before explained, is calculated from the formula of Prony, which for measures in English inches becomes $U = V \left(\dfrac{93\cdot39 + V}{124\cdot14 + V} \right)$ . . . (11)

According to this latter table the mean velocity corresponding to a surface velocity of 5 feet per minute, is 3·75 feet, whereas Mr. Beardmore makes it 2·5. For 600 feet per minute, the highest surface velocity in the table, the mean is 524 feet per minute, Mr. Beardmore's corresponding velocity being 548.

On the whole, it must be confessed this is a subject on which no very striking agreement in opinion is to be expected. Not only do the experiments and formulæ of separate authors differ from each other, but even those of the same authors differ among themselves, and vary to an unusual extent.

Although I have thought it necessary to bring into juxtaposition a few of the most prominent and leading facts connected with this subject of surface and mean velocities, it must be observed that this method of gauging rivers is by no means a desirable one, and should never be adopted where the more certain modes of gauging by means of sluices or weirs can be followed. In large rivers, however, it will often be impracticable to gauge in any other way than by taking velocities; and when the extreme surface velocity has been ascertained, with proper precautions, as accurately as possible, it must be reduced to the mean by using one or other of the coefficients or values of V; when this mean, so reduced, is multiplied by the area of water-way, taken in the same dimensions as V, the result will be the discharge of the river in terms of V, in the same unit of time for which the velocity V is taken.

When no other mode of gauging is practicable, except that by means of surface velocity, a part of the river should be selected where the channel is most straight, and the section most uniform and regular. Several accurate sections across the river should then be made, from which the area of water-way must be accurately computed. Stakes are then to be driven down at fixed distances of 10 or 20 feet apart; and if the river is very wide, it is desirable to have corresponding stakes on each side of the river, and a pole erected at each stake, in order that an observer, standing at each pole and looking across to the one directly opposite, may note the exact time at which the float crosses the imaginary line joining the corresponding poles. The float used should be a leaded cork, an orange, or some other body only a little lighter than water, so that it will be nearly all submerged, and present very little surface to the action of the wind.* A number of observations should be taken at a season of perfect calm, in which there is no wind to disturb either the water or the float; and an average surface velocity having been obtained from these, the mean is to be calculated from one of the formulæ already given, and multiplied by the area of the river, to find the discharge as before explained. Mr. Beardmore observes: "The most useful instrument for getting velocities where a float is not applicable, and where an under current is probable, is the current meter formed by a vane in the Archimedean form, carrying an endless screw, which turns a wheel divided on the circumference. In gauging by velocities care should be taken to ascertain that the current does not underrun at the place of observation."

* Some of the French experimenters have used common wafers for the floats in very delicate experiments, while others have made use of small cubical pieces of oak, with a cavity in one side for putting in lead to ballast or weight the wood; others have used small balls of wax ballasted with lead. They recommend several precautions, which are not commonly attended to in practice. Some of these have reference to the extent to which the float should be submerged; and for this they recommend very shallow floats, in order to attain a true surface velocity, and not one at a small depth below the surface. Other precautions refer to the buoys to be fixed in large rivers, or the stakes to be fixed on the banks at right angles to the current of the river. They insist further, that a body floating on the surface of a current has at first a velocity greater than that of the water, and that the velocity ought not to be taken for the purpose of calculation until the float is observed to pass through equal spaces in equal times.

3. *Gauging of Water passing through Sluices or Orifices.*

Dr. Hutton and other writers on mechanics have shown that independently of friction, 1. The velocity of water passing through a sluice is equal to that acquired by a heavy body falling from the surface of the water to the centre of the orifice. 2. That in any single unit of space, as a foot or an inch, the velocity per second expressed in similar units is equal to twice the square root of the force of gravity. 3. That the velocity acquired in falling through any height is as the square root of the height fallen through. Now let h represent the height or space fallen through, v the velocity acquired in falling through that space, and g the force of gravity or the space fallen through by a heavy body in one second (this, in the latitude of London, is equal to 16·08 feet). Then the preceding propositions may be thus expressed algebraically :

$$v = 2 \sqrt{g} = 2 \sqrt{16\cdot08} = 8\cdot02 \quad . \quad . \quad (12),$$

where h, the height fallen through, is equal to unity ; and where h is any height whatever,

$$v = 2 \sqrt{16\cdot08\, h} = 8\cdot02 \sqrt{h} \quad . \quad . \quad (13).$$

This last is justly called by French writers the fundamental equation in hydraulics. It is the one most commonly used. It is the one which, when modified in each case by the proper coefficient, forms the foundation of formulæ for the flow of water, not only through sluices and over weirs, but also its flow in canals and pipes. We shall see hereafter how invariably this simple equation plays its part in all determinations of this kind.

Since the time of Dr. Hutton and his contemporaries a somewhat more accurate value has been assigned to g, the force of gravity. In accordance with the most recent experiments the value of g has been somewhat increased, and it is more accurate to write $v = 8\cdot03 \sqrt{h}$.

If it were not for the friction of the particles of water amongst themselves and against the sides of the orifice, and for the contraction of the stream in issuing out, the mean velocity would be something more than 8 times the square root of the height. It is evident that the discharge through a sluice will always be equal to the mean velocity multiplied by the area of the opening. Hence, if we put A = the area, we shall have the theoretical discharge, independently of friction, equal to $8 \cdot 03$ A $\sqrt{h}$ per second . . . (14)

and $481 \cdot 8$ A $\sqrt{h}$ per minute . . . (15).

The water which issues through any orifice, however, is diminished by friction and by the contraction of the fluid vein. The relation between the theoretical discharge and that due to the real sectional area of the vein has never yet been determined by mathematical investigation; we must therefore be guided solely by experiment. These experiments have determined that for very small orifices the above expressions must be multiplied by a coefficient which differs little from $0 \cdot 62$. This would reduce the coefficient $481 \cdot 8$ to 300, this being the multiplier Mr. Beardmore uses in his tables for ordinary sluices.

The principal experiments on which formulæ for discharge by sluices have been founded are those by Poncelet and Lesbros. These are far from satisfactory, because, although they were made with heads of water varying from 0 to 10 feet, yet the orifices were extremely small. The orifices used in the experiments were all rectangular, with an invariable breadth of $7\frac{3}{4}$ inches, and varying in height from $\frac{1}{3}$ of an inch to $7\frac{3}{4}$ inches. These were evidently much smaller than the sluices in practice for the regulation of supplies of water, for the discharge of surplus water, &c. The fact most worthy of observation in these experiments appears to be, that the coefficient was always greatest for the smallest orifice, and *vice versâ;* which goes to prove that the effect of the *vena contracta* is much more sensibly felt as the size of the orifice increases. Looking carefully to the tables of Poncelet and Lesbros' experiments, which were published in their 'Expériences Hydrauliques sur les Lois de l'Ecoulement de l'Eau' (Paris, 1832), and which have since been copied by Genieys, Boileau, Dupuit, and every other French writer on hydraulics, there seems no reason to alter the coefficient ·62. This appears, in the present state of our knowledge, to be the one best adapted to give the discharge through well-constructed sluices in lock-gates, mill-dams, and other situations of the kind.   Hence we have

$$481\cdot8 \times \cdot62 \times A \sqrt{h} = 298\cdot7\, A \sqrt{h} = \text{in round numbers to}$$
$$300\, A \sqrt{h} \quad . \quad . \quad . \quad (16),$$

the discharge per minute through an orifice whose area is A, h being the height from the centre of the orifice to the surface of the water.   "Where the orifice of the sluice is covered, as in locks and river sluices, the height h is the difference of level between the respective surfaces." *

The following is a brief *résumé* of the other writers who have experimented on the subject, with the results which they arrived at, condensed from Neville's Hydraulic Formulæ :—

	Resulting Coefficients.
Dr. Bryan Robinson, in 1739. On circular orifices of one-tenth to eight-tenths of an inch diameter, with heads of 2 to 4 feet	·728 to ·774
Dr. Mathew Young, in 1788. On circular orifices one-fifth of an inch in diameter, with a mean head of 14 inches .	·623
Michelotti. On circular orifices 1 to 3 inches diameter, and square orifices 1 to 9 inches in area, with heads from 6 to 23 inches	·609 to ·64
The Abbé Bossut. On small circular and rectangular orifices from half-inch diameter to an area of 4 square inches, under heads of 4 feet and 15 feet . . .	·613 to ·619
Brindley and Smeaton. On an orifice 1 inch square, with head varying from 1 to 6 feet	·632 to ·639
Rennie. On circular orifices from ¼ to 1 inch diameter, with head varying from 1 to 4 feet	·584 to ·671
Rennie. On rectangular and triangular orifices, each with an area of 1 square inch, and head varying from 1 to 4 feet	·572 to ·668

None of the preceding experiments are on so extensive a scale as those made at Metz by Messrs. Poncelet and Lesbros.

* Beardmore's Hydraulic Tables, where some general rules are given to meet the most ordinary cases in practice relating to discharge through vertical and horizontal sluices.

These have been already alluded to. They appear generally to confirm the conclusion arrived at by the French engineers and others, that the coefficient for practical purposes may be taken at ·62. Some writers have attempted, as it were, to expand the experimental results of Poncelet and Lesbros for sluices of proportionate dimensions but much larger size. This has been done by Mr. Neville in his Hydraulic Formulæ, but it really seems a waste of time and an attempt to refine where the data are not sufficient for the purpose.

It will be observed in the preceding list of experiments that Dr. Bryan Robinson's coefficients are much the highest. It is probable that most of the other experiments were made with orifices in a thin plate, while the aperture in Dr. Robinson's experiments was probably in a plate of some thickness, with the inner edge rounded off. This would occasion the coefficients to be higher, and is probably the reason why they are so.[*]

Experiments on a much larger scale are still very desirable on this subject, and I perfectly coincide in the following observation, which occurs in Mr. Neville's book :—

"It is almost needless to observe, that all these coefficients are only applicable to orifices in thin plates, or those having the outside arrises chamfered......Very little dependence can be placed on calculations of the quantities of water discharged from other orifices, unless where the coefficients have been already obtained by experiment for them. If the inner arris next the water be rounded, the coefficient will be increased."

Some carefully-conducted experiments on the time of filling and emptying lock-chambers or canals would, after all, be much more valuable than any of those which have been cited. Where the culverts or sluices from the upper level into the lock-chamber, and between the latter and the tail of the lock, are in good order, these experiments might be made with considerable accuracy, while the lock-chamber itself would afford an excellent measure for the volume of water passing through the sluices in given spaces of time, with certain parts of their area open.

[*] Neville.

4. Gauging by means of Current Meters, and other Instru-
ments for observing Velocities at various depths.

On Hydrometers, or Instruments for measuring Velocities.—
These are of two kinds, namely those which act on the principle of Pitot's tube, and those which act as dynamometers, by
opposing a solid resistance to the action of the current, and
indicating the force of this action by means of a dial or graduated bar.

The Pitot tube, as originally proposed more than a century
ago by the philosopher whose name it bears, is a hollow tube
bent at a right angle, and so used that one part of the tube is
horizontal in the line of the current, while the other part is
vertical. The open end of the horizontal branch being opposed to the action of the current, this causes the water to rise
above the surface of the stream in the vertical tube, and attempts have been made to determine velocities at various depths
according to the amount of this rise or difference of level.
The simple tube of Pitot, however, is liable to many objections
as a hydrometrical instrument. Among these are, its small
degree of sensibility (as considerable increase of velocity produces very slight variations of height), the oscillations of the
water in the vertical part of the tube, and the errors arising
from capillary action. These defects conspire to render the
Pitot tube in its original form a very uncertain and imperfect
instrument for ascertaining velocities.

In his recent Treatise, which has been already mentioned,
Boileau describes a very ingenious modification of the Pitot
tube, which appears capable of being applied with very good
effect to the measurement of velocities. The tube in this
arrangement consists of a horizontal part and of an inclined
part, the tube being flexible, and so arranged that the horizontal part can be adjusted at any required depth in the current or stream. The upper end of the tube is surmounted
with a metallic or glass cylinder, five or six times as much in
diameter as that of the tube, and by means of valves the com-

munication can be closed at any moment between this cylinder and the tube, while the contents of the cylinder can be discharged by another small tube. The cylinder is provided with a metallic float which rises as the cylinder fills, and acts on a needle or pointer which indicates its exact height. The graduated scale to which the needle is applied points out the velocity of the current, but of course this requires very accurate graduation, and many preparatory experiments would have to be made for this purpose. The following appear to be the principal advantages that would attend the use of this modification of Pitot's tube :—

1. It can be adapted without difficulty to the deepest and largest as well as to the shallowest rivers.

2. It dispenses with the use of an accurate chronometer, which, on the other hand, is essentially necessary where either floats or other kinds of current metres are employed.

3. When once fixed at any particular station, it need not be removed until the whole series of observations shall have been made which are necessary for that particular vertical line.

There is another beautiful adaptation of a tube to the measurement of velocities which deserves notice. The tube is of glass, having one extremity fitted with a smaller adjutage, which can be varied according to the current. Before being used the tube is prepared by closing the larger end and filling the tube with water, so as to leave a small bubble of air, similar to the bubble in the glass tube of the common spirit-level. The tube is immersed in the direction of the current at the required depth, and the velocity observed by means of the passage of this bubble of air from a mark at one extremity of the tube to the other end as soon as the latter is opened. Although a very beautiful philosophical toy, this is almost too delicate an instrument for practical purposes.

The Pitot tube described by Genieys is made of tin, is about 6 feet in length, and 2 inches diameter, with a horizontal arm about a foot in length. The extremity of the horizontal arm is contracted conically, leaving only a very small orifice to receive the pressure of the water. The open end of the tube is fitted with a float and wooden rod, graduated to show the rise of water in the tube.

In using the Pitot tube it is necessary to have a strong stake or small pile fixed in the river, to which the tube can be adjusted at any required depth. Except in very minute investigations, it is unnecessary to take more than two observations at each place, namely one at the surface and one at the bottom, the arithmetical mean between these being the mean velocity in that vertical line. Care must be taken to place the horizontal part of the tube exactly in the line of the current, for which purpose it must be turned about, and read off in that position when the water rises highest in the tube. The graduation of the tube or wooden bar may be calculated from the formula $U = 8 \cdot 03 \sqrt{h}$, where U = velocity in feet per second, and h = height in feet of water in the tube above the surface in the river. Thus, a rise of one foot in the tube will mark a velocity of 8 feet per second, and so for any intermediate height the velocity will be as the square root of the height.

Pitot's tube has been much used by Mr. Scott Russell and others, in making experiments on the speed of vessels passing through water.

Current Meters acting on the principle of Dynamometers.— The first current meter of this kind, as introduced by Woltmann in 1790, consisted of a small light waterwheel, with narrow pallets or float-boards whose breadth was about one-fourth the diameter of the wheel. On the axis of the wheel is an endless screw which works one or more toothed wheels, the revolution of which marks the velocity of the current. To this apparatus is attached a vane or rudder, of about the same depth as the diameter of the waterwheel, and this vane, acted on by the current, causes the waterwheel to rotate exactly in the true axis of the stream. The whole apparatus is movable on a vertical shaft, to which it can be clamped, so as to act at any required depth. In the earlier current meters the pallets or blades of the waterwheel were fixed radially around the circumference, similar to those in the simpler forms of the common waterwheel. The pallets have been constructed either of flat boards of wood, or metal, or with one surface flat and the other spherical, in order to oppose less resistance in rising out of the water.

These earlier forms of wheels for measuring velocities of currents had many objections. They were made in too clumsy a manner, and had so much friction in the axle, in the endless screw and the toothed wheels, that they were incapable of recording small velocities, and not possessed of sufficient delicacy for experimental purposes. Besides this, the toothed wheels, being immersed in the water without any protection or covering, were liable to be affected by sand, gravel, weeds, &c. getting between the teeth and preventing them from working properly. These and other objections and inconveniences arising from the use of the current meters, led to several improvements by Laignel, Boileau, and others. In one of the most improved forms of current meters the wheel is made with helical blades, and great care is taken to have as little friction as possible in the axle and the bushes in which it works. In place of the endless screw and toothed wheels of Woltmann's meter, M. Laignel substitutes a simple screw, on which works a movable nut. The nut does not revolve with the screw, but travels along it, and carries a pointer, which marks by the distance on a bar the revolutions of the helical-bladed wheel. The screw on which the nut traverses is made of copper, and very accurately turned, with a pitch of $\frac{1}{25}$ of an inch, and is $\frac{1}{8}$ of an inch in diameter. The screwed part of the axle, the travelling nut, and the graduated bar on which the revolutions are read off, are all enclosed in a box, so that the delicate mechanism is protected from that kind of injury to which the meter of Woltmann is so subject. The screw is about a foot or 15 inches in length, and the extremity of the box in which it is enclosed is provided with a vane or rudder, to keep the machine in the centre of the current. The helical wheel and box carrying the screw are supported on a bracket, which slides on a vertical pillar, and can be clamped at any required height. There is a contrivance for stopping the revolution of the wheel at any required moment. The length of the screw is sufficient to allow the travelling nut and marker to record the revolutions of the wheel during 25 or 30 seconds, when it is stopped, and the machine taken out of the water, and the distance corresponding with the number of revolutions read off on the graduated bar. This instrument, like all others of the same kind, requires the employment of a very accurate chronometer, and has the inconvenience of causing much loss of time, by the necessity for taking the instrument out of water to read off the number of revolutions at every separate experiment.

M. Boileau has proposed an extremely ingenious and delicate instrument, acting on the principle of measuring the velocity of a current without the employment of a wheel at all. A small copper plate is opposed to the action of the current at any required depth, and this, by means of a vertical bar, is made to act on a delicate and very sensitive elliptical spring. The compression of the spring is measured with extreme accuracy by means of a micrometer screw. For the purpose of minute and delicate observations and experiments this instrument appears to be extremely well adapted, and is free from many of the defects which attach to all current meters acting by means of wheels.

Having obtained in a large river, either by means of the Pitot tube or some kind of current meter, the mean velocities of the water in several vertical lines across the stream, we have then to find the mean velocity of the whole river. There are various ways of doing this, but the geometrical method described by D'Aubuisson, in his 'Traité d'Hydraulique à l'usage des Ingénieurs,' is perhaps entitled to the preference on account of its extreme simplicity. "Having fixed on the station where the cross section of a large river is to be taken, and the velocities ascertained, take a number of soundings across the stream, at 8, 10, or 12 points, according to the breadth. These lines of sounding divide the section into a number of trapezia, and the area of each of these is to be calculated. Then at a point half-way between each of the two lines of sounding is to be fixed a small boat containing the current meter, such as Woltmann's wheel, the Pitot tube, or other instrument by means of which 5, 6, or 7 velocities are to be determined in the same vertical line. The arithmetical mean of these is then to be multiplied by the area of the trapezium to which they apply. The sum of all these products is evidently the discharge of the river,—it is equivalent to the total sectional area multiplied by the mean velocity."

5. *Gauging by means of Weirs.*

Referring again to what has been called the fundamental theorem of hydraulics, namely $v = 8 \cdot 03 \sqrt{h}$, it may be at once applied to the case of weirs, v being the theoretical velocity corresponding to the height h of the notch or weir. This height must be the difference of level between the surface of the notch over which the water flows and the still water in the pond above. This height is not to be confounded with the thickness of the sheet of water flowing over the weir, from which it is quite distinct. As in the case of orifices, the theoretical expression $8 \cdot 03 \sqrt{h}$ for the velocity per second, or $481 \cdot 8 \sqrt{h}$ for the velocity per minute, must be multiplied by a coefficient to give the true velocity, as this in a similar manner is affected by friction and by contraction of the sheet of water flowing over the weir.

Hence the velocity per minute of water flowing over a weir is equal to $481 \cdot 8 \, C \sqrt{h}$ . . . (17),
where C is a coefficient to be determined by experiment.

The value of C varies somewhat, not only according to the shape of the weir, but also varies for different depths and widths. The limits of this variation will be seen when we come to describe the experiments more particularly.

The principal observers who have made experiments on the flow of water over weirs are the Chevalier Du Buat, Eytelwein, MM. D'Aubuisson and Castel, and MM. Poncelet and Lesbros.

In our own country experiments have been made by Messrs. Smeaton and Brindley, by Dr. Robinson, Messrs. Leslie, Blackwell, Hawksley, and others.

The experiments of Du Buat were made in 1779 on weirs 18⅓ inches wide and an extreme depth of 6¾ inches.

MM. Poncelet and Lesbros made an extensive series of experiments at Metz in the years 1827 and 1828. Their weir was of the constant width of 7¾ inches, and various heads were tried from ¾ inch up to 8 inches. They found the coefficient varied considerably for different heads. The experiments of MM. D'Aubuisson and Castel were made at the Toulouse Waterworks in 1834, with various widths of weir, and with a head ranging from 1 to 8 inches. Messrs. Smeaton and Brindley made their experiments with a constant width of 6 inches, and with depths varying from 1 to 6 inches. For Dr. Robinson's experiments see the 'Encyclopædia Britannica.'

Mr. Blackwell's experiments were made in 1850, and consisted for the most part of observations on the discharge of water by weirs out of a large pond on the Kennet and Avon Canal, which afforded a perfectly uniform head of still water.

These experiments were 264 in number, and embraced observations on various forms of overfall, from a very thin plate of sheet iron to a weir 3 feet in thickness ; the length or breadth of the weir extending from 3 feet to 10 feet.

If we substitute m for C 481·9 in the equation for the velocity over the weir, we have the discharge of water for each foot in width $= m \sqrt{h} \times h = m \sqrt{h^3} = m h^{\frac{3}{2}}$.

The value adopted for m by Mr. Beardmore and by most English engineers, is 214 where h is in feet ; and where h is in inches the value adopted is 5·1. These two values are in fact identical, because $\dfrac{214}{12^{\frac{3}{2}}} = \dfrac{214}{41·6} = 5·1$

Hence the formula given by Mr. Beardmore, where h is in feet, is $214 \sqrt{h^3}$ . . . (18)
for the discharge in cubic feet per minute, and the rule deduced

TABLE SHOWING THE VARIATIONS OF THE COEFFICIENTS FOR DIF-
FERENT HEADS OF WATER, ACCORDING TO MR. BLACKWELL'S EX-
PERIMENTS.

Number of Experiments.	SPECIES OF OVERFALL.	Head.	Mean coefficient for formula $m \sqrt{h^3}$ when h is in feet.	Mean coefficient for formula $m \sqrt{h^3}$ when h is in inches.
		in. in.		
6	Thin plate, 3 feet long.	1 to 3	212	5·10
		3 ,, 6	194	4·68
11	Thin plate, 10 feet long.	1 to 3	241	5·76
		3 ,, 6	210	5·16
		6 ,, 9	178	4·32
23	Plank, 2 inches thick, 3 feet long.	1 to 3	165	3·96
		3 ,, 6	185	4·44
		6 ,, 9	196	4·62
56	Plank, 2 inches thick, 6 feet long.	1 to 3	173	4·14
		3 ,, 6	191	4·60
		6 ,, 9	189	4·44
		9 ,, 14	173	4·14
40	Plank, 2 inches thick, 10 feet long.	1 to 3	167	4·08
		3 ,, 6	191	4·60
		6 ,, 7	180	4·32
		9 ,, 12	172	4·14
4	Plank, 2 inches thick, and 10 feet wide, with wings.	1 to 2	230	5·52
		4 ,, 5	213	5·22
7	Overfall, with crest 3 feet wide, 3 feet long, sloping 1 in 12.	1 to 3	165	3·96
		3 ,, 6	158	3·78
		6 ,, 9	163	3·96
9	Overfall, with crest 3 feet wide, 3 feet long, sloping 1 in 18.	1 to 3	175	4·20
		3 ,, 6	152	3·66
		6 ,, 9	160	3·84
6	Overfall, with crest 3 feet wide, 10 feet long, sloping 1 in 18.	1 to 4	158	3·78
		4 ,, 8	169	4·08
14	Overfall, with crest 3 feet wide, level, 3 feet long.	1 to 3	147	3·54
		3 ,, 6	150	3·60
		6 ,, 9	153	3·66
15	Overfall, with crest 3 feet wide, level, 6 feet long.	3 to 7	159	3·72
		7 ,, 12	149	3·60
12	Overfall, with crest 3 feet wide, level, 10 feet long.	1 to 5	147	3·54
		5 ,, 8	158	3·78
		8 ,, 10	151	3·66
61	Chew Magna experiments, overfall bar 2 inches thick, 10 feet long.	1 to 3	210	5·06
		3 ,, 6	240	5·78
		6 ,, 9	243	5·85

R 3

from Eytelwein, where h is in inches, is $5 \cdot 1 \ \sqrt{h^3}$ . . (19)$=$ discharge in cubic feet per minute.

As these two formulæ are far more generally employed by engineers than any others, I have thought it would be useful to adapt the results arrived at by Mr. Blackwell to each of them. The table on the preceding page (p. 369) is compiled from an abstract given by Mr. Blackwell himself, with the substitution by me of the last two columns.

The results shown in the preceding table are very valuable. All the experiments except the last (61), apply to still ponds without any current in the water, and we shall first consider these.

Over a thin plate 3 feet long, the experiments agree with the usual coefficients 214 and 5·1 up to 3 inches of head, above which they fall below them.

Over a thin plate 10 feet long, the experiments show a higher coefficient than the common ones; between 3 and 6 inches the coefficients are nearly the same, and above 6 inches the coefficients are less.

Over planks 2 inches thick an opposite law seems to prevail, for the coefficients increase with the depth instead of becoming less as the latter increases. This is the case with all the experiments on the 3 feet plank, the coefficient up to 3 inches being 165, and increasing up to 196 for heights between 6 and 9 inches. All the experimental coefficients are below the usual ones in these twenty-three experiments. In the next ninety-six experiments, over 6 feet and 10 feet planks, the maximum coefficient is for a depth of 5 or 6 inches, beyond which it diminishes as far as the experiments extend.

The four experiments made over a plank with wing boards show, as might be expected, a better discharge than the former ones. The coefficient here is highest at 1 inch, and gradually diminishes as the head increases.

In the experiments on weirs with inclined crests the lowest coefficient is in each case at 4 inches, an increase taking place when the depth is diminished below this, and also when the depth is increased.

The experiments on level crests 3 feet wide and of different lengths present the greatest anomalies. The crest 3 feet long has the lowest coefficient at 4 inches. The one 6 feet long has the lowest at 10 inches, and the one 10 feet long has the lowest at an inch, from which it increases to 5 inches, and then again diminishes.

The sixty-one experiments at Chew Magna are applicable to weirs across rivers in which the pond is small in proportion to the breadth of the weir, and in which it is difficult to avoid a slight current in the water as it approaches the weir.

In these experiments the coefficients are below the mean up to 3 inches, then they rise almost to a maximum at 5 inches, then fall slightly to $6\frac{1}{2}$ inches, when they ascend to a maximum at 8 inches, below which they again descend.

The mean coefficients determined by Mr. Blackwell for this kind of weir, when the depth exceeds 3 inches, are all higher than those in common use, so that the ordinary formulæ may be used with safety for plank weirs of this description across rivers and streams, but appear somewhat too high for plank weirs where the pond is large and the water entirely without current. There can be no doubt that Mr. Blackwell's experiments are entitled to the highest confidence, as they were made with great care, and afford an admirable model for that kind of experiment which is so necessary to furnish facts for the engineer.

Recurring now to the formula at page 368, where $m\sqrt{h^3} =$ discharge for each foot in width, it is evident if b be the width of the weir in feet, $mb\sqrt{h^3}$ will be the whole discharge per minute. Hence we have only to employ the corresponding coefficient in the table in order to obtain the discharge in cubic feet per minute. If h be in feet, the value of m must be taken from the fourth column, and if in inches, it must be taken from the fifth column.

Example: Required the discharge over each foot in width of a plank 2 inches thick and 10 feet long, with a still pond above it, and a head of ·25 feet or 3 inches. Using ·25 for the height, the corresponding coefficient in the table will be 191.   Hence

$191 \times \cdot25^{\frac{3}{2}} = 23\cdot9$ cubic feet per minute, or using 3 for the height in inches, the coefficient is 4·6.   Hence

$4\cdot6 \times 3^{\frac{3}{2}} = 23\cdot9$ cubic feet as before.

The quantity calculated from Mr. Beardmore's coefficient 214 would be $214 \times \cdot25^{\frac{3}{2}} = 26\cdot7$ cubic feet.

EXPERIMENTS OF DU BUAT AND OTHERS.

Mr. Blackwell gives a diagram showing the mean results arrived at by other experimenters, and from this the following Table of Coefficients has been prepared.

Depth over weir.	Du Buat's experiments on a weir 18¼ inches wide.		Brindley and Smeaton, on a weir 6 in. wide.		Poncelet and Lesbros, on a weir 7¼ in. wide.		Castel and D'Aubuisson on a weir 7·87 in. wide.	
Inches.	Mean coefficient when h is in feet.	Mean coefficient when h is in inches.	Mean coefficient when h is in feet.	Mean coefficient when h is in inches.	Mean coefficient when h is in feet.	Mean coefficient when h is in inches.	Mean coefficient when h is in feet.	Mean coefficient when h is in inches.
1			227	5·47	200	4·82	200	4·82
2	202	4·87	204	4·91	197	4·75	196	4·72
3	169	4·07	194	4·67	193	4·65	193	4·65
4	168	4·05	194	4·67	190	4·58	190	4·58
5	168	4·05	194	4·67	190	4·58	190	4·58
6			178	4·29	190	4·58	190	4·58
7	202	4·87	201	4·84	188	4·53	191	4·60
8					184	4·43	191	4·60
Mean.	186	4·48	204	4·91	192	4·63	193	4·65

Some engineers are in the habit of gauging the depth over a weir by observing the height to which the water rises on the face of a common 2-foot rule held flatwise to the stream, and at the outer edge of the overfall bar or crest of the weir.   Mr. Blackwell made many trials to ascertain the value of this method, and he observes as a result, "that the head of water above an overfall may be ascertained approximately, but only so, by the insertion of a 2-foot rule held against the stream on the overfall bar, and observing the height to which the water rises, as the total head above the crest."

There can be no doubt that this is a difficult and uncertain method of observation, as the water coming with considerable velocity against the face of the rule, does not form a level and uniform head there, but is liable sometimes to be depressed and sometimes elevated above its true level. The method recommended by Mr. Beardmore is therefore much to be preferred. He says a stake should be driven in the still pond above the weir, until the top is exactly level with the top of the weir or overfall bar. The depth of the water over the stake can then be dipped with the 2-foot rule or any other instrument for reading it, and recorded either in inches or in hundredths of a foot. One great advantage of this method is this, that when the stake or mark in the still pond is once accurately fixed—and this should be done by means of the spirit-level in the ordinary manner of putting in level stakes—the dipping on the head of the stake may afterwards be entrusted to any careful person who can read and write; whereas the method of dipping on the weir with the flat side of the rule requires extreme caution, and should not be lightly entrusted in the hands of any but an experienced assistant. It is important to observe that the water in the still pond has frequently a slight inclination for several yards above the weir, so that the stake should be driven down where the water is perfectly level and has no perceptible motion. The best formulæ for calculating the flow over weirs are all based on the *greatest* head or depth being observed.

M. Du Buat had assumed in the theory on which his formula was founded, that the thickness of the blade of water was equal to half the total depth from the crest to the top of the water. Mr. Blackwell made experiments on this subject by immersing the thin brass slide of the rule and reading the depth when held edgewise to the course of the stream. Mr. Blackwell found that the assumption of Du Buat was by no means correct. In the case of the overfall plank 2 inches wide the thickness varied from six-tenths to eight-tenths of the total depth, following the law of increase, as the total head increased. In the case of weirs with long crests the thickness is generally much less, varying from one-fourth to one-half of the total depth.

Mr. Blackwell observes truly, as the result of his experiments, that no formula with a constant coefficient will give the true discharge of water by a weir.

On the Importance of accurate Gauging over Weirs.

There is no subject more deserving of attention by the waterworks engineer than the accurate determination of the discharge over weirs. In cases of impounding reservoirs the embankment is commonly furnished with a weir, over which the waste water passes for the use of millers on the stream below. The discharge over the weir is frequently regulated by Act of Parliament, and the Company is bound to allow a certain quantity of water to flow off for the use of the mills. Hence the extreme importance of determining the exact discharge of a weir in proportion to the head of water flowing over it.

In laying out or designing waterworks also, the accurate gauging of streams is of the utmost consequence, in order to show the quantity of water which may be depended on at all seasons of the year, and to show by comparison with future observations the quantity abstracted from mills, &c. In cases of compensation to millers for loss of water-power, the accurate gauging of the various weirs and sluices, in order to show the comparative quantity of water used and wasted by the millers at different seasons, becomes of great importance.

Probably the most convenient method of gauging weirs would be to have a rule graduated, to show the discharge in cubic feet per minute for each foot in width, so that the discharge could be read off at once by inspection, without refer ence to a table or calculation. The scale below shows how a rule of this description should be graduated, where the usual coefficient 5·1 is adopted in the formula $C\, h^{\frac{3}{2}} = D$:—

				Cubic feet per minute.
Thus at 1 inch on the rule should be marked				5·1
1¼ ,,	,,	,,		7·14
1½ ,,	,,	,,		9·23
1¾ ,,	,,	,,		11·78
2 ,,	,,	,,		14·43

and so on.

A rule of this description would enable an observer, by a single trial, to pronounce on the spot what is the approximate discharge of a weir, and this without the aid of a level or any instrument for putting in a stake, as it would be near enough for this purpose to gauge over the lower edge of the weir, with the flat side of the rule opposed to the current.

Where greater accuracy is required than that derived from a coefficient, such as 5 or 5·1 assumed for all depths, the rule might have divisions corresponding to the 1·5 powers of the depths, or one of the faces might be so divided, in which case the observer might employ his own coefficient.

It may be worth noticing that the discharge in cubic feet per minute is readily converted into gallons per day by multiplying by 9000: thus, 35 cubic feet per minute = 35 × 9000 = 315,000 gallons per day of 24 hours. This is an operation so simple, that it can readily be performed in the head without any great effort of mind.

On the Employment of the Coefficient for calculating the Discharge over Weirs.

Having settled the general formula for the discharge over weirs as $C\, h^{\frac{3}{2}}$, it is obvious that all we have to determine from experiment is the value of C, without any reference to the expression $2g$ or the force of gravity. In fact, there is considerable ambiguity in the use of this expression and the meaning attached to it.

Dr. Hutton makes g to represent $16\frac{1}{12}$, but Mr. Blackwell, Mr. Neville, and others who have written on the subject of weirs, have taken the value of g at double, or $32\frac{1}{6}$. Hence, as the expression is only necessary to elucidate the theory of the subject, it is better to get rid of it altogether in practical calculations.

Mr. Blackwell uses two coefficients, one of which he terms m and the other k.

The first, m, is merely a factor, by which $8·03$ is to be multiplied to give the discharge in cubic feet per second, or by which $481·8$ is to be multiplied, to give the discharge in cubic feet per minute. His other coefficient, k, is to be used when the depth is taken in inches, and really does dispense with the factor $\sqrt{2g}$, and simply means the factor by which $h^{\frac{3}{2}}$ is to be multiplied to give the discharge in cubic feet per second.

Hence it follows that my coefficient* for feet is equal to Mr. Blackwell's m multiplied by $481·8$, and my coefficient for inches is equal to Mr. Blackwell's k multiplied by 60; or algebraically,

My coefficient when h is in feet $= 481·8\, m$

,,   ,, h is in inches $= 60\, k$

Mr. Neville, in his tables of the discharge over weirs, has termed the theoretical discharge $321\, h^{\frac{3}{2}}$ instead of $481·8\, h^{\frac{3}{2}}$ in cubic feet per minute, and his coefficients form factors by which 321 is to be multiplied, in order to give the discharge in cubic feet per minute. Hence, collecting all the corresponding values,

	In terms of Mr. Blackwell's coefficient.	In terms of Mr. Neville's coefficient.
My coefficient when h is in feet	$= 481·8\, m$	$321\, m$
,, ,, h is in inches	$= 60\, k$	$7·722\, m$

* See Tables at pages 369 and 3 2.

On the Velocity at which Water should flow in Channels.

It is the opinion of M. Genieys and some other hydraulic authorities, that in order to preserve the salubrity and freshness of water flowing in open channels, the surface velocity should not be less than about 80 feet per minute. Dupuit, although not disputing the advantage of rapid motion in open channels, is of opinion that no very extraordinary sacrifice of economy ought to be made to secure such a velocity as 80 feet per minute, and considers that 50 feet per minute would be sufficient, especially in aqueducts and conduits of brick or masonry.

Mr. Beardmore gives a table of the characteristics of rivers, compiled from a paper in the 'Philosophical Transactions.'

From this paper and the table which precedes it in Mr. Beardmore's book, the following brief notes are taken.

1. The artificial canals in the Dutch and Austrian Netherlands have $I* = 2$ to $9{\cdot}05$ and $V* = 30$ to 40.

2. Rivers in low flat countries, with many turns and windings, having a very slow current, and being subject to frequent and lasting floods; as, the Nene below Peterborough, where $I = 2$ and $V = 66$, section of river in feet $44 \times 5{\cdot}5$; the Thames below Staines, where $I = 1{\cdot}50$ to $3{\cdot}73$ and $V = 101$ to 130. When the volume and depth are greater, however, the velocity is much increased, with only a small addition to the rate of fall. Thus, the Severn between Worcester and Gloucester, with a section of 160×16, has $I = 5$ and $V = 190$.

Canals in Flanders are said to have $I = 6{\cdot}33$ and $V = 275$, but this is probably a mistake, as the velocity is much too great for the fall. The Neva at St. Petersburg, with a section of 900×63, is said to have $I = 1{\cdot}7$ only, and $V = 156$. The Ganges, of which the volume is much greater, with $I = 4$ has $V = 264$. When the Nile is low, the section at Cairo is about 900 feet wide $\times$ 14 feet deep, and $I = 3\frac{1}{4}$, $V = 110$;

* I means the fall in inches per mile, and V means the surface velocity in feet per minute.

when the river is high, the section at Cairo being 1100×40, $I = 5$ to 5_2 and $V = 300$.

Mr. Beardmore says, however, there are rivers of such a character that $I = 12 \cdot 18$ and $V = 60$. (This proportion between the inclination and velocity could scarcely obtain unless the river were much impeded by aquatic plants or other obstructions. Scarcely any river can have a section so disadvantageous as to produce an hydraulic mean depth $= 1$; and yet an hydraulic mean depth of 1, with an inclination of 1 foot per mile, will give a velocity, according to Mr. Beardmore's own formula, of nearly 80 feet per minute.)

3. The upper parts of rivers in low countries, and rivers generally in districts having a mean inclination between flat and hilly. Examples of such rivers may be found in the Dee above Chester, where $I = 11$, the Forth near Stirling, where $I = 13$, the Seine from Paris to Hâvre, where $I = 12 \cdot 4$ and $V = 125$,* the Shannon below Lough Allen, where $I = 12$. Such rivers may be taken in most countries to have a mean inclination of nearly 16 inches, and $V = 90$, although this is often exceeded in rivers of great depth and volume.

4. Rivers in hilly countries, with a strong current, or nearly straight course, and rarely overflowing. Examples:—the Lune above Lancaster, where $I = 23$; the Thames below Oxford, where $I = 21$ and $V = 176$; the Rhone from Besançon to the Mediterranean, where $I = 24 \cdot 18$; † the Rhine between Strasburg and Schenckenschantz, where $I = 24$; ‡ the Tiber at Rome, where $V = 197$; ‡ the Loire, where $I = 242$ and $V = 256$. Mr. Beardmore gives as the type of such rivers $I = 24 \cdot 37$ and $V = 180$, but is surely not correct in saying that they rarely overflow —witness the Thames below Oxford and the Nene between

---

* From an observation made by M. de Chezy on the Seine below Paris, namely between Surênes and Neuilly, the fall appears to be only 7·92 English inches per mile, and the velocity 150 feet per minute.

† The Rhone at Arles, according to Dupuit, has a velocity at low water = 287 feet per minute, and at Bamaire $V = 511$.

‡ Dupuit.

Wansford and Peterborough, where the valley has I = 21·8 Both these rivers are remarkable for their extensive and devastating floods, which however are doubtless due in a great measure to the weirs built across them for purposes of navigation.

5. Rivers in their descent from among mountains down into the plains below, in which plains they run torrent-wise. Mr. Beardmore gives as the type of such rivers, I = 31·68 and V = 300, and for rivers which are absolute torrents he gives I = 37·27, V = 480. His table of actual rivers contains only two examples of such currents, namely the Nene in the oolitic district above Wansford, where I = 38·7, and the Rhine between Schaffhausen and Strasburg, where I = 48.

Artificial Canals and Aqueducts.

The Canal de l'Ourcq, with a fall of 4 inches per mile, has a calculated velocity of 75 feet per minute, while the New River, according to Mr. Beardmore, with a velocity between the sluices of 2½ inches per mile and of 5 inches per mile including the sluices, has an actual velocity of 50 to 60 feet per minute.

The following is a list of some celebrated Aqueducts, taken from Dupuit's work, showing the inclination and velocities.

NAME OF AQUEDUCT.	Fall in feet per mile.	Velocity in feet per minute.
Pont du Gard at Nîmes (Section of water 4′ × 3′·4″)*	2·11	120
Aqueduct of Pont Pyla, Lyons (Section of water 1′·10″ × 1′·10″) *	8·80	177
,, ,, Metz (Section of water 3′·2″ × 2′·2″)* .	5·30	167
,, ,, Arcueil	2·20	
,, ,, Trappes	0 66	106
,, ,, Roquencourt	1·54	
,, ,, Maintenon	1·11	
,, ,, Caserte, Naples, (Section of water 3′·11″ × 2′·7″)*	1·10	80
,, ,, Montpellier (Section of water 12″ × 6″)	1·52	43

* Velocity calculated. All the velocities in this table are mean velocities, which being multiplied by the area of the water, will give the discharge per minute.

The four first aqueducts in the preceding table are works of the Romans, and are remarkable for the high velocity of the water which flowed in them. That of Nîmes in particular when full must have been capable of conveying about eighteen million gallons in twenty-four hours.

The aqueducts of Trappes, Roquencourt, and Maintenon were formed for conveying the waters from various streams and impounding reservoirs to supply the magnificent lakes and fountains in the gardens of Versailles.

The aqueduct of Caserte was constructed by Charles III., King of Naples, to supply a palace which he built in the plain of Capua, near Naples. The aqueduct of Montpellier, which has a slower velocity than any of the others, supplies water to the town of Montpellier, consisting of 33,000 inhabitants.

ON THE FLOW OF WATER THROUGH PIPES.

This subject is intimately related to the flow of water in uniform channels, which has been already treated at some length; in fact, the one subject may be termed a modification of the other. A pipe is simply a channel in which the wetted perimeter is the inner periphery of the pipe, and it follows, from the relation between the circumference and the area of a circle, that the hydraulic mean depth of a pipe when filled with water is simply one-fourth of the diameter, thus:

$$\frac{\text{Area of pipe}}{\text{Circumference}} = \frac{\cdot 7854\, d^2}{3 \cdot 1416\, d} = \frac{d}{4}.$$

The hydraulic mean depth of a pipe half full is also $= \dfrac{d}{4}$.

The same eminent men who have investigated the flow of water in open channels, have also applied their distinguished attainments to the subject of pipes. Du Buat, Bossut, Couplet, and Coulomb have shown that the same equation holds good with respect to water flowing in pipes as in open channels, and that $R\,I = a U + b U^2$.

The values of the coefficients a and b in the case of pipes, however, are different from those which apply to open channels.

Thus Prony determined, from fifty-one experiments made by Du Buat, Bossut, and Couplet, with pipes from 1 to 5 inches in diameter and from 30 to 1700 feet in length, and one pipe 19 inches in diameter and nearly 4000 feet long, that $a = \cdot0000173314$ and $b = \cdot000348259$; and substituting these values in the above equation, it follows that $U = (2871\cdot09\ R\,S + \cdot0006129)^{\frac{1}{2}} - \cdot0249$ for measures in metres, and for measures in English feet

$$U = (9419\cdot75\ R\,S + \cdot00665)^{\frac{1}{2}} - \cdot0816 \quad . \quad . \quad (20).$$

M. Prony has further simplified this expression for the case of pipes in actual practice, and replaces the preceding equation by the following, in which the value of R is ex·pressed in terms of the diameter: $\frac{1}{4}\,D\,S = \cdot0003483\,U^2$, or

$$D\,S = \cdot0013932\,U^2 \quad . \quad . \quad . \quad (21).$$

Hence $U = 26\cdot79\sqrt{D\,S}$ for measures in metres; and when the measures are in English feet the expression becomes

$$U = 48\cdot49\sqrt{D\,S} \quad . \quad . \quad . \quad . \quad (22).$$

Now if for S, the sine of the inclination, $\left(= \frac{h}{l} \right)$, we substitute H, the head or fall in feet per mile, we obtain

$$\frac{48\cdot49}{\sqrt{5280}}\sqrt{D\,H} = U \quad . \quad . \quad . \quad (23)$$

$$= \frac{\sqrt{D\,H}}{1\cdot5} \text{ and } H = \frac{2.25\,U^2}{D} \quad . \quad . \quad . \quad (24).$$

Eytelwein, who followed the same mode of investigation as Prony, also determined the value of a and b from fifty-one experiments of Du Buat, Bossut, and Couplet. He found[*] that for pipes $a = \cdot0000223$ and $b = \cdot0002803$, from which, when the measures are in metres, we derive

$$U = (3567\cdot29\ R\,I + \cdot00157)^{\frac{1}{2}} - \cdot0397,$$

and this, reduced for measures in English feet, becomes

$$U = (11704\ R\,I + \cdot01698)^{\frac{1}{2}} - \cdot1303 \quad . \quad . \quad (25).$$

[*] Neville.

The decimal ·01698 may be neglected in this equation ; and if we convert it as before into terms of the diameter and of the fall per mile, it becomes

$$U = \left(\frac{11704 \, D \, H}{5280 \times 4}\right)^{\frac{1}{2}} - ·1303 = \left(\frac{D \, H}{1·80}\right)^{\frac{1}{2}} - ·1303 \ . \ . \ (26).$$

The common form of Eytelwein's rule for pipes is,

$$U = 50 \left(\frac{d \, h}{l + 50 \, d}\right)^{\frac{1}{2}}$$

when h is the head and l the length, both in feet.

The expression 50 d, which is added to the denominator in the above equation, may be entirely neglected in very long pipes. Putting H for the fall per mile as before, and rejecting 50 d, Eytelwein's form becomes

$$U = \left(\frac{5280 \, D \, H}{2500}\right)^{\frac{1}{2}} = \left(\frac{D \, H}{2·112}\right)^{\frac{1}{2}} \ . \ . \ . \ (27)$$

and

$$H = \frac{2·112 \, U^2}{D} \ . \ . \ . \ . \ . \ . \ . \ . \ (28).$$

The subject of the flow of water through pipes has also been investigated by Dr. Thomas Young, Sir John Leslie, and others. Julius Weisbach, in a recent work ('Ingenieur- und Maschinen-Mechanik'), takes a somewhat different view from other writers, and proposes this formula :—

$$H = \left\{ ·0144 + \frac{·01716}{U^{\frac{1}{2}}} \right\} \frac{l}{D} \times \frac{U^2}{64·4} \ . \ . \ (29),$$

where l is the length in feet, H the height in feet required to overcome friction in that length, and the other letters represent the same as before.

If we now take $l = 5280$ feet in one mile, H becomes =

$$\frac{76·032 \, U^2}{64·4 \, D} + \frac{90·6 \, U^2}{64·4 \, D \, \sqrt{U}} = \frac{1·18 \, U^2}{D} + \frac{1·41 \, U^{\frac{3}{2}}}{D} \ . \ . \ (30).$$

This formula is not materially different from Prony's and Eytelwein's, where the velocities are small, but differs considerably for high velocities. A very excellent table, calculated from the formula of Weisbach, is inserted in Weale's 'Engineer and Contractor's Pocket-book for 1855-6.' This table has been calculated in English measures by Messrs. Thomson and

Fuller, Civil Engineers, of Belfast, and shows the head required
to overcome friction in 100 lineal feet of pipe varying in diameter from 3 inches up to 30, and at various velocities, increasing by one-fifth of a foot per second from 2 feet up to 7.
In this table the head multiplied by 52·8 gives the fall per
mile necessary to overcome friction.   The table also gives the
quantity of water delivered in cubic feet per minute, corresponding with each amount of head and rate of velocity.   This discharge in feet per minute, when multiplied by 9000, becomes
gallons in twenty-four hours.

   This table gives in all cases a greater velocity, and consequently a less head for friction, than Prony's formula.   For
example: with a velocity of 2 feet per second the following
is the head in feet per mile required to overcome friction :—

	According to Weisbach's table.	According to formula $H = \dfrac{2\cdot 25\ U^2}{D}$
3 inch pipe	34·79	36·00
6 inch ,, ...	17·37	18·00
12 inch ,, ...	8·71	9·00
24 inch ,, ...	4·33	4·50
For Velocities of 3 feet per second.		
3 inch pipe	71·28	81·00
6 inch ,, ...	35·85	40·50
12 inch ,, ...	17·89	20·25
24 inch ,, ...	8·98	10 12
For Velocities of 7 feet per second, the highest in the table.		
3 inch pipe	335·28	441·00
6 inch ,, ...	197·90	220·50
12 inch ,, ...	83·95	110·25
24 inch ,, ...	41·92	55·12

   It will be observed that for small velocities, such as 2 and
3 feet per second (those which are most common in the pipes
of waterworks), the results in the table of Weisbach differ only
slightly from those given by the more simple formulæ.   For
high velocities, such as 7 feet per second, the difference is much
greater; but this is a velocity far too great for water flowing

through pipes, and one which occasions so much loss by friction as to be seldom used.

For long pipes we have seen (Equation 24) that according to Prony,

$H = \dfrac{2\cdot25\ U^2}{D}$, and according to Eytelwein (Equation 28),

$H = \dfrac{2\cdot112\ U^2}{D}$.

In order to allow for the effects of bends and other irregularities in pipes, Mr. Blackwell proposes to use 2·3 in the above equation; and this allowance is one which certainly errs on the safe side, as for a given fall and diameter it shows a somewhat less velocity than that which Prony establishes.

I must here acknowledge my obligations to Mr. Blackwell for his valuable suggestion of this very simple English form of Prony's equation. In calculating the allowance for friction in pipes of considerable length and diameter, it was found an exceedingly ready and convenient method, although suggested by Mr. Blackwell without any pretence to extreme accuracy. Having lately however had the opportunity of comparing it and deducing it directly from Prony's equation, which must be considered the parent of it, I am induced to reproduce it here with more confidence.

As this formula will be found of immense practical value in solving a great many questions relating to the dimensions of pipes, the velocity of water flowing through them, and the head to be given to overcome friction, I shall briefly recapitulate the several values of U, D, and H.

Let U be the velocity in feet per second.

D = diameter of pipe in feet.

H = inclination of pipe in feet per mile.

Then $U = \sqrt{\dfrac{D\,H}{2\cdot3}}$ . . . (31)

$\qquad D = \dfrac{2\cdot3\ U^2}{H}$ . . (32)

$\qquad U = \dfrac{2\cdot3\ U^2}{D}$ . . (33)

EXAMPLES.—1. Required the velocity of water issuing from a pipe 2 feet diameter 4 miles in length, connecting two reservoirs, one of which is 30 feet above the other.

$$\text{Here } D = 2, \; H = \frac{30}{4} = 7.5,$$

$$\text{then } U = \sqrt{\frac{2 \times 7.5}{2.3}} = \sqrt{6.52} = 2.55,$$

the velocity in feet per second.

A 2-feet pipe has an area of 3·1416 feet, and, with a velocity of one foot per second, will deliver 188·5 cubic feet per minute. Hence, with a velocity of 2·55, it will deliver $188.5 \times 2.55 = 480.675$ cubic feet per minute, or 4,326,075 gallons in 24 hours.

2. Required the diameter of a pipe having a fall of 10 feet per mile, capable of delivering water with a velocity of 3 feet per second.

$$\text{Here } D = \frac{2.3 \times 3^7}{10} = 2.07 \text{ feet, the diameter required.}$$

3. Required the head or fall per mile necessary to overcome friction in a pipe 3 feet diameter, discharging 6 million gallons in 12 hours.

$$\text{Here } D = 3 \text{ and } U = \frac{6,000,000}{60 \times 4500 \times 3^2 \times .7854}$$

$$= \frac{6,000,000}{1,908,522} = 3.14,$$

$$\text{then } H = \frac{2.3 \times (3.14)^2}{3} = 7.56.$$

Hence a gravitating main, 3 feet diameter, must have a fall of 7·57 feet per mile, or say in round numbers 8 feet per mile, in order to discharge 6 million gallons in 12 hours.

If the pipe be a pumping main, the same height of 8 feet per mile must be added to the pumping lift in calculating the power of the engine required to perform the work. See examples at page 279, where the height for friction is there added in computing the work to be performed by pumping engines.

S

Let Q be the discharge in cubic feet per second. Then from Equation 24 (Prony's) we derive

$$Q = (\cdot 274 \ D^5 H)^{\frac{1}{2}} \quad . \quad . \quad . \quad (34);$$

also from Equation 28 (Eytelwein's) we derive

$$Q = (\cdot 292 \ D^5 H)^{\frac{1}{2}} \quad . \quad . \quad . \quad (35);$$

and from Equation 31 (Blackwell's)

$$Q = (\cdot 268 \ D^5 H)^{\frac{1}{2}} \quad . \quad . \quad . \quad (36).$$

For short lengths of pipe the following formula will be found useful, where d is the diameter in inches, h the height or fall of the pipe in feet, and l the length in feet. Then the discharge in cubic feet per minute is

$$= \left\{ \frac{h \, d^5}{(\cdot 0448 \ (l + 4 \cdot 2 \, d)} \right\}^{\frac{1}{2}}.$$

Example: Required the discharge of a 2-inch pipe 100 feet long, with a fall of 4 feet.

Here $\left(\dfrac{4 \times 32}{\cdot 0448 \ (100 + 8 \cdot 4)} \right)^{\frac{1}{2}} = \left(\dfrac{128}{4.86} \right)^{\frac{1}{2}} = 5 \cdot 23$ cubic feet.

Let Q' be the discharge in cubic feet per minute. Then we have from Prony's equation (20)

$$Q' = (986 \ D^5 H)^{\frac{1}{2}} = 31 \cdot 4 \ (D^5 H)^{\frac{1}{2}} \quad . \quad . \quad (37);$$

from Eytelwein's equation (24),

$$Q' = (1051 \ D^5 H)^{\frac{1}{2}} = 32 \cdot 42 \ (D^5 H)^{\frac{1}{2}} \quad . \quad (38);$$

and from Blackwell's equation (27),

$$Q' = (965 \ D^5 H)^{\frac{1}{2}} = 31 \ (D^5 H)^{\frac{1}{2}} \quad . \quad . \quad (39).$$

If for H we substitute h and l, both in feet, we have only to multiply the coefficient in each of the above equations by 72·66, the square root of 5280, the number of feet in a mile. Thus the value of Q',

according to Prony, is $2282 \dfrac{(D^5 h)^{\frac{1}{2}}}{\sqrt{l}} \quad . \quad . \quad . \quad (40);$

according to Eytelwein, $2356 \dfrac{(D^5 h)^{\frac{1}{2}}}{\sqrt{l}} \quad . \quad . \quad . \quad (41):$

according to Blackwell, $2252 \dfrac{(D^5 h)^{\frac{1}{3}}}{\sqrt{l}}$ . . . (42).

Formula 41 is identical with the one which Mr. Beardmore has used to calculate his Table of Discharges (see page xvii., Beardmore's 'Hydraulic Tables,' 2nd edition).

If d, the diameter of the pipe, be expressed in inches, as is frequently the case, all the other dimensions being in feet, we have only to divide the coefficient by $12^{\frac{5}{2}} = 498\cdot8$. We then obtain the following values of Q' :—

By Prony's formula, $4\cdot57 \dfrac{d^5 h)^{\frac{1}{3}}}{\sqrt{l}} = Q'$ . . (43)

Eytelwein's, ,, $4\cdot72 \dfrac{(d^5 h)^{\frac{1}{2}}}{\sqrt{l}} = Q'$ . . (44)

Blackwell's ,, $4\cdot51 \dfrac{(d^5 h)^{\frac{1}{2}}}{\sqrt{l}} = Q'$ . . (45)

Neville's ,, $4\,81 \dfrac{(d^5 h)^{\frac{1}{2}}}{\sqrt{l}} = Q'$ . . (46)

Formula 44 is the same as that which has been derived from Eytelwein's equation for short lengths, in which the diameter exceeds $\frac{1}{100}$th part of the length, namely

$$Q' = \left\{ \frac{d^5 h}{\cdot0448\,(l + 4\cdot2\,d)} \right\}^{\frac{1}{2}}.$$

Here $\cdot0448 = \dfrac{1}{(4\cdot72)^2}$, or the reciprocal of the square of Eytelwein's coefficient.

Eytelwein's formula 44 is also identical with that given by Mr. Pole, where l, the length, is expressed in yards, d in inches, and h in feet as before. (Paper in 'Journal of Gas-Lighting,' 10th June, 1852.)

Mr. Pole's form is,

$$Q' = 2\cdot72\,d^2 \sqrt{\frac{h\,d}{l + d}}.$$

Here $2\cdot72 = 4\cdot72 \div \sqrt{3}$, or equal to Eytelwein's coefficient divided by the square root of 3.

APPENDIX.

TABLE I.

Table of Horse-Power, showing the amount required to raise from 50,000 to 10,000,000 gallons 1 foot high in 24 hours.

Gallons lifted 1 foot high in 24 hours.	Horse Power required.	Gallons lifted 1 foot high in 24 hours.	Horse Power required.	Gallons lifted 1 foot high in 24 hours.	Horse Power required.
50,000	·0105	1,400,000	·2946	2,750,000	·5787
100,000	·0210	1,450,000	·3051	2,800,000	·5892
150,000	·0316	1,500,000	·3157	2,850,000	·5998
200,000	·0421	1,550,000	·3262	2,900,000	·6103
250,000	·0526	1,600,000	·3367	2,950,000	·6208
300,000	·0631	1,650,000	·3472	3,000,000	·6313
350,000	·0737	1,700,000	·3577	3,050,000	·6418
400,000	·0842	1,750,000	·3683	3,100,000	·6524
450,000	·0947	1,800,000	·3788	3,150,000	·6629
500,000	·1052	1,850,000	·3893	3,200,000	·6734
550,000	·1157	1,900,000	·3998	3,250,000	·6839
600,000	·1263	1,950,000	·4104	3,300,000	·6945
650,000	·1368	2,000,000	·4209	3,350,000	·7050
700,000	·1473	2,050,000	·4314	3,400,000	·7155
750,000	·1578	2,100,000	·4419	3,450,000	·7260
800,000	·1684	2,150,000	·4524	3,500,000	·7365
850,000	·1789	2,200,000	·4630	3,550,000	·7471
900,000	·1894	2,250,000	·4735	3,600,000	·7576
950,000	·1999	2,300,000	·4840	3,650,000	·7681
1,000,000	·2104	2,350,000	·4945	3,700,000	·7786
1,050,000	·2200	2,400,000	·5051	3,750,000	·7892
1,100,000	·2315	2,450,000	·5156	3,800,000	·7997
1,150,000	·2420	2,500,000	·5261	3,850,000	·8102
1,200,000	·2525	2,550,000	·5366	3,900,000	·8207
1,250,000	·2631	2,600,000	·5471	3,950,000	·8312
1,300,000	·2736	2,650,000	·5577	4,000,000	·8418
1,350,000	·2841	2,700,000	·5682	4,050,000	·8523

TABLE I.—*Continued.*

Gallons lifted 1 foot high in 24 hours.	Horse Power required.	Gallons lifted 1 foot high in 24 hours.	Horse Power required.	Gallons lifted 1 foot high in 24 hours.	Horse Power required.
4,100,000	·8628	6,100,000	1·2837	8,100,000	1·7046
4,150,000	·8733	6,150,000	1·2942	8,150,000	1·7151
4,200,000	·8838	6,200,000	1·3047	8,200,000	1·7256
4,250,000	·8944	6,250,000	1·3153	8,250,000	1·7361
4,300,000	·9049	6,300,000	1·3258	8,300,000	1·7467
4,350,000	·9154	6,350,000	1·3363	8,350,000	1·7572
4,400,000	·9259	6,400,000	1·3468	8,400,000	1·7677
4,450,000	·9365	6,450,000	1·3573	8,450,000	1·7782
4,500,000	·9470	6,500,000	1·3679	8,500,000	1·7887
4,550,000	·9575	6,550,000	1·3784	8,550,000	1·7993
4,600,000	·9680	6,600,000	1·3888	8,600,000	1·8098
4,650,000	·9785	6,650,000	1·3993	8,650,000	1·8203
4,700,000	·9891	6,700,000	1·4008	8,700,000	1·8308
4,750,000	·9996	6,750,000	1·4204	8,750,000	1·8414
4,800,000	1·0101	6,800,000	1·4309	8,800,000	1·8519
4,850,000	1·0206	6,850,000	1·4414	8,850,000	1·8624
4,900,000	1·0312	6,900,000	1·4519	8,900,000	1·8729
4,950,000	1·0417	6,950,000	1·4625	8,950,000	1·8834
5,000,000	1·0522	7,000,000	1·4731	9,000,000	1·8940
5,050,000	1·0627	7,050,000	1·4836	9,050,000	1·9045
5,100,000	1·0732	7,100,000	1·4941	9,100,000	1·9150
5,150,000	1·0838	7,150,000	1·5046	9,150,000	1·9255
5,200,000	1·0943	7,200,000	1·5152	9,200,000	1·9360
5,250,000	1·1048	7,250,000	1·5257	9,250,000	1·9466
5,300,000	1·1153	7,300,000	1·5362	9,300,000	1·9571
5,350,000	1·1259	7,350,000	1·5467	9,350,000	1·9676
5,400,000	1·1364	7,400,000	1·5573	9,400,000	1·9781
5,450,000	1·1469	7,450,000	1·5678	9,450,000	1·9887
5,500,000	1·1574	7,500,000	1·5783	9,500,000	1·9992
5,550,000	1·1679	7,550,000	1·5888	9,550,000	2·0097
5,600,000	1·1785	7,600,000	1·5993	9,600,000	2·0202
5,650,000	1·1890	7,650,000	1·6099	9,650,000	2·0307
5,700,000	1·1995	7,700,000	1·6204	9,700,000	2·0413
5,750,000	1·2100	7,750,000	1·6309	9,750,000	2·0518
5,800,000	1·2206	7,800,000	1·6414	9,800,000	2.0623
5,850,000	1·2311	7,850,000	1·6520	9,850,000	2·0728
5,900,000	1·2416	7,900,000	1·6625	9,900,000	2·0834
5,950,000	1·2521	7,950,000	1·6730	9,950,000	2·0939
6,000,000	1·2626	8,000,000	1·6835	10,000,000	2·1044
6,050,000	1·2732	8,050,000	1·6940		

TABLE II.

Showing the Power of Cornish Engines working with a load of 18 lbs. per square inch on the Piston, and an effective Velocity of 110 feet per minute.

Diameter of Cylinder in inches.	Area of Cylinder in square inches = a.	Horse Power of Engine $\frac{a \times 18 \times 11}{3300}$	Diameter of Cylinder in inches.	Area of Cylinder in square inches = a.	Horse Power of Engine $\frac{a \times 18 \times 11}{3300}$
15	177	11	23	415	25
16	201	12	24	452	27
17	227	14	25	491	29
18	254	15	26	531	32
19	284	17	27	573	34
20	314	19	28	616	37
21	346	21	29	661	40
22	380	23			

TABLE III.

Showing the Power of Cornish Engines working with a load of 17 lbs. per square inch on the Piston, and an effective Velocity of 110 feet per minute.

Diameter of Cylinder in inches.	Area of Cylinder in square inches = a.	Horse Power of Engine $\frac{a \times 17 \times 11}{3300}$	Diameter of Cylinder in inches.	Area of Cylinder in square inches = a.	Horse Power $\frac{a \times 17 \times 11}{3300}$
30	707	40	38	1134	64
31	755	42	39	1195	68
32	804	45	40	1257	71
33	855	48	41	1320	75
34	908	51	42	1385	79
35	962	54	43	1452	82
36	1018	58	44	1521	86
37	1075	61	45	1590	90

TABLE IV.

Showing the Power of Cornish Engines working with a load of 16 lbs. per square inch on the Piston, and an effective Velocity of 110 feet.

Diameter of Cylinder in inches.	Area of Piston in square inches = a.	Horse Power $\frac{a \times 16 \times 11}{3300}$	Diameter of Cylinder in inches.	Area of Piston in square inches = a.	Horse Power $\frac{a \times 16 \times 11}{3300}$
45	1590	85	53	2206	118
46	1662	88	54	2290	122
47	1735	92	55	2376	127
48	1810	97	56	2463	131
49	1886	100	57	2552	136
50	1963	103	58	2642	141
51	2043	108	59	2734	146
52	2124	113	60	2827	151

TABLE V.

Showing the Power of Cornish Engines working with a load of 15 lbs. per square inch on the Piston, and an effective Velocity of 110 feet per minute.

Diameter of Cylinder in inches.	Area of Piston in square inches = a.	Horse Power $\dfrac{a \times 15 \times 11}{3300}$	Diameter of Cylinder in inches.	Area of Piston in square inches = a.	Horse Power $\dfrac{a \times 15 \times 11}{3300}$
60	2827	141	81	5153	259
61	2922	146	82	5281	264
62	3019	151	83	5411	271
63	3117	156	84	5542	277
64	3217	161	85	5675	283
65	3318	166	86	5809	290
66	3421	171	87	5945	297
67	3525	176	88	6082	304
68	3632	182	89	6221	311
69	3739	187	90	6362	318
70	3848	192	91	6504	325
71	3959	198	92	6648	332
72	4071	204	93	6793	339
73	4185	209	94	6940	347
74	4301	215	95	7088	354
75	4418	221	96	7238	361
76	4536	227	97	7390	369
77	4657	233	98	7543	377
78	4778	239	99	7698	384
79	4902	245	100	7854	392
80	5026	251			

TABLE VI.

HORSE-POWER OF CORNISH STEAM-ENGINES.

By Mr. John Darlington.

The following Table has been compiled with the object of furnishing an approximate value of the power in horses rendered by Cornish Pumping Engines having cylinders from 15 to 100 inches diameter. The elements employed for the calculations are those most usual with Cornish engineers; and the effective horse-power per stroke is given, that the inquirer may ascertain the total value of horse-power resulting from working any given number of strokes per minute. The steam in most of the Cornish Pumping Engines is only permitted to act on one side of the piston; hence such mode of working is technically termed "single acting." Recently however it has been considered that equal economy is obtained by introducing the steam on both sides of the piston, and a few engines are in operation on this principle. The horse-power of such (double-acting) engines may be found by doubling the results given in the Table.

Horse-Power, Load in Pounds, and Speed per Minute of Cornish "Single Acting" Expansive Steam Pumping Engines, having Cylinders from 15 inches to 100 inches diameter.

Initial Pressure of Steam, 30 lbs. per sq. inch. Temp. 251·6°. Full Pressure of Steam, 1-4th of stroke.—Mean Pressure of Steam, 17·8 lbs. less 1-5th friction = 14·24 lbs.

Diameter of Cylinder.	Length of Stroke in Cylinder.	Area of Cylinder.	Load in pounds, less 1-5th for Friction.	Strokes per Minute.		Speed per Minute in Feet.		Horse Power.		Effective Horse Power per Stroke.
				Economical Working.	Safe Working.	Economical Working.	Safe Working.	Economical Working.	Safe Working.	
In.	Ft.	In.	lbs.			Ft.	Ft.			
15	8	176·71	2,516	5	14	80	224	3·04	8·53	·609
16	8	201·06	2,863	5	14	80	224	3·47	9·71	·694
17	8	226·98	3,232	5	14	80	224	3·91	10·96	·783
18	8	254·46	3,623	5	14	80	224	4·39	12·29	·878
19	8	283·52	4,037	5	14	80	224	4·89	13·70	·978
20	9	314·16	4,473	4½	12	81	216	5·48	14·63	1·219
21	9	346·36	4,932	4½	12	81	216	6·05	16·14	1·345
22	9	380·13	5,413	4½	12	81	216	6·62	17·71	1·476
23	9	415·47	5,916	4½	12	81	216	7·26	19·36	1·613
24	9	452·39	6,442	4½	12	81	216	7·90	21·08	1·756
25	9·5	490·87	6,989	4½	10	85½	200	9·05	21·17	2·012
26	9·5	530·93	7,560	4½	10	85½	200	9·79	22·90	2·176

TABLE VI.—*Continued.*

Diameter of Cylinder.	Length of Stroke in Cylinder.	Area of Cylinder.	Load in pounds, less 1-5th for Friction.	Strokes per Minute.		Speed per Minute in Feet.		Horse Power.		Effective Horse Power per Stroke.
				Economical Working.	Safe Working.	Economical Working.	Safe Working.	Economical Working.	Safe Working.	
In.	Ft.	In.	lbs.			Ft.	Ft.			
27	9·5	572·55	8,153	4½	10	85½	200	10·56	24·70	2·347
28	9·5	615·75	8,768	4½	10	85½	200	11·35	26·57	2·524
29	9·5	660·52	9,405	4½	10	85½	200	12·18	28·50	2·707
30	10	706·86	10,065	4	10	80	200	12·20	30·50	3·050
31	10	754·76	10,747	4	10	80	200	13·02	32·56	3·256
32	10	804·24	11,452	4	10	80	200	13·88	34·70	3·470
33	10	855·30	12,179	4	10	80	200	14·76	36·90	3·690
34	10	907·92	12,928	4	10	80	200	15·67	39·17	3·917
35	10	962·11	13,700	4	10	80	200	16·60	41·51	4·151
36	10	1017·8	14,492	4	10	80	200	17·56	43·91	4·391
37	10	1075·2	15,310	4	10	80	200	18·55	46·39	4·639
38	10	1154·1	16,148	4	10	80	200	19·57	48·93	4·893
39	10	1194·5	17,009	4	10	80	200	20·61	51·54	5·154
40	10	1256·5	17,894	4	10	80	200	21·68	54·22	5·422
41	10	1320·2	18,799	4	10	80	200	22·78	56·96	5·696
42	10	1385·4	19,728	4	10	80	200	23·91	59·78	5·978
43	10	1452·2	20,679	4	10	80	200	25·06	62·66	6·266
44	10	1520·5	21,652	4	10	80	200	26·24	65·61	6·561
45	10	1590·4	22,647	4	10	80	200	27·45	68·62	6·862
46	10	1661·9	23,664	4	10	80	200	28·68	71·71	7·171
47	10	1734·9	24,705	4	10	80	200	29·94	74·86	7·486
48	10	1809·5	25,768	4	10	80	200	31·23	78·08	7·808
49	10	1885·7	26,852	4	10	80	200	32·54	81·37	8·137
50	10	1963·5	27,956	4	10	80	200	33·88	84·71	8·471
51	10	2042·8	29,089	4	10	80	200	35·25	88·14	8·814
52	10	2123·7	30,240	4	10	80	200	36·65	91·63	9·163
53	10	2206·1	31,414	4	10	80	200	38·07	95·19	9·519
54	10	2290·2	32,612	4	10	80	200	39·52	98·82	9·882
55	10	2375·8	33,831	4	10	80	200	41·00	102·51	10·251
56	10	2463·0	35,072	4	10	80	200	42·50	106·27	10·627
57	10	2551·7	36,336	4	10	80	200	44·04	110·10	11·010
58	10	2642·0	37,620	4	10	80	200	45·60	114·00	11·400
59	10	2733·9	38,930	4	10	80	200	47·18	117·97	11·797
60	10·5	2827·4	40,260	4	9½	84	200	51·24	122·00	12·810
61	10·5	2922·4	41,614	4	9½	84	200	52·96	126·10	13·240
62	10·5	3019·0	42,988	4	9½	84	200	54·71	130·26	13·678
63	10·5	3117·2	44,388	4	9½	84	200	56·49	134·50	14·123
64	10·5	3216·9	45,808	4	9½	84	200	58·30	138·81	14·575
65	10·5	3318·3	47,252	4	9½	84	200	60·13	143·18	15·034

TABLE VI.—*Continued.*

Diameter of Cylinder.	Length of Stroke in Cylinder.	Area of Cylinder.	Load in pounds, less 1-5th for Friction.	Strokes per Minute.		Speed per Minute in Feet.		Horse Power.		Effective Horse Power per Stroke.
				Economical Working.	Safe Working.	Economical Working.	Safe Working.	Economical Working.	Safe Working.	
In.	Ft.	In.	lbs.			Ft.	Ft.			
66	10·5	3421·2	48,716	4	9½	84	200	62·00	147·62	15·500
67	10·5	3525·6	50,204	4	9½	84	200	63·89	152·13	15·974
68	10·5	3631·6	51,712	4	9½	84	200	65·81	156·70	16·453
69	10·5	3739·2	53,246	4	9½	84	200	67·76	161·35	16·941
70	11	3848·4	54,800	4	9	88	198	73·06	164·40	18·266
71	11	3959·2	56,379	4	9	88	198	75·17	169·13	18·793
72	11	4071·5	57,978	4	9	88	198	77·30	173·93	19·326
73	11	4185·3	59,598	4	9	88	198	79·46	178·79	19·866
74	11	4300·8	61,240	4	9	88	198	81·65	183·72	20·413
75	11	4417·8	62,909	4	9	88	198	83·87	188·72	20·969
76	11	4536·4	64,592	4	9	88	198	86·12	193·77	21·530
77	11	4656·6	66,310	4	9	88	198	88·41	198·93	22·103
78	11	4778·3	68,036	4	9	88	198	90·71	204·10	22·670
79	11	4901·6	69,798	4	9	88	198	93·06	209·39	23·206
80	11.5	5026·5	71,578	4	8½	92	196	99·77	212·56	24·943
81	11·5	5153·0	73,378	4	8½	92	196	102·28	217·91	25·571
82	11·5	5281·0	75,201	4	8½	92	196	104·82	223·32	26·206
83	11·5	5410·6	77,046	4	8½	92	196	107·39	228·80	26·849
84	11·5	5541·7	78,913	4	8½	92	196	110·00	234·04	27·499
85	12	5674·5	80,804	4	8	96	192	117·53	235·06	29·383
86	12	5808·8	82,717	4	8	96	192	120·31	240·63	30·078
87	12	5944·6	84,651	4	8	96	192	123·12	246·25	30·782
88	12	6082·1	86,609	4	8	96	192	125·97	251·95	?1·494
89	12	6221·1	88,588	4	8	96	192	128·85	257·71	32·213
90	12	6361·7	90,590	4	8	96	192	131·76	263·53	32·941
91	12	6503·8	92,614	4	8	96	192	134·71	269·42	33·677
92	12	6647·6	94,661	4	8	96	192	137·68	275·37	34·422
93	12	6792·9	96,730	4	8	96	192	140·69	281·39	35·174
94	12	6939·7	98,821	4	8	96	192	143·73	287·47	35·935
95	12	7088·2	100,925	4	8	96	192	146·80	293·60	36·700
96	12	7238·2	103,071	4	8	96	192	149·92	299·84	37·480
97	12	7389·8	105,230	4	8	96	192	153·06	306·12	38·265
98	12	7542·9	107,410	4	8	96	192	156·23	312·46	39·058
99	12	7697·7	109,615	4	8	96	192	159·44	318·88	39·860
100	12	7854·0	111,840	4	8	96	192	162·67	325·35	40·669

Table of the Yield of Chalk Wells.

Situation.	Yield in gallons per day.	Authority.
Amwell Hill Well . .	2,400,000	Evidence of J. Muir, Esq., before the Royal Commission on Water supply.
Amwell End Well . .	2,500,000	R. W. Mylne, Esq.
Cheshunt Well . .	702,000	Ditto.
Tottenham Court Road*	630,000	Ditto.
Southampton . . .	288,000	'Hampshire Independent,' May, 1841.
Camden Station† . .	300,000	Mr. Paton.
Brighton . . .	232,000	Mr. R. Stephenson.
Plumstead Common .	600,000	Mr. Homersham.
Reid's Brewery . .	277,200	Mr. Braithwaite.
Truman and Hanbury .	166,320	Mr. Davidson.
Experimental Well at Watford . . .	1,800,000	Mr. Stephenson & Mr. Paton.

List of some of the principal Chalk Springs in England.

1. *Situate on the Long Slope of the Chalk.*

Leatherhead, close to Guildford Road.

Croydon, near the church.

Carshalton.

Orpington.

Birchington, Isle of Thanet.

Bedhampton, near Portsmouth.

Chadwell, near Hertford, yielding from 2,700,000 gallons up to 4½ million gallons per day.

Woolmers, in the valley of the Lea, yielding 2,700,000 gallons per day.

River Lea, above Luton, chiefly spring water, yielding 5,400,000 gallons per day.

Bourne Stream, Riddlesdown, 2,025,000 gallons per day.

Grays Thurrock springs, now pumped up for the supply of Brentwood, Romford, &c., capable of yielding 7,000,000 gallons a-day.

2. *Springs situate on the Escarpment side, or Short Slope of the Chalk.*

Bourne Mill, near Farnham.

The Holy Well at Kempering, on the south side of the North Downs.

---

* In 1843 Mr. Mylne stated the yield of this well was 423,360 gallons per day. The spring was met with 234 feet below the surface.

† Well sunk 180 feet deep down to the chalk, then bored 200 feet deep in chalk.

Lydden Spout, near Folkestone.

The Holy Well, Beachey Head Cliff.

Nine Wells, near Cambridge, yielding 423,000 gallons per day according to my gaugings in October, 1854.

Cherry Hinton, near Cambridge, yielding 707,000 gallons per day according to my gaugings in October, 1854.

Godstone, Surrey.

Cheriton, near Folkestone.

SOUTH STAFFORDSHIRE WATERWORKS.

SPECIFICATION OF ENGINES, BOILERS, AND PUMPS.

This Contract comprises the making, erecting, and setting to work of one pair of engines, with boilers and pumps complete, capable of delivering in twelve hours' work 2,500,000 gallons of water through the main pipe, under a head on the pumps, including the friction of the water in the pipes, of 355 feet.

The engines are to be erected on a certain plot of land called Sandgate, adjoining the South Staffordshire Railway, and lying about halfway between the Hammerwich and Lichfield Stations of that Railway, and bounded on one side thereof by the Lichfield Branch of the Birmingham Canal.

The general arrangement and design of the engine, &c. is shown on the drawing attached to this Specification.

The engines are to be condensing and expansive double-acting beam engines, coupled together and working with cranks on the same crank-shaft, with one fly-wheel between them.

The connecting-rods are to be made so that either engine can be easily disconnected at the crank-pin; and either engine must be able to work alone as well as in conjunction with the other; and when working alone must be able to perform one-half of the work specified for the pair.

Each engine to have one steam-cylinder without jacket. The steam-valves to be double-beat gun-metal, with proper nozzles complete, and throttle and expansive gearing, so as to cut off the steam at any required portion of the stroke.

The air-pump valves to be formed of vulcanized india-rubber flaps, working on gun-metal gratings.

The principal dimensions of the engines are to be as under:—

Steam-cylinders 4½ inches diameter each; length of stroke of ditto, 8 feet; extreme length of main beam from cylinder centre to connecting-rod centre, 26 feet 6 inches; radius of crank, 4 feet.

Height of beam centre from floor of engine-house, 21 feet 4 inches.

Width from centre of one engine to centre of the other, 14 feet 2 inches.

Diameter of fly-wheel not less than 21 feet, with large and heavy rim to regulate the motion of the engines perfectly.

The piston-rods of the cylinders must be of the best refined iron: the pistons to be fitted with the most approved metallic packing.

The cranks, crank-shafts, and beam-centres must be all of the best wrought iron.

All the bearings must be formed of the best gun-metal, of ample thickness and strength, and be fitted into their places so as to bear on the iron over their whole surface.

Every part of the engines must be made strong enough to bear without breaking ten times the maximum strain that can ever arise in working.

A cast-iron floor, supported on iron girders, is to be placed round the upper part of the cylinders, for the convenience of packing the glands and examining the pistons; to be connected with the engine-house floor below and the beam-floor above, with neat open cast-iron stairs and light handrail. Cast-iron plates are to be fixed in a neat and convenient manner over all parts of the cold-water cisterns, well, and other parts to which access is required; proper railings to be provided where required in the judgment of the engineers of the Company for the safety of the engine-workers.

A travelling-crane of the best description, capable of lifting ten tons, is to be provided and fixed over the engines.

The engines are to be finished as bright engines; all the joints are to be made metal and metal; all levers, journals, and other working bearings to be case-hardened, and joint-pins under 1½ inch diameter to be of steel; and the whole of this work throughout to be finished in the best style of modern engines.

Each engine is to be provided with one double-acting combined plunger and lift well-pump, so made as to discharge equal quantities of water in the up-and-down stroke. The plungers are to be worked by means of strong connecting-rods direct from the main beam of the engines. To be attached at the top to the fly-wheel connecting-rod centres, and at the bottom to a cross-head fitted to the top of the plunger. To be connected at each end with proper straps and bearings.

The diameter of the pumps is to be such that the engines shall raise the above specified quantity of water when they are making 15 strokes per minute.

The rising pipe of the pump to be three times the area of the plunger so far as the plunger descends into it. Proper guides to be fixed in such a manner as to secure a uniformly perpendicular motion to the plunger.

The pumps are to be placed on separate wells, in which water will stand 70 feet below the engine-house floor, and they must be provided with the most convenient ladders, stages, and means of access to all the valves and the necessary suction and delivery pipes.

Pipes are to be provided to connect the two pumps together and with the main through which both are to pump. Each pump must also be provided with a stop-valve, made perfectly tight, so as to prevent the return of the water when either or both pumps are not at work.

On each rising main, immediately above the working barrel and on the breeches-pipe connecting to the large water main, a 4-inch blow-out valve is to be placed, loaded to 150 lbs. to the square inch, and each commanded by a 4-inch double-faced screw-cock, placed between the barrel or main pipe and the blow-off valve, so as to ease the pump at starting.

An air-vessel is to be provided to each pump, the capacity of each of which must be equal to ten times the quantity of water raised by both pumps in one complete stroke of the engines. These air-vessels must be proved to be perfectly tight under a pressure of 600 feet of water.

Each air-vessel must be provided with proper discharge and filling pipes and taps, and wrought-iron diaphragm floats filling nearly the whole interior diameter of the air-vessel, so as to keep the surface of the water in the air-vessel as far as possible from contact with the contained air.

Proper double-faced stop-cocks of Nasmyth's pattern must be placed in the inlet and outlet pipe to each air-vessel and the pipe connecting the two pumps.

The pump and well-work is all to be made on the most improved plans, and the pump-valves must be Hosking's patent gutta-percha ball valves. The bucket to be geared with metallic packing.

The engines to be provided with proper balance-weights at the cylinder ends, to be fixed either in the pistons or between the cheeks of the beams, so as to balance the plunger ends.

Four boilers are to be provided, of the following dimensions, viz. cylindrical, with flat ends, and two internal tubes passing through each. Each boiler is to be 32 feet long, 7 feet diameter, and the internal tubes 2 feet 3 inches diameter above the fire-places, and 1 foot 9 inches diameter beyond.

The plates for these boilers are to be $\frac{7}{16}$ inch thick, except the end plates, which are to be $\frac{9}{16}$ inch thick, with angle iron ribs riveted on to stiffen them.

The ordinary working pressure of the steam is not to exceed 25 lbs. per square inch; but all the boilers must be proved by water pressure to be perfectly tight under a pressure of 80 lbs. per square inch before they a. fixed.

All the boiler plates are to be of the best Staffordshire iron, with the exception of those over the fire, which are to be of Lowmoor iron.

The steam-pipes are to communicate with each boiler by means of a steam-valve, and a similar steam throttle-valve is to be provided for each engine in the engine-house, so that either engine may be worked from any boiler.

Similar arrangements are to be made to enable any boiler to be fed by either engine. The blow-off hot and cold feed must be so arranged with proper stopcocks, as all to enter the boiler through one junction-pipe to be attached to the bottom of the boiler at the front end.

A wrought-iron expansion flange to be provided and riveted on the main range of steam-pipes between each boiler.

Each boiler to have two cast-iron man-hole lids riveted on over the top of the boiler, and one wrought-iron man-hole and clench plate to be fixed in part of the boiler below and between the tubes.

Each boiler to be fitted with one blow-off valve attached to one of the man-hole lids 5 inches diameter; three brass gauge-cocks, a glass guard-gauge, and float-gauge and whistle to be attached to each boiler. The ash-pit to be covered with strong smooth cast-iron plates.

Approved steam and vacuum gauges to be provided and fixed in the engine-house. A complete set of spanners to fit every sized nut in the engine and pump-work, to be arranged upon a cast-iron plate and fixed against the wall of the engine-house. Two sets of firing-irons, a set of taps and dies, with hammers, files, chisels, and vice to be provided. A set of small brass oil-cups and siphons with spring lids to be affixed to all the principal bearings. A well-gauge and float is to be fixed so as to be seen in the engine-house. An approved counter to be affixed to each engine-beam, so that it cannot be worked except the engine is worked.

The following duplicates are to be provided:—A set of main pump buckets and clacks; ditto for cold water pump; a duplicate valve of each description of steam and stop-valve throughout the engine; a dozen glass tubes for guard-gauges; twelve bolts and nuts for cylinder covers, six for pistons, twelve for clack doors, six for plunger glands.

The engine cylinders and all the steam-pipes are to be covered with 2 inches thickness of felt and canvas, to prevent the radiation of heat, lagged on the outside with 1 inch wrought and bead deal lagging well hooped round.

All the work is to be painted once before leaving the contractor's premises, and is to receive two more coats of oil-paint after being fixed,—the finishing colour being such as may be approved by the Company's engineers.

This specification is intended to describe generally the engines, pumps, and boilers required, but is to be understood that the contractor is to provide and fix complete every kind of iron-work, steam-feed, condensed and waste-water pipes, suction pipes, delivery pipes, foundation plates, washers, furnace and boiler fittings, dampers, girders, floor plates, and apparatus of every kind in the engine and boiler-house and well, that are required to make the works complete in every respect, and capable of performing the duty required, although many of the details may not be specially described in this specification.

All the work described in this specification is to be made with the best materials and workmanship, and is in all respects to be subject to the approval of the Company's engineers, and to their inspection at all times during the progress of the work.

It is to be understood that no earthwork, bricklayers' work, masons' or builders' work is included in this specification. This part of the work is to be executed by the Waterworks Company; but the contractors for the engines shall, within one month of signing the contract, supply the Company's engineer with a detail working drawing, showing the entire engines, boilers, and pump-work, together with the masonry and brickwork required for the foundations and boiler seatings, with the position of all foundation plates and holding-down bolts clearly marked on.

The engineers shall have full power to alter, vary, diminish, or increase the works, without in any way releasing the contractors from the responsibility and conditions attached to this contract; any additions to or deductions from the amount of works included in this contract which may be made by an order in writing by the engineers shall be added or deducted from the amount of the contract according to the schedule of prices attached.

The whole of these works are to be delivered, fixed, and set to work on the site before mentioned on or before September 30, 1856.

Payment to be made upon the certificate of the engineers as follows :—
30 per cent. of the contract amount immediately work to that value shall have been delivered upon the ground; an additional 30 per cent. when the whole of the work shall have been delivered; a further instalment of 30 per cent. when the engines shall have been started, and are working to the satisfaction of the engineers.

The balance by two equal instalments, one at the end of six months, and the other at the end of twelve months from the period of starting the engines, and during which period the contractors will be responsible for the engines and have to keep them in good repair; and the balance above mentioned shall only be paid provided tde contractors shall have fulfilled all the conditions of the contract to the satisfaction of the Engineers. And in case of any dispute as to any matter of account between the Company and the contractor, the same shall be referred to the engineers whose decision shall be final and binding between both parties.

INDEX.

T